The Natural History
of The Bahamas

The Natural History of The Bahamas

A FIELD GUIDE

Dave Currie, Joseph M. Wunderle Jr., Ethan Freid,
David N. Ewert, and D. Jean Lodge

Comstock Publishing Associates
An imprint of
Cornell University Press
Ithaca and London

First published 2019 by Cornell University Press

Printed in China

Library of Congress Cataloging-in-Publication Data
Names: Currie, Dave 1967– author. | Wunderle, Joseph M., author. |
    Freid, Ethan, author. | Ewert, David N., author. | Lodge, Deborah Jean,
    1953– author.
Title: The natural history of the Bahamas : a field guide / Dave Currie, Joe M.
    Wunderle Jr., Ethan Freid, David N. Ewert, and D. Jean Lodge.
Description: Ithaca [New York] : Comstock Publishing Associates, an imprint
    of Cornell University Press, 2019. | Includes bibliographical references and
    index.
Identifiers: LCCN 2018035574 (print) | LCCN 2018035782 (ebook) |
    ISBN 9781501738029 (pdf) | ISBN 9781501738036 (epub/mobi) |
    ISBN 9781501713675 | ISBN 9781501713675 (pbk. ; alk. paper)
Subjects: LCSH: Natural history—Bahamas—Guidebooks. |
    Bahamas—Guidebooks.
Classification: LCC QH109.B3 (ebook) | LCC QH109.B3 C87 2019 (print) |
    DDC 508.7296—dc23
LC record available at https://lccn.loc.gov/2018035574

Front cover photos: (top) *Cyclura rileyi rileyi*, by Dave Currie; (bottom left)
Melocactus intortus, by Ethan Freid; (bottom middle) *Coereba flaveola
bahamensis*, by Bruce Hallett; (bottom right) *Papilio andraemon*, by Dave
Currie. Spine photo: *Setophaga flavescens*, by Bruce Hallett.
Cover design and book design and composition by Julie Allred, BW&A Books

The Bahamas National Trust (BNT) was created by an act of Parliament in 1959 to build and manage the national parks system of The Bahamas. The BNT is a science-based organization dedicated to effectively managing national parks to conserve and protect Bahamian natural resources. This comprehensive network of effectively managed national parks and protected areas is important for biodiversity conservation, environmental education, and green spaces for public recreation. Today, 32 national parks are designated to protect more than 2 million acres of The Bahamas. The Trust is the only known nongovernmental organization in the world with the mandate to manage a country's entire national park system.

THE BAHAMAS NATIONAL TRUST
P.O. Box N-4105, Nassau, The Bahamas
242-393-1317 • bnt@bnt.bs • www.bnt.bs

Contents

Amphibians 241

Reptiles 249

Birds 291

Mammals 387

The Natural History of The Bahamas

Introduction

Two political entities—The Commonwealth of The Bahamas (henceforth **The Bahamas**), an independent country, and the British Overseas Territory of The Turks and Caicos Islands (henceforth **TCI**)—comprise The Bahamas Archipelago, a chain of islands that share a common subtropical or tropical climate, geography, geology, ecology, and human culture as well as many plant and animal species. The archipelago is in the western Atlantic Ocean, east of Florida where it runs southeasterly to just north of the Greater Antilles, namely the islands of Cuba and Hispaniola. The islands extend over an enormous expanse of ocean (c. 215,000 km^2 [83,000 mi^2]) from Walker's Cay, east of Ft. Pierce, Florida, for more than 1000 km (621 mi) to Salt Cay, just south of Grand Turk Island north of the Dominican Republic. The archipelago comprises low-lying limestone islands.

The Bahamas consists of 29 large islands, 661 cays, and 2387 islets (total land area: 13,878 km^2 [5358 mi^2]). The highest point in the archipelago is 63 m (206 ft) on Cat Island. The Bahamas' closest islands to the United States (the Biminis) are just 85 km (53 mi) from Miami, and the southernmost island in the chain (Great Inagua) is located 80 km (50 mi) from Cuba and Hispaniola. Spread over The Bahamas on 32 inhabited islands is a human population of about 393,000 (2016) of which 70% reside on New Providence Island (site of the capital, Nassau) and 18% on Grand Bahama. The other 30 inhabited islands are collectively referred to as the "Family Islands." The TCI are located southeast of The Bahamas and east of Cuba. The TCI include 8 inhabited main islands (total land area: 616 km^2 [238 mi^2]), with a population of approximately 35,500 people, most of whom live on the island of Providenciales.

Geology

Limestone: The Bedrock of the Archipelago

Compared with other islands in the region and most of continental North and South America, the archipelago is geologically young with a relatively simple geology derived from calcium carbonate (limestone) sediments. The archipelago's geology differs from the Greater Antilles, which is derived from continental, volcanic, and igneous rock and limestone sediments, and the Lesser Antilles islands that are mostly of volcanic origin. The Bahamas Archipelago lies off the North American continent on a 6000 m (19,685 ft) shelf or bank consisting of multiple layers of shallow water limestone sediments. The weight of these limestone sediments has depressed the crust underlying the archipelago.

The archipelago's limestone sediments are derived primarily from the abundant marine life of the banks. Owing to its primarily biogenic origins, limestones are typically excellent sources of fossils. Some limestones are composed entirely of fossil reefs. Corals, particularly those along the edge of the banks where substantial walls have developed, together with other marine animals and algae have contributed sediments to the banks. Rapidly growing marine algae, especially the green calcareous algae, extract great quantities of calcium carbonate from the seawater and deposit it as sand or mud. The sediments of the banks are derived mostly from these algae.

Another major sedimentary source is the oolitic sands, especially on the edge of the banks. Oolitic sands are derived from dissolved calcium carbonate, which precipitates out as ooliths (small and near spherical grains of calcium carbonate) when the waters warm while flowing over the shallow banks. The oolitic sands of the archipelago have played an important role in island formation. This process started about two million years ago during the most recent ice age when much water was bound up in ice, and global sea levels were 100 m (330 ft) lower than present day. As a result of the lower sea levels, the round oolitic grains were exposed to the powerful easterly trade winds and blown into lines of tall sand dunes. Over time, these oolitic sand dunes hardened into rock ridges. Sea levels rose following the end of the ice age, and ridges became islands. Islands formed in this manner are always located along the edge of banks, as expected for wind-derived sand dunes (Figure 1). Oolitic sand dunes are not the only source of islands in the archipelago, as some islands or parts of islands consist of limestone rocklands, exposed seabed from an earlier time of higher sea level (6 m [20 ft]; mainly from 125,000 BP [before present] when sea level was highest). The rocklands, originally formed from the seabed, undergo erosion by karst processes (see below) once exposed by sea level decline. These rocklands are evident in the broader islands such as Andros and Grand Bahama, whereas the long, thin islands, such as Acklins and Long Island, are composed mostly of ridges derived from oolitic sand dunes. Some islands, such as New Providence, are a mixture of both ridges and rocklands.

The carbonate banks of the archipelago are believed to have played an important role in facilitating dispersal of terrestrial organisms among islands on the same bank, at least during the ice ages when the sea levels were lower and banks were exposed. Movement of terrestrial organisms between the banks has been constrained, even at the lowest sea levels of the past, because of deep water trenches separating the individual banks. Eight significant carbonate banks are found within The Bahamas, most of which are now partially submerged. These include the Little Bahama Bank (encompasses Grand Bahama and Great Abaco); the Great Bahama Bank (Andros, the Biminis, the Berrys, New Providence, Eleuthera, Exuma, Cat Island, Long Island,

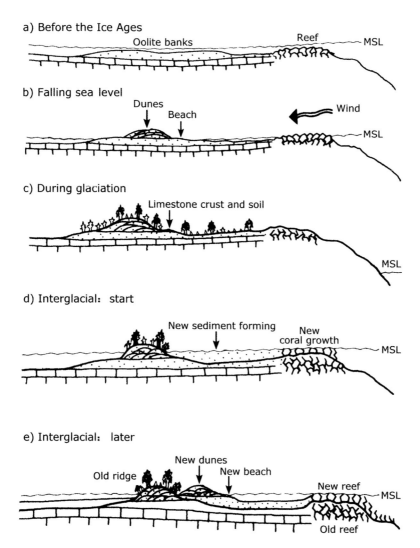

a) Before the Ice Ages

Oolite banks   Reef   MSL

b) Falling sea level

Dunes   Beach   Wind   MSL

c) During glaciation

Limestone crust and soil   MSL

d) Interglacial: start

New sediment forming   New coral growth   MSL

e) Interglacial: later

Old ridge   New dunes   New beach   New reef   MSL   Old reef

Figure 1. Creation of ridges and dunes in the archipelago occurs as a consequence of alternating glacial expansion and retreat as sea level (MSL) changes in the archipelago as illustrated here by Sealey (2006). Before the ice ages is portrayed in (a). As sea level falls, oolite sands from the beach are blown into dunes (b), which are subsequently colonized and stabilized by vegetation (c) during glaciation. When sea level rises and returns to normal at the start of the interglacial period, new limestone particles are washed up on new beaches (d) from which winds blow the particles into new dunes (e). During subsequent glaciation events (lowering of sea levels) and interglacial periods (accompanied by a rise in sea level), the cycle will be repeated.

Figure 2. An example of a rock from rockland surface that has been pitted by solution weathering of the solid rock. Rainfall becomes slightly acidic when it passes through the atmosphere and soil, and as a result it dissolves limestone. As weathering proceeds, small holes connect and enlarge and eventually the larger rock breaks into smaller pieces.

and Ragged Island); the Cay Sal Bank; the Crooked-Acklins Bank; Mayaguana; and the Inaguan Bank (see inside cover for islands and the major banks). Within the TCI are found three carbonate banks, two of which are currently emergent (Turks Bank and Caicos Bank) and another, which is now submerged (Mouchoir Bank). An additional two submerged carbonate banks lie farther southeast: the Silver Bank and Navidad, which are politically part of the Dominican Republic, although geologically they are a continuation of The Bahamas Archipelago. Other smaller or minor banks are found in the archipelago and are indicated for certain species in the species accounts.

A different karst type. Limestone is easily eroded by water, especially freshwater in the form of rainfall and surface runoff. Carbon dioxide in the atmosphere and in the soil dissolves in water and creates carbonic acid, a mild acid that dissolves limestone (Figure 2). This freshwater erosion or "solution weathering" results in surface and subterranean features, such as caves, characteristic of limestone areas. These formations are collectively referred to as "karst," after the limestone area first studied in Europe. The archipelago, however, lacks many of the features of typical karst landscapes from which it differs in terms of the purity of its limestone (i.e., 98% calcium carbonate), its formation from shallow warm water sediments, its young age, and a relatively high water table.

Continual erosion. The highly porous nature of young Bahamian limestone allows rainwater to permeate rapidly through the rock. Consequently, no permanent freshwater streams or rivers occur in the archipelago. Over time, rainwater and its runoff erode the limestone surface and produce a latticework of pits or sinkholes

Figure 3. Sinkholes are common surface features in the archipelago. They are formed when subterranean chambers collapse or when surface chemical weathering results in the creation of shallow depressions or holes, such as the example here from San Salvador. These holes accumulate soil, vegetation debris, and rainwater, providing an excellent site for growth of trees as well as a location for growing crops, such as bananas and pineapples from which they derive their local names. Sinkholes also provide microclimates for some mesic plant species, such as the Maidenhair Ferns seen here growing on the walls of this sinkhole.

ranging from a few centimeters to several meters across (Figure 3). These sinkholes are variously termed "solution holes" and "potholes" and are known locally as "banana holes" (or "pineapple holes" for the smaller ones). Banana holes are circular to oval chambers, 5 to 10 m (16–33 ft) in diameter, and 1 to 3 m (3.3–9.8 ft) deep. They are a common surface feature found throughout the archipelago and are formed when subterranean caves and chambers collapse, resulting in the creation of shallow pits. Following collapse, accumulation of soil, vegetative debris, and water in the hole provides an excellent location for growing crops, including bananas or pineapples from which they derive their local names. Rainwater ultimately permeates through the limestone and collects on top of the subterranean saltwater. Here, because of its lower density, the subterranean freshwater floats on the saltwater and forms a convex layer or a freshwater lens. Saltwater or saline wetlands or marshes occur when low elevation land meets the underlying water table.

As the limestone surface dissolves in rainwater, the runoff flows off or down various pores, cracks, and sinkholes and increases erosion of the rock. Left behind is a honeycombed surface of sharp, jagged ridges with occasional upright projections or "castles" of hard limestone. The honeycombed limestone surface continues to dissolve in rainwater, causing the remaining pieces of limestone to fall in on one another. Thus, the surface may become littered with loose rocks, boulders, and flat rocks or plates that may rest loosely on more solid rock below.

Rainwater runoff not only dissolves limestone, it also carries the dissolved calcium carbonate to other locations where it may be redeposited and form a hard crust due to (water) evaporation. This limestone crust, locally referred to as "flint rock"

Figure 4. Blue holes, such as this one on San Salvador, are formed from rainwater erosion of sinkholes that reach the water table. In the foreground and surrounding this blue hole is a dense growth of the Golden Leather Fern, which is found in wet areas including saltwater associated with mangroves.

because of its hardness, is deposited in layers coating the surfaces of rocks and edges of sinkholes. Crusts are most frequently formed beneath the soil where they can prevent erosion of the underlying softer limestone and crusts may also block plant roots from reaching water contained in softer limestone below.

Sinkholes, blue holes, and caves. Sinkholes grow in width as rainwater erodes the limestone surface between neighboring holes, eventually causing them to connect, enlarging the overall width of the sinkhole. This process of sinkhole enlargement can continue for decades or centuries. Although some channels may be formed below the water table, their rate of formation is slow because of slow carbonate dissolution rates under water, and large sinkholes can occur only in sites with very deep water tables. By contrast, the rocklands, which have shallow water tables, do not have large sinkholes.

The deepest sinkholes in the archipelago are known as blue holes. In the present day they are filled with water, but they were created during the four Pleistocene Ice Ages (2.5 million to 11,700 years BP) during each of which the sea level fell. In the last of these ice ages, the sea level was approximately 122 m (400 ft) lower than present day. As glaciers advanced and sea level fell, the islands' water tables were also correspondingly lower. Blue holes can be very deep; the deepest in the archipelago is Dean's Blue Hole on Long Island (depth 201 m [663 ft]). All the major islands of The Bahamas have blue holes (Figure 4). They are most prevalent on Andros, which has

Figure 5. The Conch Bar Caves of Middle Caicos Island are the largest caves in the Turks and Caicos Islands and are protected by the Turks and Caicos National Trust. These caves were mined for bat "guana" or guano for ten years in the 1880s, when bat guano was used as fertilizer and exported overseas. During the mining period, evidence of Lucayan habitation of the caves was uncovered, suggesting that aboriginal people used the caves as places of worship and as shelter from hurricanes.

178 on land and about 50 in the nearby ocean (on the currently submerged carbonate banks). Those found in the ocean are believed to have been formed by the same rainwater erosion process when the banks were exposed during the ice ages. Terrestrial blue holes are connected to the ocean and are tidal; when the tidal response is marked, they are locally known as ocean holes. Although most of the water moving through blue or ocean holes is saltwater, some blue or ocean holes penetrate through a freshwater lens and therefore have a freshwater layer on top.

Caves are characteristic of karst regions and are widely distributed throughout the archipelago. Flooded caves are found below the water table as exemplified by the caves associated with blue holes. Above the water table are the dry caves, which are accessible on some islands (Figure 5). Usually the dry caves consist of a vertical section formed by rainwater dissolution of the limestone as it flows downward and a horizontal section as the rainwater continues to dissolve the limestone as it flows laterally or along fissures, often toward the sea. Some caves may have multiple levels resulting from changes in the water table associated with sea level changes during glacial and interglacial periods. The freshwater lens, both at its surface where acidic rainwater accumulates and at its bottom layer where it meets the underlying saltwater, provides the chemical conditions for increased limestone dissolution resulting in cave formation. These dissolution conditions are especially potent where the top and bottom layers come together at the edge of a freshwater lens. Because the freshwater lenses often diminish at the coast, this is where many caves are formed, but not all (e.g., Salt Pond, Cartwright's, Hatchet Bay, and Conch Bar caves). Caves initiated where the top and bottom layers of the freshwater lens meet have been termed by some as "flank margin caves". Most archipelago caves were initiated in this manner, and flank margin caves are found along the coast and provide an indication of previous sea levels.

The Ephemeral Islands (Campbell 1978). Since the archipelago's islands are low lying, as a consequence of the slow processes involved in their creation (slow growth of corals, slow rate of sedimentation, and the fact that corals can never be higher than sea level at a given time), they are highly susceptible to variation in sea level. Changes in sea level have played an influential role in the creation of these islands and their landforms and subterranean traits. The principal cause of these sea level changes, as mentioned previously, was large-scale glaciation events. During past ice ages, sea levels fell as freshwater was taken up in ice sheets, which subsequently thawed in the interglacial periods releasing the water and causing sea levels to rise.

During the Pliocene (5 million to 2.5 million years BP), sea levels were, on average, 25 m (82 ft) higher than present-day levels. More recently in the Pleistocene (2.5 million to 11,700 years BP), sea levels varied by up to 140 m (459 ft). As recently as 125,000 years BP, sea levels were approximately 6 m (20 ft) higher than present day, and the vast majority of the archipelago was submerged. By contrast, during the height of the last significant ice age (the Wisconsin Ice Age; c. 16,000 years BP), sea levels were as much as 122 m (400 ft) lower than now. This fall in sea level exposed the carbonate banks thereby consolidating the present-day Bahamas into five major islands and several smaller islands with a resulting tenfold increase in land area from the present day by c. 13,900 km^2 (5366 mi^2 to approximately 124,716 km^2 [48,149 mi^2]). Similarly, in the TCI the Caicos group was consolidated into a single island c. 1500 km^2 (579 mi^2) while the Turks island group was consolidated into an island of c. 300 km^2 (116 mi^2). The three currently submerged carbonate banks southeast of the Turks Bank were formed as separate islands including the Mouchoir Bank, Silver Bank, and Navidad Bank. With the lower sea level of the Wisconsin Ice Age, the highest points on the archipelago were more than 200 m (656 ft) higher above sea level than present-day levels. At the end of the Wisconsin Ice Age, 12,000 to 13,000 years BP, sea levels rose and much of the low-lying land was submerged, re-fragmenting the large islands into smaller ones. The archipelago achieved its present size around 6000 years ago, and the shallow seas (3–30 m [10–98 ft]) around the modern archipelago are the submerged remains of these larger islands.

Lithified Sand Dunes. During the Ice Ages, the sea level fell and exposed the carbonate banks at least four times. During each interglacial period, as the sea rose and covered the banks, fresh oolitic sands were produced, which on subsequent exposure to the wind created new sand dunes as the sea level fell during each Ice Age. Thus,

Figure 6. Cross-bedding in the exposed ridge, as shown here in southern Eleuthera, is typical of oolitic sands deposited by wind and subsequently cemented into solid rock (lithification).

the supply of oolitic sands was replenished during each interglacial period and was then exposed to the winds, producing a new row of sand dunes on the islands. With time, the oolitic sand dunes became solidified (lithified) and formed the ridges of the modern archipelago's islands. Layers of lithified soils, some running in different directions (termed "cross bedding") and with fossil origins, make up the oolitic ridges, some of which may be partially eroded (Figure 6). Although the ridges on most islands tend to run north–south as a result of the northeastern trade winds, some islands, such as New Providence, have ridges running east–west as a result of the north and northwesterly winds associated with cold fronts from North America. On islands on small banks, such as San Salvador, all of the surface and ridges have been built up from drifting sands. Southerly winds played an important role in ridge formation in some of the southeastern islands of the archipelago.

Soils. Soils include a layered mix of organic and inorganic materials resulting from the combined interaction of parent material or rock with climate and vegetation. Given that the archipelago's soils are primarily derived from the weathering of limestone, the soils are alkaline. These soils are young, reflecting the geologically young age of the limestone parent material. This time frame is evident in soil texture, which is dominated by stones and sand, whereas smaller-sized particles such as clay are usually rare. The lack of clay in most soil types limits soil water-holding capacity, which is further exacerbated by the rarity of humus in some of the archipelago's soils. Because of their immaturity, the soils are thin with one or two discontinuous layers or horizons above the bedrock in contrast to mature soils, which are deeper and have several distinct layers present, as found in the Greater Antilles and on the continents. Soil is absent or patchy in distribution on some islands, especially on the southeastern islands and on small cays, where plant cover is sparse due to inadequate rainfall or wind exposure. Three major soil types are generally recognized and include:

Organic soils. Often called "black soils" because of their color, these soils are characteristic of rocklands (sometimes termed "blacklands") in the forested northern and central islands of The Bahamas. The soil contains a mix of decomposing vegetable matter, humus, and scattered pieces of varying-sized limestones. In texture, black soils contain particles that range in size from sand to gravel. Although black soils are generally shallow (< 15 cm [< 6 in]), they can be relatively deep (~ 30 cm [1 ft]) in some depressions where farmers may place their crops, as in "pothole farming." Organic soils are the most common soil type in the archipelago and are highly susceptible to erosion when the vegetation is removed.

Red clay soils. Red clay or pineapple soils are less common than organic soils and are usually restricted to the lower slopes of ridges or in hollows (vales) between ridges. Red clays are produced in the Sahara Desert and are blown across the Atlantic as dust. Although Sahara dust continues to periodically fall out on the archipelago, most of the red clay soils were formed primarily from dust deposited during the last glacial period, when the Sahara was exceptionally dry. As wind deposited soils, the red clays likely were deposited on ridgelands and vales alike, but later were washed from the ridges into the vales. These soils are rich in insoluble iron and aluminum oxides (termed "lateritic") but are nutrient-poor and do not readily release water to plants. When dry, the red clay soils become rock hard. Despite these deficiencies, red clay soils are used for cultivation, especially for pineapples, when mixed with sand, sedimentary soils, and fertilizers.

Sedimentary soils. This group of soils is mostly associated with sand dunes that consist of a mixture of sand and varying amounts of organic material or humus. The amount of humus is dependent on the amount of vegetation growing on the soil. In its simplest form, sedimentary soils are found on newly formed sand dunes with little vegetation cover. Here the soils are based on geologically young (Holocene) marine calcareous sands with an upper layer of gray or grayish- to reddish-brown coloration, which is usually less than 30 cm (12 in) above the pure mineral sand below. Referred to locally as "whiteland," it is a coastal soil most commonly encountered on the windward side of islands between the beach and the lithified or oolitic dunes. Here the sand dunes may vary from almost flat to gently rolling to short and steep.

Climate and Natural Disturbance Regime

Climate: "The Isles of perpetual June" (Campbell 1978)

The climate of the archipelago reflects its location in the border of the tropical and subtropical zones where frosts are unknown. The tropics are defined as the region of the earth surrounding the equator, north to the Tropic of Cancer (23°26'N) and south to the Tropic of Capricorn (23°26'S). The subtropics are those regions immediately adjacent to the tropics, usually between approximately 23.5° and 35° latitude in both the southern and northern hemispheres, respectively. The archipelago lies between latitudes 20° and 28°N and is bisected by the Tropic of Cancer, which runs through the southern Exumas and Long Island in the central Bahamas. Thus, the archipelago's position within the subtropical and tropical belts, combined with its location in the western Atlantic near the North American continent, dictates much of its climate and seasonal variation. Overall, the archipelago has a maritime climate that is largely dominated by the northeast trade winds, which in the winter are periodically disrupted by cold fronts from North America, and in late summer and fall temporarily disrupted by tropical storms from the south.

The weather system of the archipelago is strongly influenced by the Bermuda-Azores High, a large semi-permanent area of high atmospheric pressure located in the north Atlantic that influences weather patterns along the East Coast of North America. This high pressure system moves westward during the summer and fall, when it is known as the Bermuda High and is associated with warm humid weather. During the winter and early spring, it is primarily centered near the Azores (an archipelago in eastern Atlantic, c. 1500 km [930 mi], west of Lisbon, Portugal), when it is termed the Azores High. This condition brings warmer weather to Europe and allows cooler air from the Arctic to move southward along the eastern seaboard of North America. As a result, the archipelago has a subtropical climate with two distinct seasons: a hot, wet summer from May to October and a warm but drier winter season from November to April. Winds are predominantly easterly throughout the year but tend to become northeasterly from October to April and southeasterly from May to September. During the summer, the climate is dominated by moisture-rich winds from the east. Midsummer temperatures range from 23.3 to 31.7 °C (74–89 °F) with a relative humidity of 60 to 100%. In winter, the climate is affected by the movement of cold polar air masses from North America, the effects of which are moderated by the warming effects of the Gulf Stream, a warm water current flowing north from the tip of Florida between the U.S. mainland and the archipelago; winter temperatures range from 16.7 to 25 °C (62–77 °F). Although frost has never been recorded in the archipelago, light snow flurries have occurred on Grand Bahama. Whereas the southern islands are slightly hotter than the northern ones in the summer, they are markedly warmer than the northern islands in the winter.

Rainfall in the archipelago is seasonal and heaviest in the summer months (May–October) as a consequence of a northward shift of the northeasterly trade wind belt in contrast to the winter dry season when the trade wind belt shifts southward. When rain does occur in the winter, it is usually associated with cold fronts moving south from the nearby continental United States, whereas most summer rainfall results from easterly waves from the Atlantic Ocean causing convection rain (from local thunderstorms) or tropical storms and hurricanes. These seasonal differences in rainfall can be quite marked: in Nassau, winter rainfall averages 50.8 mm (2 in) per month, but 203 mm (6 in) per month in summer. In addition to the seasonality of rainfall, a distinct north–south gradient in rainfall occurs, in which the northern islands are much wetter than the southern islands. Annual rainfall in the northern islands averages 1549 mm (61 in), whereas annual rainfall in the TCI is 605 mm (24 in). The southern islands have not only lower rainfall but also higher average temperatures than do the northern islands, resulting in higher evaporation rates; therefore, the southern half of the archipelago is semi-arid and water deficient. The latitudinal gradients in rainfall and evaporation are reflected in the vegetation; the northern wetter islands have pine forest and/or broadleaf vegetation (locally known as "coppice"), whereas the drier southern islands are characterized by shorter broad-leaf vegetation and scrub, although scattered Caribbean Pine is also known from the typically scrubby TCI.

Hurricanes and Fire: Disturbance Regimes in the Archipelago

Two primary natural sources of disturbance occur in the archipelago: hurricanes and fire, which in conjunction with rainfall and soil or edaphic traits are the primary factors determining vegetation structure and composition.

Hurricanes are a frequent disturbance. Tropical cyclones, such as hurricanes, are an annual occurrence in the Caribbean between June and November, during which time they typically originate over the warm waters off the west coast of Africa, the Caribbean Sea, and the Gulf of Mexico before moving through the region. Cyclones start when rising warm air creates low-pressure systems associated with rain showers and shifting winds. As the systems pass over warm water, the winds may become more organized, initially forming a wave but eventually, under the right conditions, the winds become organized into a circular motion around the low-pressure area and produce bands of showers that radiate out like spiral arms from the low-pressure area. As the winds spiral in a counterclockwise direction around the lower pressure area, they may pick up speed. Once the winds reach speeds of 64 kmh (40 mph), the cyclone is officially designated as a tropical storm until winds reach 118 kmh (73 mph) or higher, when the cyclone is classified as a hurricane. Although some defoliation can occur in hurricanes with winds of 119 to 152 kmh (74–95 mph), severe structural damage to large trees and buildings are associated with winds of 210 kmh (131 mph) or higher.

The Bahamas is one of the most hurricane-prone countries in the Caribbean, receiving an average of one hurricane every 1.4 years (e.g., 66 hurricanes were recorded in the archipelago from 1899 to 1989). The northern islands have more hurri-

Figure 7. Shown here is a storm beach or berm composed of a mix of beach rock slabs and coral pushed up on the right side of the beach by the storm surge from Hurricane Joaquin (2015) in southeastern San Salvador. Beach rock is a sedimentary rock that forms along a coastline and consists of a mix of gravel-, sand-, and silt-sized sediments cemented with carbonate minerals. Here the hurricane swept accumulated nearshore material inland where it was deposited and will likely remain permanently, unless disturbed by humans.

canes than do the southern islands; larger islands are more frequently hit than small islands; and islands that lie in a north–south orientation are more frequently struck than are islands in a west–east orientation. Given their low-lying stature, islands of the archipelago are especially vulnerable to the high winds and associated coastal flooding (storm surges) of cyclonic storms. Some storm surges have been observed to push seawater as far as 1.7 km (1 mi) inland, and hurricanes with winds of 178 to 209 kmh (111–130 mph) can produce a surge of 2.5 to 3.7 m (8–12 ft) above sea level (asl) in the archipelago (Figure 7).

Tropical cyclones with winds of tropical storm or hurricane force occur often enough in the archipelago to influence the structure and composition of ecological communities. These storms affect the biota both directly and indirectly, with the indirect effects having the most substantial and longest lasting impact. Direct effects of hurricanes include mortality caused by high winds and exposure to heavy rains, particularly for organisms that have limited shelter options. For example, aquatic birds, such as pelicans, herons, egrets, flamingos, gulls, terns, and some shorebirds are more prone to mortality from the direct effects of hurricane winds and rain than are land birds or bats. Storm surges have been found to cause immediate significant declines in populations of non-volant (non-flying) terrestrial animals, such as lizards and insects, as well as loss of vegetation cover and the degradation and loss of soil and leaf litter. Species in the archipelago are adapted to these sporadic disturbances, however, and local populations are usually able to recover surprisingly quickly, often aided by immigration of individuals from less damaged sites or islands. Even on small, low-lying islands, in which storm surges have washed over the islands, some populations of lizards have been found to recover after 3 to 10 years depending on their ability to re-colonize by way of immigration or from eggs (e.g., anoles; see index for scientific names) that have survived the saltwater inundation.

Indirect effects of hurricanes on wildlife populations are evident almost immediately after a storm's passage and include loss of food supplies or foraging substrates, as well as loss of cover for hiding, nesting, or roosting. Loss of vegetation cover can change the microclimate required by certain organisms, such as frogs and lizards, resulting in their population declines. Hurricane winds that strip leaves from vegetation also remove flowers, fruits, and seeds, and therefore it is not surprising that nectarivore, frugivore, and seedeater populations decline after storms. Parrots, because of their fruit and seed diet, are especially sensitive to hurricanes, and post-storm mortality has been attributed to loss of their food resources in The Bahamas and elsewhere. Not all post-hurricane population declines are attributable to mortality, however, as many animals shift from sites where resources have been lost to locations where resources still remain abundant. White-crowned Pigeons, typically strong fliers, may leave islands in the aftermath of hurricanes and move to other islands where seeds and fruit remain. As animals wander through defoliated areas in search of food, they may become vulnerable to predation, especially if food-stressed, and not surprisingly, some predators benefit from increased prey availability in a storm's aftermath.

Plants differ in their vulnerability to storm damage—some species are sensitive to saltwater flooding as a result of storm surges or from the wind-borne salt spray that can be carried well inland and can result in leaf loss and mortality. Susceptibility to wind damage can vary with plant age, with large old trees more susceptible to branch breaks, trunk snaps, and blow downs than are small younger trees and shrubs; hence, tall coppice is more vulnerable to structural damage from storms than is short coppice. Caribbean Pines are sensitive to saltwater flooding (Figure 8) and older pine trees are susceptible to trunk snaps and uprooting (wind throw), which contributes to mortality. Nonetheless, storm-produced snags and branch breaks eventually benefit cavity nesting and roosting animals. Similarly, as vegetation recovers, insectivorous animals take advantage of outbreaks of herbivorous insects, such as moth and butterfly larvae, associated with production of new leaves. Increased light levels in forested areas stimulate understory growth and result in increased flowers, fruits, and seeds, allowing nectarivore and fruit- and seed-eater populations to recover following initial declines after storm passage. Thus, despite the short-term detrimental effects of hurricanes on some small populations, most

Figure 8. As shown here on Grand Bahama, numerous Caribbean Pines died as a result of the storm surge from Hurricane Wilma (2005), which drove saltwater inland and inundated some of the island's pine forests. Pines are sensitive to saltwater, which could threaten pine forests as sea level continues to rise in the archipelago.

plant and animal populations with adequate population sizes survive, and some even benefit from habitat changes caused by these storms.

    Fires. Fire is another important natural disturbance factor in the archipelago. The frequency and intensity of fires have been modified by humans, however, as first became evident with the arrival of the Lucayans (Amerindians who first colonized the archipelago) when fire frequency increased. Recent evidence suggests fires were extensive at the time of Loyalist (Americans loyal to the Crown after the American War of Independence) arrivals and the establishment of slave-based agriculture in the 18th century. This may have resulted in sedimentation of lagoons and ultimately created some of the archipelago's present-day coastal ponds and lakes. Fire naturally occurs in the archipelago in Caribbean Pine and hardwood (coppice) habitats due to lightning strikes and also anthropogenically when fire escapes from slash-and-burn agriculture, from burning rubbish, or when fire is set to clear the understory to improve hunting opportunities (especially for feral pigs and land crabs). The ecological and conservation effects of fire may be beneficial or detrimental based on the seasonality, intensity, duration, spatial coverage, and weather events associated with any particular fire.

Figure 9. Fire is common in pine forests of the archipelago, as shown here after a recent mild fire on Abaco. Although the Bracken Fern fronds were burned, the fern resprouts vigorously, and the larger palmettos and pines also survive ground fires. Fire is of natural occurrence in pine forests of the archipelago, but most fires now are set by humans clearing brush or opening the understory to facilitate hunting.

The fire-adapted Caribbean Pine forests of the archipelago, which cover 23% of the landmass, are best developed on Andros, Grand Bahama, Abaco, and New Providence, but with some outliers in the TCI. Other imbedded or adjacent fire-dependent systems (i.e., fire is required to maintain the communities) include freshwater marshes, wet prairies, palm savannas, and salt marshes, the latter of which would develop into mangroves in the absence of fire. These systems burn naturally in the dry season at intervals of approximately 1 to 7 years. Fires following hurricanes, however, are more intensive and extensive when fallen trees, caused by blow-downs or tree mortality from saltwater intrusion, provide more fuel for fires. Lightning-initiated fires occur most commonly at the end of the November–April dry season when the vegetation is driest and just before the transition to the May–October wet season. Once burned, the ash from the accumulated debris (e.g., needles, dried grasses, twigs, and branches) release nutrients. Seeds falling on the newly burned areas flourish, which are also free of shading, and thus germinate and continue the cycle of the plant communities, especially where gaps in the canopy favor growth of the young pine. Pines are especially well adapted to fire, as their buds are protected by dense clusters of needles, and older trees have thick bark, to protect them from fire (Figure 9). Without fires, the islands' pine forests would be replaced by coppice as pine seedlings are light-demanding and cannot compete with broadleaf species in the shade.

The archipelago's broadleaf or coppice forests, including low-stature forests or shrublands in the dry southern islands, cover most of the archipelago and, historically, have probably burned less than Caribbean pine forests. The many hardwood species that make up these forests have few adaptations to fire, such as protected buds and thick bark, so when fire occurs, many individuals of these plant species may die, or aboveground parts may be burned back to the surface. Some plants resprout after fires because of underground reserves in the root system, but the species composition of the burned area often changes to an early succession community with scattered resprouting of trees and shrubs characteristic of more mature mid- to late coppice habitat. Fires in coppice, as with pine forests, may be more frequent, intensive, and extensive following particularly dry periods, especially post-hurricane, when downed branches and trunks and increased leaf litter can readily carry fire. At these times, fire burns deep into the leaf litter, creating a slow-moving fire in which the heat is transferred to plant roots, stems, and trunks over a long time, resulting in more damage and mortality in plants. Consequently, coppice is more susceptible to fire and does not exhibit the regenerative effects as observed in the pine forests. Today most fires in coppice are set by humans and are more common than in the past, but coppice fires are still relatively infrequent compared to the occurrence of fires in pine forests. Under moist conditions, most pineyard fires stop on reaching coppice stands.

Habitat Types

Comparatively few major habitats are found in the archipelago, in part because the limestone substrate and associated young soils vary little among islands; climate differs little among islands (except for the NW–SE rainfall–temperature gradient); variation in elevation is limited among islands (e.g., some estimates suggest that 80% of land area is less than 1.5 m [4.9 ft] above MSL; maximum elevation is 63 m [206 ft]); and surface freshwater is scarce given the absence of streams or rivers. In addition, many different habitats share the same species because island organisms tend to be generalists and occupy a diverse array of habitats in contrast to continents where habitat specialization is common resulting in habitats with unique assemblages of species. Here we describe major terrestrial and wetland habitats and plant communities, but we caution readers that many habitat types blend into each other, as habitat edges or ecotones, and edges of roads contain a mix of species shared among habitats.

Coastal Terrestrial and Wetland Habitats
Coastal habitats include rocky shores and cliffs, beaches and dunes, coastal coppice, and wetlands, such as mangroves and mudflats. These resilient habitats are often lashed by high winds and associated wave action, storm surges, and salt spray. Species in coastal areas are tolerant or even dependent on tides, salt spray, and periodic and frequent wind damage to persist.

Rocky shores. The native plants in these communities tend to hug the ground, especially those closest to the shoreline, but become taller, even if wind-pruned or stunted, as distance from the shoreline increases. Especially close to the shoreline, bonsai-shaped plants root in small pockets in the limestone where water may collect and a bit of soil may form. Little or no leaf litter is seen on the exposed bedrock and the plants here, of necessity, are salt tolerant. Typical plants include Sea Ox-eye, Mosquito Bush, Wild Thyme, Bay Lavender, Buttonwood, and Seven-year Apple.

Coastal beaches and dunes. Sand dunes, most common along the eastern shores of the islands, are zones of stress for plants and animals as a result of the high winds that concentrate salt; the salt saturates the sands through which rainwater readily percolates. The winds sweep the sands off the beach and pile it on to the dunes, which are anchored in place by various plants, thereby preventing the inexorable migration of the sands inland. These sand dunes have been formed relatively recently, at least in the Holocene (past 11,700 years), whereas inland, just beyond these sand dunes are found the older Pleistocene dunes now consolidated into rock. Dune vegetation can be dominated either by herbs and vines or by low shrubs, such as Sea Grapes, Bay Cedar, and Coco Plum. Typical of the herb- and vine-dominated sand dunes are Sea Oats, which are among the most important plants for stabilizing the dunes along with others restricted to only these sites, including Spider Lily, Sandbur, Bay Gerana, and Rail Road Vine. Also colonizing the dunes are invasive nonnative plants, such as White Ink Berry, Coconut Palms, Jumbay, and Australian Pine (Figure 10), which contribute to the demise of the dune and its community (described in section of invasive nonnative species). Curly-tailed Lizards are most abundant on dunes and sand cliffs and Whiptail Lizards occur on dunes as well as Rock Iguanas on isolated cays. The ground-nesting Antillean Nighthawk is one of the few birds that nests on dunes and beaches, although Wilson's and Snowy Plovers nest on beaches, which are also visited by migrant shorebirds. Nocturnal Ghost Crabs are common on the beaches and the occasional Caribbean Hermit Crab can be encountered there. Diurnal-feeding sandflies are noteworthy pests, which breed in wet sands of beaches, whereas the nocturnal-feeding sandflies breed in muds of inland salt ponds (or salinas), mangroves, and ponds.

Mangroves, mud flats, and salinas. Where nearshore waters have a gradual slope and protection from high-intensity waves, oolitic sands precipitate from the warm ocean waters and fine-grained muds accumulate to form mud or sand flats, usually found on the western side of islands. These shallow warm water flats are colonized by a group of taxonomically unrelated trees known as mangroves because of their tolerance of saline or brackish water. The most salt-tolerant species and first to colonize the shallow water flats are the stilt-rooted Red Mangroves, which trap sediments and provide shelter and nurseries for seemingly myriads of fish, larvae, and other marine life. Immediately inland from the Red Mangroves are the Black Mangroves with their upright aerial roots (pneumatophores), specialized for gaseous exchange, which also trap sediments. If this process continues without a major disturbance, the

Figure 10. Sea Oats and other native plant species can stabilize a dune by holding the sand particles in place, as is evident on this dune on San Salvador. Here the fore dune has been eroded by recent wave action, but beyond the eroded section are Sea Oats, which anchor the dune in place. A row of Australian Pines or Casuarinas lines the back part of the dune. As they colonize the dune, native plants will be lost, and winds will blow the sands through the Casuarinas. Casuarinas do not bind the sand in place as well as the native species, however, so eventually the dune will disappear and the Casuarinas will fall over. They will subsequently re-colonize the site again once the dune is gone.

nearshore flats are slowly converted to a mangrove flat that is eventually colonized by White Mangroves and Buttonwood on the semi-dry land.

Coastal ponds, marshes, and wetlands. Small pools with limited open water to large ponds of fresh or saline water are often found behind or between dunes. Other sites, such as Sabal Palm flats or savannas are temporarily flooded, and such areas may also contain White-head Rush and Saw Grass. Sites with more permanent freshwater may also support cattails. Saline wetlands may support halophytic (salt-tolerant) plants, such as Saltwort, Glasswort, and Sea Purslane, and may be lined by scattered White Mangroves and Buttonwoods. In sites with deeper water, Red Mangroves fringe inland saline wetlands. Often abundant in small bodies of water of various salinities are Bahamian Mosquito Fish. Mosquito fish can tolerate a wide range of salinities and survive in temperatures that range from 4 to 35 °C (low 40s–95 °F), which facilitates their survival in ponds, creeks, salt ponds, ocean holes, and even cisterns. As their name suggests, mosquito fish readily consume mosquito larvae and thus public health officials have spread mosquito fish throughout the archipelago. Also found in water bodies with a wide range of salinities are pupfish of several species, which are capable of self-fertilization, thereby facilitating colonization of water bodies lacking the opposite sex. Together with mosquito fish, pupfish serve as a food source for a variety of wetland birds, including herons and egrets. Resident Least Grebes and White-cheeked Pintails occur in some wetlands, which may also be occupied by winter resident Blue-winged Teal, Lesser Scaup, and American Coots.

Terrestrial Habitats

Pine forests, pineyards, or pine barrens. Pine forests, many of which cover extensive tracts of land, are restricted to the northern islands of Grand Bahama, Abaco, Andros, and New Providence, and some of the TCI. Pines are capable of colonizing nutrient-poor and droughty soils thanks to their obligate fungal partners (symbionts). The mutually beneficial fungus–root associations known as mycorrhizae facilitate nutrient uptake by plants such as pines. Although pines may dominate the canopy and reach heights of 15 m (50 ft) in the northern islands, they are relatively short and scattered in the TCI. Commercial harvest of pines began on Abaco in 1905 and subsequently spread to the other northern islands; by the 1970s most of the virgin or old growth pines had been removed. Today, old growth pine can be found only in a small patch on Abaco (Figure 11) and western Andros, isolated in wetlands on hummocks above the limit of storm floods. Large areas of nearly pure, even-aged pines are common in the northern islands, a legacy of past clear-cut harvests, and storm blow-downs. As fire frequency increases, pine forest area expands whereas the less fire-tolerant coppice retreats. By contrast, saltwater intrusion from storm surges kills pine, which is replaced by the more salt-tolerant coppice vegetation.

Pineyards situated above shallow water tables or frequently flooded with freshwater often have an understory of palms (e.g., Sabal and Thatch Palm), whereas on slightly higher ground and ridges, the Silver Top Palm locally dominates the understory. The broadleaf understory varies in height with fire frequency and ranges in stature from ankle height shortly after a fire to head high or greater when fire is infrequent. Poison Wood can be especially abundant in some pine understories, although other plant species also occur including Blolly, Pain-in-Back, Ram's Horn, White Sage, Wild Guava, Golden Dew Drop, and various sedges and grasses. Golden Creeper is often common in the understory and its flowers are especially attractive to hummingbirds such as the Cuban Emerald. On some sites, the pine understory may be covered in a thick impenetrable cover of Bracken Fern, which inhibits colonization by other plant species. Although the reason for the Bracken Fern dominance in some pine understories is poorly understood, it has been attributed to a combination of fires during the dormant season and soil conditions. In open understory, the orchid known as Pineyard Pink can be readily identified by its bright pink flowers when blooming in early winter. Birds such as the Pine Warbler are restricted to the pine forest and are found on all the northern islands, whereas others such as the Bahama and Olive-capped Warblers occur only in pines on Grand

Figure 11. Mature Caribbean Pine stands that have not been logged in the past fifty years, such as this stand of old pines on Great Abaco, are rare in the archipelago. Although this stand has been burned repeatedly over the years, as evidenced by the fire scar at the base of the center pine, the shrubby understory suggests it has not been burned recently. Typical plants of the understory here include Poison Wood, Bracken Fern, Wild Guava, Wild Coffee, Five Finger, and Thatch Palm.

Bahama and Abaco, and the rare Brown-headed Nuthatch is found only in pines on Grand Bahama.

Dry broadleaf evergreen formations or coppice. Most broadleaf trees and shrubs in this habitat maintain their foliage cover throughout the year, and hence the evergreen designation. The prevention of water loss from the plants during the dry season requires the leaves to have various drought adaptations, such as leathery leaves (e.g., leaves of Sea Grape). Some plants, however, such as Gammalamme (Gum Elemi) or Gumbo Limbo, readily lose their leaves to reduce water loss during droughts. The hardwood trees and/or shrubs are often found in dense stands, locally known as coppice, and are similar to hammocks of south Florida. The term "coppice" is derived from the forestry term used for trees that readily re-sprout from the base when the main trunk is broken or cut. This re-sprouting is a common response by many hardwood species to disturbances such as hurricanes. Coppice vegetation can vary in stature from high or tall coppice, usually occurring on the more nutrient-rich organic upland soils, to the low coppice or scrublands that occur on the more nutrient-poor coastal sand soils. A distinct north–south gradient based on rainfall is common, with taller, larger canopy forest in the northern wetter islands and dwarf shrublands in the drier southern islands.

Figure 12. Profile of Inland Limestone Coppice or rockland exposed when a bulldozer cut through the coppice. Note that the bulldozer not only removed the coppice vegetation but also scraped off most of the leaf litter and organic soil at this site.

Inland Limestone Coppice. Inland Limestone Coppice is characterized by dark (black or red), loamy soils, which are relatively nutrient-rich and have a high water-holding capacity, and hence, can support tall coppice and the highest biodiversity of any terrestrial habitat in the archipelago. No tree species currently dominate the present-day communities, but species historically once common in the canopy of coppice include Mahogany, Horseflesh, Mastic, and Cedar; all were extensively harvested and therefore, when abundant, are usually of small stature. Gammalamme is often common in the canopy where it occurs with other species such as Pigeon Plum, Poison Wood, Darling Plum, Short-leaf Fig, Willow Bustic, Blolly, Wild Tamarind, Lignum Vitae, Five Finger, Velvetseed, and Butterbough (Figure 12). In undisturbed tall coppice with closed canopy, the absence of direct sunlight restricts understory growth and consequently the understory can be quite open. Storm or human disturbance can open up the canopy facilitating growth of understory trees and shrubs such as Wild Coffee, Snow Berry, and various stopper species (*Eugenia* spp.). At the end of the dry season, associated with the early rains, flowering commences and attracts insect pollinators including bees and wasps and a suite of butterflies, such as Florida Purplewing, Cuban Daggerwing, and Leaf Wing Butterfly. Bahama Woodstar, an endemic hummingbird, occurs in coppice, and is most readily found where plants with red or orange flowers are present. Typical of undisturbed coppice are epiphytes, such as various orchids and bromeliads, which may be abundant on tree trunks and branches and on rocks. Trunks and stems here may harbor lizards, such as anoles, which remain alert for predators including snakes, such as boas and Brown Racers as well as avian predators including Great Lizard-cuckoos, Red-legged Thrushes, Bahama Mockingbirds, and La Sagra's Flycatchers, which are common in

Figure 13. Coastal Sand Coppice, such as this site on Middle Caicos island, has plants such as the Silver Top Palm, Sea Grape, Black Torch, Joe Wood, Bay Lavender, and others.

coppice on some islands. Key West Quail-doves forage for seeds and fruit on the ground under closed canopies and are more often heard than seen, as are the Greater Antillean Bullfinches. In the winter, migrant songbirds (from North America), particularly warblers of various species and Gray Catbirds, may be common in coppice.

Coastal Sand Coppice. These communities, more typical of coastal areas and the southern islands, are named for the white limestone or sand substrates of various origins, which are low in nutrients and water-holding capacity and have a sparse and patchy leaf litter layer (Figure 13). Often found behind beaches, dunes, and headlands, the Coastal Sand Coppice may be thick and scrubby, less than 2 m (6 ft) tall, but becomes taller as distance from the shoreline increases. Often designated as coastal coppice when near the shore, this and other Coastal Sand Coppice may have Sea Grape, Sabal Palm, Seven-year Apple, Horse Bush, Cinnecord, Granny Bush, Black Torch, White Sage, Darling Plum, and Strong Back. In many locations this community is dominated by the Silver Top Palm. Although sand coppice may be dense, small openings exposing the sandy substrate are common. As in dune communities, lizards such as curly-tails and whiptails also occur here. Land birds, such as Bananaquits, Greater Antillean Bullfinches, and Bahama Woodstars, are especially common in Coastal Sand Coppice communities where flowers and fruits are plentiful. Agaves can be common here, and when flowering, the nectar attracts a variety of resident and migrant bird species as well as Carpenter Bees and the nectarivorous Tiger Moths (often mistaken for wasps).

Figure 14. The endemic New Providence Cusk-eel or Bahama Cavefish (*Lucifuga spelaeotes*) illustrates an instance of "use or lose" evolution in which a dweller of dark, flooded caves has, over time, lost its eyes after many years of inhabiting a habitat that has insufficient light for vision. These fish evolved from ancestors that originally lived in habitats with light and had eyes.

Subsurface Habitats

Caves. In the absence of light and plants to provide energy, cave ecosystems are dependent on bats to provide the necessary nutrients to support the cave food web. By foraging outside the cave and returning to roost, bats release feces and urine on the cave floor, which along with discarded fruit seeds as well as dead bats provide nutrient-rich organic material or guano for a plethora of decomposers. These decomposers of the bat guano—fungi and bacteria—break down the organic material into simple nutrients. In turn, these nutrients and some of the decomposers are consumed by millipedes and small crustaceans. Bigger arthropods, such as beetles, feed on these organisms as well as the eggs of crickets. Centipedes, spiders, and whip scorpions feed higher up on the cave food web by preying on the smaller arthropods. Nutrient-rich bat guano was commonly harvested by the early settlers for fertilizer (see Figure 5). Care should be used when exploring poorly vented caves with dried guano deposits, because disturbance of the guano can release spores of the fungus *Histoplasma capsulatum*, which causes the disease histoplasmosis. The severity of histoplasmosis depends on the number of spores inhaled, and fortunately most infections are mild and go unnoticed. Flu-like symptoms, however, occur in about 10% of the cases, and severe cases can be fatal. Thus, use of a respirator with a filter capable of eliminating spore particles is recommended for extended stays in guano-filled caves.

Because of their geology and water chemistry, the water-filled caves and blue holes of the islands provide specialized habitats for a diverse array of bizarre organisms, many of them endemic to the archipelago and some endemic to a single island. As with many obligate cave-dwelling organisms, some have lost their eyes or at least their vision, and others have also lost their surface pigmentation. Also typical of this "use or lose" evolution are blind cave fish (Figure 14) that originated from ancestors that had dwelt in sunlit environments and had eyes. Many cave-adapted aquatic species are shrimp or closely related crustaceans, and some resemble those preserved in the fossil record, and hence have been designated as "living fossils."

Biogeography

The Banks and Biogeographic Regions

The terrestrial biota of the archipelago is probably younger than 65 million years, perhaps a consequence of widespread extinction resulting from a large meteor striking the Yucatan Peninsula of Mexico at the end of the Cretaceous (66 million years BP). This event had both regional and global catastrophic effects on the flora and fauna.

The most recent colonization of the archipelago was possible once the islands re-emerged in the Pleistocene ice ages (2.5 million to 11,700 years BP) after their Pliocene submergence. Colonization was facilitated by lowered sea levels during the ice ages, which exposed the carbonate banks, consolidating present-day islands into larger "mega-islands." These mega-islands were closer to adjacent land areas (e.g., continent, other islands) than they are now, providing larger targets and more habitat for colonists departing from source areas. Included among the separate mega-islands were the Little Bahama Bank (Grand Bahama, Great Abaco), the Great Bahama Bank (the Biminis, Andros, New Providence, Eleuthera, the Exumas, Long and Cat islands), and several smaller consolidated islands in the southeast including Crooked and Acklins, the TCI, and Great Inagua and Mayaguana (see map of the islands and the banks on the inside cover). In the southeast, the conjoined islands provided larger "stepping stones" into the archipelago for colonists from Hispaniola. Because of these differences in proximity to various colonization sources, the archipelago's modern flora and fauna has a strong West Indian influence (via Cuba and Hispaniola) in the southern Bahamas. A North American influence is evident for birds in the western and northern islands. Not all colonists originated from these sources as some originated earlier in South America and may have come north by island hopping through the Lesser Antilles into the Greater Antilles, or they may have passed through Central America and the Yucatan Peninsula to the Greater Antilles, or along the Gulf Coast of North America to Florida.

Natural colonization of the archipelago has occurred only by water or over-water dispersal, as there have been no land bridges connecting the archipelago with either North America or the Greater Antilles. Dispersal by water or over-water both act as filters because organisms differ greatly in their dispersal abilities using these modes, which is evident in the composition of the biota of oceanic islands. For instance, winged or volant animals (insects, birds, and bats) are good over-water colonizers and, consequently, are better represented in island faunas than non-volant species (amphibians, reptiles, and mammals other than bats). Non-volant species are reliant on rafts of vegetation, making them highly dependent on prevailing winds and ocean currents. For plants, passive dispersal occurs by wind and water, although the former mode is limited to small seeds and the latter is limited by saltwater tolerance. Seeds and fungal spores can also be carried by animals in their guts or unintentionally attached to their skin, feathers, or fur and thus plants with animal-dispersed seeds are well represented in island floras.

Colonizations of the archipelago were followed by extinctions in a cyclic pattern as Pleistocene ice sheets expanded (i.e., sea levels fell) and retreated (i.e., sea levels rose) during or between the various ice ages. Falling sea levels exposed more land area and reduced oceanic distances between islands. This cycle facilitated dispersal and increased species colonizations. As sea levels rose again, land surface area decreased, causing some species to become extinct. Some of the species losses associated with diminished island area are attributable to the basic biogeographic principle that large islands host more species than small islands, all else being equal (elevation, distance to nearby sources, latitude, longitude, climate, and so forth). All else was not equal, however, because the Pleistocene climate was generally cooler (by approximately 4°) and drier than present day; consequently, prairie grassland habitats were extensive. Thus, at the end of the last ice age, postglacial warming of the climate combined with the subsequent reduction in island land area, and the expansion of more mesic (wetter) environments led to the extinction and range fragmentation of many species in the archipelago.

The geographical isolation of archipelagos provides conditions in which species may, over time, diverge evolutionarily from each other. The divergence may begin with the small group of colonists that were blown or rafted to an island and that initially departed with only a small subset of the genetic variation from their ancestral or source population (founder effect). Over time, the isolated colonists diverge further from their source population because of selective pressures, mutations, and/or genetic drift, at least when the founding population is small. Given adequate time and geographic isolation on their "own" island, the colonists or founders may genetically diverge sufficiently that they no longer breed with their ancestral population, even if they subsequently co-occur, and hence, they are recognized as separate species. The Bahamas Archipelago has a suite of endemic species, although the proportion of endemics is lower than what is found in more isolated archipelagos primarily owing to its closer proximity to potential source populations. The archipelago is situated only 85 km (53 mi) from a continental source (Florida) in the northwest, and 80 km (50 mi) from two large island sources (Cuba and Hispaniola) in the south. Nonetheless, recent analyses of the flora indicate that of the 1371 vascular plant species known from The Bahamas Archipelago, 89 (or c. 6%) are endemic. Mobile species, such as birds, have only 6 endemic species (c. 5% endemic out of 120 breeding species), but at least 24 endemic subspecies. By contrast, less mobile species can exhibit higher degrees of endemicity as evident in reptiles in which approximately 68% of native species are endemic. Other less mobile taxa can be especially species rich (e.g., more than 300 species of peanut snails in the archipelago), but these and other taxa are in need of additional studies before the numbers of species and proportions of endemics in the archipelago will be known.

Fossil Record Indicates Extinctions and Extirpations

Fossils show that a much richer fauna existed in the Pleistocene and early Holocene than is now found in the archipelago. Included among the fossils are bird species that no longer occur in the archipelago and that are typical of xeric (dry) grassland and savanna habitats, such as meadowlarks (fossils from New Providence) and Eastern Bluebirds (Abaco) and a caracara (a now extinct falcon relative; fossils from New Providence and Great Abaco). During this period, the pine savannas of Abaco harbored Hispaniolan Crossbills, a pine seed specialist that now survives only in pine woodlands of Hispaniola. Other fossil birds include a suite of crow species now absent in the present day Bahamas (a remnant remains on the TCI); several owl species, the most conspicuous being a giant owl that is related to the Barn Owl (c. 1.0 m [3 ft] tall); and a flightless rail species (endemic, now extinct; fossils from Abaco). Climate change coupled with loss of roosting caves from a sea level rise may have also contributed to the loss of at least eight bat species from the archipelago and at least six local extirpations of existing species. The archipelago also once supported several large species of reptiles including giant tortoises, crocodiles, and large iguanas more than 2 m (6 ft) long. In fact, crocodiles, either the American Crocodile and/ or the Cuban Crocodile, were recorded in the southern Bahamas as recently as the early 1800s.

Recent fossil discoveries, from blue holes and elsewhere in the archipelago, demonstrate that the terrestrial food web of the late Pleistocene and Holocene (c. 4200 to 1000 yr BP) was dominated by reptiles prior to human occupation. Remains of Cuban Crocodiles and two extinct species of tortoises in a blue hole on Great Abaco indicate that the latter were the largest browsing herbivores (46 cm [18 in] carapace length) while crocodiles were the largest predators in the grassy pinelands at the time. Cuban crocodilians are believed to have preyed mostly on terrestrial species, such as birds, hutias, and tortoises based on isotope studies of their bones and observations of remnant populations on Cuba. In addition, a tortoise carapace with crocodile bite marks was found with the crocodile fossils at the same site, corroborating crocodile predation on the tortoises.

As observed globally, especially on islands, waves of extinctions and extirpations followed shortly after the first humans arrived in the archipelago. This appeared to be the case at two sites on Great Abaco where late Holocene fossils occurred concurrently with evidence of the earliest Lucayans (Amerindians). Of the seventeen species of terrestrial vertebrates found at a Holocene site on Great Abaco with evidence of Lucayan presence, seven species no longer occur on Abaco. Two of these species, Cuban Crocodile and Albury's Tortoise, disappeared from Great Abaco during Lucayan occupation before the arrival of the Europeans in the 15th century, whereas the crocodile and remaining species survive elsewhere. Although it is unknown what exactly led to these species' disappearances coincident with Lucayan occupation, the changes in climate and sea levels during the Holocene were minor relative to those of the Pleistocene when many species were lost from the islands. Lucayan arrival in the archipelago coincided with an increase in the frequency of wildland fires on several islands, suggesting that perhaps the earliest inhabitants altered the natural fire regime, which had negative consequences for some species. Whatever the exact cause(s), it is apparent that island extirpations and extinctions coincided with Lucayan occupation, suggesting that human-caused loss of species was already underway in the region before European arrival, which subsequently caused an even greater loss of species.

Climate Change

The islands of the archipelago have already begun to experience the effects of current global climate change as evident in an increase in the incidence of extreme weather events as well as more subtle changes in temperature, rainfall patterns, and sea level rise. These changes are consistent with model projections, which predict increases in average atmospheric and sea surface temperatures, decreases in annual rainfall, and increases in the frequency of the most powerful hurricanes (Classes IV and V).

The archipelago's islands are especially vulnerable to climate change given their low-lying terrain and lack of topographic relief. Rising sea levels, which have occurred over the past 4000 years since the last glacial retreat, are of special concern given the accelerating loss of the ice caps from global warming. During the past 4000 years, the sea level in the archipelago has been rising at a rate of 0.4 mm/yr (0.02 in/yr), although in the past century it has been rising at a rate of 1.0 to 2.5 mm/yr (0.04–0.10 in/yr), which is not far below the rate of 4.0 mm/yr (0.16 in/yr) projected for the upcoming years. Scientists have estimated that 15% of the wetlands of The Bahamas will be affected by a 1 m (3.3 ft) sea level rise, placing freshwater resources at risk. Sea level rise also exacerbates the negative effects of hurricane storm surges on coastal erosion and salt-intolerant pine and coppice plant communities, which in turn can have negative consequences for organisms dependent on these habitats. This is to say nothing of the increased risk to humans and their property and infrastructure, much of which is concentrated in coastal zones. Thus, as sea level continues to rise, those living in the archipelago will be challenged to devise and implement adaptation and mitigation strategies for the continued use, development, and conservation of coastal areas.

If the pace of global warming continues unabated at the current global average of 0.07 °C (0.13 °F) per year, the archipelago's organisms and ecosystems will face challenging consequences. As surrounding waters warm, the likelihood of coral bleaching increases as coral polyps respond to heat stress by ejecting their symbiotic algae (zooxanthellae, which is responsible for providing the coral animals [polyps] with carbohydrates derived from photosynthesis) and causing polyp mortality. With the expulsion of the zooxanthellae, polyps die and the coral loses its color or "bleaches." Because the polyps are the living portion of the coral, the coral dies and the bleached reef may eventually be covered with algae or may disintegrate. Further stressing corals and other marine life dependent on calcium carbonate for shells, exoskeletons, and other structures is the increased acidity of seawater as atmospheric CO_2 increases, thereby increasing dissolved CO_2 in seawater (and increasing acidity). Increased temperatures will challenge organisms with the inability to adapt or disperse to new locations, which may be impossible for some organisms without warm temperature adaptations.

Some species may have sufficient genetic variation to enable evolution of genetically based adaptations to the new environmental conditions, which fall outside their existing physiological tolerances. Some organisms may respond to the environmental changes by dispersing to other locations where their required environmental conditions still occur. Adaptation on islands may not be an option for some terrestrial species with small population sizes and limited ability to adapt to changing conditions. Populations of species that cannot adapt to climate change, and that have weak dispersal abilities, may be especially vulnerable to climate change. Moreover, species inhabiting the archipelago's low-lying islands do not have the option to respond to rising temperature as some continental species have by shifting their ranges far to the north or up mountains to escape increasing temperatures. Unless humans can substantially decrease their release of greenhouse gases into the atmosphere, rising temperatures will continue to directly or indirectly challenge the survival of many of the archipelago's species that cannot adapt to or escape from the changing climate conditions.

Human History

A Brief History of The Bahamas

As early as 2400 to 2500 years BP, the Tainos peoples of the Greater Antilles, who were related to the Arawaks of South America, began to colonize the southern part of the archipelago. Once in the archipelago, the Tainos were referred to as Lucayans, after the Spanish *Lucayos* (probably derived from the Taino *Lukku-Cairi* meaning "people of the islands"), and subsequently spread north to colonize the rest of the archipelago. Although they subsisted primarily on seafood (conch and other mollusks, crabs, fish, sea turtles), the Lucayans also consumed terrestrial animals and used rudimentary slash-and-burn agriculture to grow cassava, sweet potatoes, corn, beans, squash, and tobacco.

The Spanish arrived in The Bahamas in 1492. The name "Bahamas" may derive from the Spanish *baja mar* ("shallow seas") or perhaps from the Lucayan word for Grand Bahama Island, *ba-ha-ma* ("large upper middle land"). Although the Spanish never officially colonized the archipelago, they exported the majority of the Lucayans to slavery to facilitate development of Hispaniola and Cuba. Within 20 years of Spanish arrival, the Lucayans had been virtually extirpated from the archipelago as a result of slavery and European diseases, for which they had no immunity. Afterward, the archipelago remained mostly uninhabited for 130 years.

There was little European interest in the archipelago until the French attempted to settle in 1625. In 1629 the archipelago was claimed by Great Britain, but not permanently settled until 1649 when English Puritans known as "Eleutheran Adventurers" from the British colony of Bermuda settled on the island of Eleuthera in search of sanctuary from civil and religious unrest. The first arrivals eked out a meager existence through harvesting native tree species, collecting ambergris (a waxy substance originating from the intestines of the Sperm Whale and used as a spice or as a fixative in perfumes), and salvaging shipwrecks. In 1666, the island of New Providence was formally settled by the British and by 1671, census data indicated the presence of 1000 permanent residents, of which 40% were slaves.

Following the American War of Independence (1775–1783), thousands of pro-British loyalists and their slaves moved to the archipelago and established plantations. Growth and prosperity increased with the arrival of loyalists; the population grew from 2300 in 1740 to about 11,000 in 1789 with the initiation of permanent populations on islands other than New Providence and Eleuthera. Plantation agriculture, however, failed due to a combination of factors including climate, limited soil fertility, and pestilence; despite early promise, plantation production of cotton, sugar, cocoa, and ginger all failed.

The abolition of slavery in 1834 ended plantation agriculture in the archipelago. When emancipated, many of the freed slaves turned to slash-and-burn agriculture for subsistence farming on abandoned plantations and adjacent lands. Commercial production of several fruit and vegetable crops, including tomatoes and pineapples, and several cash crops, including tobacco and sisal (an agave species used for rope production), was initially profitable in the mid-19th century, but production ultimately failed to compete with foreign competition. In the same period, sponge fishing, a prominent feature of the Bahamian economy from the mid-1800s, ultimately failed because of disease that destroyed an estimated 90% of the crop, and from which the industry never fully recovered.

As recently as the 1930s, development of infrastructure and amenities was still limited primarily to New Providence. Infrastructure developed by the United Kingdom and the United States during the Second World War, however, facilitated the rapid expansion of postwar tourism within The Bahamas. Coupled with this increased economic prosperity was an increase in housing development and an increased standard of living for many on the "family islands" and not just for those in Nassau and Grand Bahama.

In 1973, The Bahamas gained independence from the United Kingdom and became a free and sovereign country. The continued development of a large and stable tourism industry and the development of offshore banking brought unprecedented prosperity to the islands. Currently, most of the Gross Domestic Product is generated from tourism and financial services, and the remainder is from industry (diverse industries mostly on Grand Bahama and salt production on Great Inagua), agriculture, and fisheries (primarily lobster, conch, and grouper).

A Brief History of the Turks and Caicos Islands

The TCI, probably so named after the common Turk's Head Cactus and also the Lucayan term *caya hico*, meaning string of islands, were discovered in approximately 1512. Post-discovery, the Lucayans experienced the same fate as those in The Bahamas, and scholars estimate that within 20 years they were extirpated. At various times the islands were controlled by Spanish, French, and British. No permanent settlements on the TCI were created, however, until the 1680s when salt collectors from the British colony of Bermuda built the first permanent settlement on Grand Turk. The initial attraction was the naturally produced salt in *salinas* (salt ponds) on Grand Turk and Salt Cay. Salt in the 17th and 18th centuries was especially important as it eliminated the dependence on the seasonal availability of food and thus allowed travel over long distances, especially in the developing New World. Its importance in food preservation as well as in the textile and tanning industries persisted to the end of the 19th century.

The natural salt ponds on the Turks islands were expanded and improved by the Bermudians to allow salt production on a larger commercially viable scale. Despite the highly lucrative nature of the salt industry, however, it was also labor intensive and required the use of slave labor to prosper. From a trade perspective, salt production was limited and seasonal, occurring between spring and fall (hurricane season). Therefore, to supplement income, other industries were developed, such as salvaging wrecks and harvesting sponge, turtle shell, and conch. In addition, beginning in 1890, sisal (an agave) was extensively planted for fiber production on the Caicos islands, but the enterprise ultimately failed.

As the lucrative salt industry developed in the TCI during the 1700s, tensions developed between the crown authorities in The Bahamas and the salt rakers (most of Bermudian heritage) on the Turk Islands, with the crown authorities in The Bahamas attempting to tax and influence their salt production. The Bermudians on the TCI were largely self-governing and controlled the salt industry prior to 1799 when the TCI were incorporated into The Bahamas parliament. Their assimilation with The Bahamas was unpopular with island residents, and in 1848 they successfully petitioned for and subsequently ceded from The Bahamas. They became a separate colony under the supervision of the governor of Jamaica and were later annexed to Jamaica. In 1959, the islands received their own governor, and when Jamaica was granted independence in 1962, the TCI became a crown colony. From 1965, the governor of The Bahamas was also governor of the TCI and oversaw their affairs for the islands. When The Bahamas gained independence in 1973, the islands received their own governor.

Ultimately the importance of the TCI as salt producers dwindled as the international demand for salt expanded. The TCI were unable to increase supply because of the lack of deep water anchorages, limited areas of salinas, and the seasonality of production. In combination with the development of domestic sources in the United States and other international suppliers, commercial salt raking in the TCI ceased in the 1960s after 300 years of production. In the 1960s, tourist development was initiated on Providenciales where it successfully continues today and combined with offshore finance constitutes the TCI principal industries.

An Early Naturalist in the Archipelago

Numerous naturalists have explored the archipelago and studied its biota, but undoubtedly the most famous of the early naturalists was the British naturalist Mark Catesby. As a guest of the governor in 1725, Catesby explored The Bahamian islands of New Providence, Abaco, and Eleuthera for nine months. His visit came at the end of his surveys of the British colonies in southeastern North America, later published in his monumental work, *The Natural History of Carolina, Florida, and the Bahama Islands*. The objectives of Catesby's surveys were to collect, illustrate, and describe natural history specimens for his sponsors in England. Primarily a botanist, Catesby's attention to detail is evident in the accuracy of his sixty-five Bahamian plant illustrations, of which modern botanists have examined and identified all but one. For many of the plants and animals not illustrated in his natural history volumes, Catesby provided descriptions in the text as well as valuable observations. For example, he recounts his encounter with sap of the Manchineel tree in his eye and alerts readers to the intense prickly pain and temporary blindness he experienced. His observations include descriptions of the taste of various specimens including those of flamingos and hutias, which he consumed. His name is commemorated in the genus of lilythorns (*Catesbaea*), a gentian (*Gentiana catesbaei*), and the Bullfrog (*Rana catesbeiana*).

As a naturalist with diverse interests, Catesby described the geography of the sites he visited including topography, soils, and climate. He also described Caribbean Monk Seals and a crocodilean species, both of which are now extinct in the archipelago, as well as hutias and iguanas, which were more widespread in his day. When anchored in a boat off Andros for three nights, Catesby identified the vocalizations of Bobolinks he heard overhead in the dark and deduced that they were on route to somewhere else, becoming the first naturalist to consider the possibility of bird migration. At the time of his observations, most naturalists believed that the winter absence of certain bird species was attibutable to their movement to unlikely places, including hibernating in the bottoms of ponds. His deductions were perceptive, and in 1747 he published an important paper on bird migration. Fortunately, Catesby's two-volume natural history including text and illustrations is readily accessible on the internet for readers to explore (http://cdn.lib.unc.edu/dc/catesby/index.html).

Human Impact on Flora and Fauna after Columbus

The arrival of Europeans in the Caribbean in the 15th century heralded an era of increased deforestation and conversion of natural vegetation to agriculture, increased hunting pressure, and the introduction of nonnative plants and animals. The effects of habitat loss are typically more acute for islands, as a consequence of their relatively small size and the correspondingly small population sizes of most resident species. Moreover, island species are often sensitive to the effects of nonnative invasive species as a result of their isolation from various predators, parasites, and competitors to which they have not evolved or have lost adaptive defenses. Invasive nonnative species are often highly adaptable generalists, which can threaten certain native species through competition, damage to native habitats, as vectors of novel virulent diseases, genetic degradation, and through the direct effects of predation.

Habitat Loss and Degradation

The extent and rate of habitat loss and degradation since the arrival of Columbus has varied among islands. In the 120 to 200 years following Columbus' discovery, the islands were not immediately colonized and therefore habitat loss or degradation in the archipelago appears to have been more limited in this period than elsewhere in the Caribbean. In The Bahamas, development was initially limited to the harvesting of specific tree species including Mahogany, Satin Wood, and Caribbean Pine for shipbuilding; Brasilleto wood used in dyeing; and several other tree species used as spices and flavorings, whereas in the TCI during the initial development of the salt industry, reportedly large numbers of trees were felled to discourage rainfall that would adversely affect the salt harvesting operations. Far more widespread and significant, however, was the large-scale clearing of natural habitats that occurred in the archipelago when slave-based plantation agriculture was introduced to the islands during the influx of loyalists during and after the American War of Independence, and by the mid-19th century many of the archipelago's original habitats were lost or damaged. Abolition of slavery in 1834, in conjunction with plant diseases, pests, and the loss of soil fertility, heralded the end of plantation agriculture and a return to small-scale slash-and-burn farming (Figure 15) and the grazing of livestock, in particular goats and pigs, which resulted in a regenerating patchwork of native habitats of differing ages. Many apparently pristine present-day wooded habitats in the archipelago have at some time in their past been cleared, as evidenced by the remains of agricultural infrastructure, such as stone walls and introduced plants (Figure 16). The presence of extensive second-growth coppice and the current estimates of 51% of the land area of present-day Bahamas covered in forests suggests that forest cover in the archipelago now is more extensive than in the past couple of centuries.

In the mid-20th century, large-scale harvests of Caribbean Pine were conducted in the northern pine islands of Abaco, Grand Bahama, and Andros, and much of the native pine forest was cleared and replanted resulting in many of the even-aged stands of pines currently present on these islands. The 20th century was also characterized by the clearing of other natural habitats, in particular coastal habitats, especially dunes, coastal coppice, wetlands, and mangroves, to accommodate housing for a growing population and infrastructure to capitalize on the growth of tourism, at least in the population centers on Grand Bahama, New Providence, and Providenciales. Despite bringing financial prosperity to the islands, however, development

Figure 15. Slash-and-burn agriculture, which dates back to the Lucayans, still exists in the archipelago as depicted in this garden plot on Eleuthera. As with many garden plots, this one contains Bananas, Plantains, Papaya, Bird Peppers, West Indian Pumpkins, and Sweet Potatoes. Garden plots are usually farmed for three to four years before abandonment because of declining soil fertility and encroachment by weeds. Abandoned plots may lie fallow for many years as the farmers move to other coppice sites where they again cut, burn, and plant crops.

Figure 16. Stone walls or fences in the midst of second growth coppice woodlands, such as this wall on San Salvador, are reminders of the agriculture history of the landscape. As farmland has been abandoned throughout the archipelago many of the old pastures and croplands have reverted to second growth coppice.

has also strained local resources, including freshwater supplies (especially in southern drier islands), and highlighted the challenges of waste disposal.

Overexploitation

Evidence from archeological sites suggests that the Lucayans overexploited populations of rock iguanas, tortoises, hutias, and seabirds especially on small islands or cays. Upon European colonization, economically important hardwood trees were heavily harvested including Mahogany, Brasilleto, Mastic, Cedar, and Lignum Vitae, as is evident today in the rarity and/or small individual size of these species. To provide sustenance and to diversify island diets, inhabitants routinely harvested wildlife such as White-crowned Pigeons, which today have populations that are mere remnants of their former numbers due to both commercial hunting of adults and the collection of nestlings. Populations were not only reduced throughout their ranges by overharvesting but also some species were locally extirpated from islands, thereby reducing the number of islands on which they now occur. Seabirds, popular targets of egg harvesters, and rock iguanas and hutias, killed for their meat, have disappeared from numerous cays and small islands. Adult flamingos were hunted for food, and their breeding colonies were disrupted by hungry settlers harvesting eggs and chicks; as a result, flamingos now breed only on Great Inagua, but formerly bred on Abaco, Andros, Flamingo Cay in the Ragged Islands, Long Island, and Norman's Cay in the Exuma Cays and in the Bight of Acklins. In addition to being hunted for food, adult Cuban Parrots were shot by farmers to prevent agricultural losses, and chicks were collected for pets; as a consequence, they no longer occur on Acklins and Crooked Islands, but now breed only on Abaco and Great Inagua (although other races persist on the Cuban mainland, Isla de Juventud, and the Cayman Islands). Although the documentation of population declines or losses of plant species in the archipelago is meager or nonexistent, studies elsewhere have found that plants popular in the ornamental trade, especially orchids, bromeliads, cycads, and cacti, have suffered from overcollecting. Thus, overexploitation directly or in combination with habitat loss and degradation and the introduction of nonnative pest species has been an important factor causing reductions in population sizes and geographic ranges of some island species.

With extensive coral reefs and the close proximity of deep water, exploitation of the archipelago's marine resources has always played a significant role in the economy and culture of the region. Cetaceans (e.g., dolphins and whales, including Humpback and Sperm Whales) were hunted during their migration to and from their southern breeding grounds. Caribbean Monk Seals were hunted in the archipelago throughout the 18th and early 19th century until it was no longer financially viable to do so, and the species subsequently went extinct in the mid-20th century. Sea turtles were harvested for their meat, eggs, and shells; the shell industry peaked in The Bahamas in the 1860s and subsequently declined as local populations were diminished. Queen Conch and Caribbean Spiny Lobster (or Crawfish) have been

Figure 17. Goat overgrazing or overbrowsing of coppice is characterized by loss of herbaceous plants in the understory, often leaving the ground devoid of herbaceous plants. All but the leaves of the most unpalatable trees, shrubs, and vines are consumed in the understory from the ground up to a browse line (~2.5 m [8 ft]), beyond which foliage remains unconsumed by the goats. This photo was taken on Long Island where free-ranging goats are common.

harvested historically through to the present day. Exploitation of resources continues today with some marine species under threat from overfishing, including staples such as the Nassau Grouper and Queen Conch.

Introduction of Nonnative Species

The earliest human-aided introductions of organisms began with the Lucayans who introduced cotton, indigo, and Sapodilly to the islands as well as dogs, which likely contributed to the loss of iguanas and tortoises on some islands. The European colonists accelerated the introduction of nonnative species, both accidently and intentionally. These introductions included invertebrates, such as insects (cockroaches, termites, and ants), reptiles, amphibians, and small mammals (rats and mice). Nonnative species were also deliberately introduced by both Europeans and North Americans as "a memory from home," in a misdirected effort to "enrich" the local fauna or to provide sport and/or an alternative source of food (e.g., Raccoons and game birds). Various nonnative plant species were introduced for use in agriculture, landscaping, and erosion prevention, often to the detriment of native species. A vari-

Figure 18. Hunting or trapping of feral pigs is a popular pastime and food source for some island residents, as evident in the trapping success of these two hunters on Abaco. Columbus brought eight pigs or hogs with him on his second voyage to the West Indies in 1493/94, and there have been subsequent introductions or re-introductions on various islands. Feral pigs can damage wetland and understory vegetation by rooting, wallowing, and trampling and as predators can destroy nests of ground-nesting birds, iguanas, and sea turtles.

Figure 19. Cats, both domestic and feral, are known to have substantial impacts on some wildlife populations, especially those of birds, rodents, and lizards. Illustrated here is a pet cat that has just killed an anole (*Anolis* lizard). Cats are a threat to Cuban Parrots, the females of which are depredated by cats as they incubate in their ground nests on Abaco.

ety of nonnative domestic mammal species were also introduced to the archipelago, which subsequently escaped or were deliberately released and went feral. These include cats and dogs, which accompanied the colonists, and livestock (pigs, sheep, goats, donkeys, horses, and cows). Overgrazing or overbrowsing by nonnative ungulates continues on some islands where it contributes to erosion (Figure 17). Rooting by feral pigs exposes rock iguana and sea turtle eggs, which are readily consumed by pigs (Figure 18). Feral cats (and even free-ranging pets) are known to substantially impact populations of lizards (Figure 19) and birds, and on Abaco they threaten the

Figure 20. The effects of the Pine Tortoise Scale insect (*Toumeyella parvicornis* [Hemiptera]) are evident in this young Caribbean Pine on Middle Caicos island. The pine scale, accidently introduced to the Caicos islands, secretes a sugary solution known as honeydew, which coats the pine foliage and serves as a medium for growth of Black Sooty Mold. The Black Sooty Mold (evident in the top of this pine) inhibits photosynthesis, thereby weakening and eventually killing the pine. Although the infestation has been devastating to the pine ecosystem here, an intensive conservation program holds promise for recovery of the pines on these islands.

ground-nesting parrots. Fortunately, the Small Asian Mongoose was never introduced to the archipelago as it was on most Caribbean islands to rid the sugarcane fields of rats, which it failed to do and subsequently became a threat to the native fauna of those islands.

Despite the threat of nonnative species to native ecosystems, the continued reliance of the archipelago on imported foodstuff and building materials (e.g., approximately 90% of produce is imported into The Bahamas) makes it at risk from colonization by nonnative species. Even with stricter quarantine measures implemented in the latter half of the 20th century, the islands are littered with opportunistic amphibian and reptile species in northern and western islands of the archipelago, primarily through the main ports. As recently as 2005, the Pine Tortoise Scale insect was accidentally introduced (possibly on imported Christmas trees) to the TCI, where it has decimated the endemic Caribbean Pine on those islands (Figure 20). The rise of the pet trade has also been a source of introductions due to escapees or the intentional release of unwanted individuals, including invertebrates (e.g., Giant African Land Snail), birds (e.g., Eurasian Collared Dove, Caribbean Dove, and Cuban Grassquit), reptiles (e.g., Corn Snake and various terrapin species), and fish (e.g., Lionfish).

Despite numerous plant introductions in the archipelago, many nonnative plant species have had little if any apparent detrimental effects on native species. For example, Coconut Palms introduced to the archipelago 500 years ago appear to have been one of the more benign introductions in terms of effects on native species. Although naturalized in coastal areas, where its floating seeds colonize beaches, Coconut Palm populations have been reduced due to a disease, lethal yellowing, likely introduced from Florida. Coconut Palms are valued for food, fiber, and as ornamentals by island residents and, in addition, mature Coconut Palms are important for nesting by the Bahama Oriole, a species of conservation concern and now only endemic to Andros.

Many nonnative plant species take advantage of natural (e.g., hurricanes) or human-caused habitat disturbances (e.g., agriculture and deforestation) to gain a toehold in new environments, and most nonnative plant species are described as ruderal (species that first colonize disturbed lands). For example, Jumbay, long naturalized in the archipelago, can colonize and quickly dominate disturbed sites by wind dispersal of its seeds. The presence of Jumbay on disturbed sites provides perches that attract birds and bats, which disperse seeds of native plant species, some of which are shade tolerant. In the absence of re-occurring disturbance on a Jumbay-dominated site, shade-tolerant native plant species can eventually establish in the shaded understory and overtake the shade-intolerant Jumbay, which dies out once the native species have grown above them. Jumbay can persist for many years, however, if the site receives re-occurring disturbances, such as goat browsing, fire, or storm surges.

Coastal areas or shorelines naturally suffer re-occurring disturbances associated with flood tides and storms as well as from human development activities, and as a result are especially vulnerable to invasive nonnative plant colonization. For instance, a recent study of eleven Bahamian islands found that 77% of 238 randomly placed coastal plots had an abundance of invasive nonnative plants and their abundance increased over time in all re-surveyed sites. Although Jumbay, Coconut Palm, and White Ink Berry were each found in at least a third of the coastal plots of the aforementioned study, the most pervasive nonnative species (in more than three-quarters of the sampled plots) was the Australian Pine or Casuarina, which is especially prominent along island shorelines (see Figure 10). As a salt-tolerant and fast-growing species, Casuarina outcompetes many native plant species for nutrients. In addition, by dropping copious amounts of small, needle-like stems on the ground and producing thick mats of stems, the Casuarinas prevent establishment of most native plants in the understory. In addition, the mats of dead Casuarina stems release toxins, which poison some plant species. Casuarina roots do not hold the sand as well as roots of most native dune plant species do; consequently, along shorelines, the wind and storm waves can sweep away the sand under the Casuarinas causing them to fall over during storms and for the dunes to disappear. After dune loss, the Casuarinas readily re-colonize these sites, but the native plant diversity declines and native dune species disappear. The presence of Casuarinas can lead to coastal retreat as Casuarinas collapse under storm conditions and the beaches narrow. Thus, not only is Casuarina a threat to native coastal vegetation, but by contributing to the loss of sand dunes, it increases the risk of storm damage to coastal property.

Conservation and Natural Areas Protection

Conservation Holds Promise to Meet Future Challenges

Despite some species losses and challenges facing the archipelago's biodiversity, conservationists have had successes and have increased public awareness, which holds promise to meet future challenges to biodiversity. Conservation efforts in the archipelago have long been in place. For example, efforts to protect breeding colonies of such iconic species as the flamingo were initiated early in the 1900s when the National Audubon Society worked with Bahamians to protect the birds at their Great Inagua breeding colonies. In the 1950s renewed attention was devoted to flamingos as numbers declined. Other pioneering conservation activities occurring at this time included the successful efforts to establish a land-and-sea park in the Exumas, initiated by the Bahamian government. Shortly thereafter, the Bahamas National Trust (BNT), a nongovernmental organization responsible for building and managing a national park system, was formed by an Act of Parliament in 1959. BNT maintains a country-wide system of marine, wetlands, and terrestrial parks and natural areas representative of the ecological diversity in the archipelago. BNT is the oldest conservation organization in The Bahamas, and its rich history includes developing education programs for Bahamian youth; managing the first land and sea park in the world, on Exuma Cays; recovering the flamingo population; fostering protection of the Bahamian race of the Cuban Parrot by protecting nests from feral cats and raccoons; maintaining White-crowned Pigeon populations through hunting regulations and protecting nesting colonies; encouraging protection of the West Indian Whistling Duck with BirdsCaribbean; monitoring seabird populations and eradicating introduced rats and other mammalian predators on some cays; collaborating with others to monitor and protect rock iguana populations on Exuma and Andros; helping to establish the Bahamas Environment, Science and Technology Commission; and many other accomplishments. Some national parks in The Bahamas and the TCI have also been designated as RAMSAR sites, wetlands of international importance. Similarly, the Turks & Caicos National Trust, founded in 1992, focuses on conservation and protection of ecosystems as well as historical and cultural sites and provides educational and other outreach programs to residents and tourists. Several sites, ranging from former plantations to caves, are protected by the Turks & Caicos National Trust.

In recent years other conservation organizations have emerged and the Bahamian government has undertaken various conservation initiatives, such as passage of the Wild Animals Protection Act of 1968; prohibiting commercial harvest of sharks; adoption of a forest management policy; participating in international conservation forums; being a party to the Convention on International Trade in

Endangered Species of Wild Fauna and Flora (CITES, as a result of the 2004 Wildlife Conservation and Trade Act); Convention on Biological Diversity; Convention on Wetlands of International Importance. Other conservation organizations have developed with a variety of foci. These include island-specific organizations such as One Eleuthera, Friends of the Environment (Abaco), Andros Conservancy & Trust, and San Salvador Living Jewels Foundation, as well as nongovernmental organizations with a topical focus: Bahamas Reef Environmental Educational Foundation, Bonefish & Tarpon Trust, Bahamas Marine Mammal Research Organization, and Bahamas Sea Turtle Conservation Group. The Cape Eleuthera Institute fosters conservation work, especially on marine systems and sustainability on Eleuthera, through research and outreach. In addition, other organizations, such as the American Museum of Natural History, BirdsCaribbean, BirdLife International, International Reptile Conservation Foundation, Island Conservation, National Audubon Society, The Nature Conservancy, Oceanic Society, Smithsonian Institute, Shedd Aquarium (Chicago, Illinois, USA), United States Forest Service, United States Fish and Wildlife Service, many universities, and field stations (Gerace Research Centre on San Salvador, Forfar Field Station on Andros) collaborate with Bahamian conservationists to implement conservation activities. In the TCI, efforts of the Turks & Caicos National Trust are complemented by those of the Turks & Caicos Museum, Turks & Caicos Reef Fund, and UK Overseas Territories Conservation Forum, Kew Botanical Garden, among others. Finally, improved enforcement of existing laws by national and international agencies (including the Royal Bahamas Defence Force) may help to limit overharvest of wildlife populations in the archipelago. These efforts are enhanced by researchers who contribute expertise and research findings that can facilitate protection, land management, and policy efforts.

Most of the protected areas in The Bahamas and the TCI are managed by either the Bahamas National Trust (BNT; http://bnt.bs/) or the Turks & Caicos National Trust (TCNT; http://tcnationaltrust.org), respectively, with a few additional areas protected by nongovernmental organizations. These parks, which are distributed throughout the archipelago, include most marine, freshwater wetlands, and terrestrial ecosystems found in The Bahamas and the TCI.

The BNT manages more than 800,000 ha (2 million ac) in more than thirty national parks, ranging from The Bahama's northernmost island, Walker's Cay, to the southernmost island, Great Inagua. The TCNT administers several national parks, including those on Middle Caicos and Providenciales, that center on the natural history of the islands. Directions to sites, entry fees, facilities, and other information about the national parks can be found at the websites for the respective trusts (see above). Visitors should note that research permits are required to collect plants or animals in the national parks, and many species are also protected outside these managed areas.

The archipelago is also blessed with an abundance of natural habitats (outside national parks) on islands other than the highly urbanized islands of New Providence and Providenciales and outside the few major urban areas on other islands. These include extensive areas of subtropical coppice and Caribbean pine forest, wetlands, mangroves, and coastal habitats, which afford numerous additional opportunities for seeing the archipelago's natural beauty and the respective plant and animal communities. Many such areas can be seen along public roads, and it can be rewarding to either drive or walk down the smaller, narrower side roads, but these often require a high-clearance vehicle because of the variable quality of the road surface. *Note:* Some of this land is privately owned, so permission should be secured before walking into unmarked off-road natural areas. Check locally about areas that permit access to natural areas.

Ecotourism and Conservation

Conservation in the archipelago faces several challenges including the uneven distribution of people throughout the archipelago, and that conservation is effective only if it has local community support. Support for conservation-related activities is facilitated when local communities benefit economically from these activities. For example, visiting sport fisherman and divers can contribute substantially to the economies of many of the family islands by engaging local guides and outfitters, as well as by staying in local hotels, eating in local restaurants, and using local transportation. The economic contributions of these visitors are recognized in local settlements and governments and have led to the implementation of nationwide conservation laws, such as a catch-and-release policy for fly fishing, protection of sharks and turtles, as well as providing the local impetus for protected areas including marine reserves. Although the contribution of ecotourism to the economies of the archipelago is small relative to the contributions of traditional tourism associated with mega-resorts and cruise ships, ecotourism is growing and is now recognized by the tourism ministries. Tourism officials are embracing ecotourism as it has the potential to disperse visitors beyond the traditional tourist centers on Grand Bahama, New Providence, and Providenciales to the less-inhabited islands. As our field guide illustrates, considerable biodiversity can be seen in the archipelago, and we hope this guide will encourage ecotourism throughout the archipelago, including to the less-populated islands. As ecotourism increases in the archipelago, so too does environmental awareness and support for conservation measures. In turn, national parks and protected areas provide a foundation for future growth of ecotourism.

How to Use This Book to Identify Fungi, Plants, and Animals

The core of this book describes and provides photographs of more than 500 of some of the most common terrestrial species in The Bahamas archipelago, including those on the seashore. This field guide complements field guides focused on marine environments (see selected references), but we include a few species of endemic freshwater fish found in inland ponds and lakes and four marine fish of economic and cultural significance because of their uniqueness or because of the widespread awareness of the importance of the species to the natural history of the archipelago.

Each taxonomic group, such as a family, is introduced by noting their characteristic or diagnostic traits. Species in each family are described and illustrated with photographs. When possible, subspecies native to the archipelago are shown. Natural history notes give additional cues for identification.

Not all taxa can be readily identified, in part because less is known about certain groups of insects or mollusks compared to plants, birds, mammals, amphibians, and reptiles. Given this disparity in knowledge, for some groups, such as fungi, insects, and other arthropods, we have highlighted families and commonly seen species, endemic species, or those species with important health or economic implications rather than attempting to describe the majority of species in a family, as is done for most vertebrates. Consult selected references at the end of the book for more comprehensive and detailed species information, especially for the comparatively little-known taxa found in the archipelago.

Species Names

We have provided the most frequently used common name for each species but recognize that common names of a species can vary within and among islands and also may change over time. Some of the common names are standardized by taxonomists (i.e., birds) but still can change with taxonomic revisions, and standardized "common" names may not be recognized by island residents. Common names for plant species can be especially variable whereas common names are often lacking for species of invertebrates and most fungi. In one instance, we created a common name for a fungus species based on its scientific name. Scientific names are provided for most species, but some organisms are identified only to the level of family (arthropods) or genus (some fungi).

Classification or Taxonomy of Organisms

Understanding how organisms are classified is especially helpful for identifying unknown organisms. Biologists have divided the world's readily visible organisms into three major kingdoms: fungi, plants, and animals. The fungi and plants are further subdivided into divisions and the animals into phyla (singular is phylum). The divisions or phyla are further divided into classes; classes are divided into orders; orders into families; families into genera (singular is genus); and genera into species. In other words, a division or a phylum consists of a group of similar classes, a class consists of similar orders, an order consists of similar families, and so on down the hierarchy to species. The basic unit of classification is the species. Scientific names for species are based on a binomial (two-name) system devised by Carl Linnaeus (1707–1778) in which the first name refers to the genus, and the second refers to the species. The names are from Latin or Greek, or words or names that have been Latinized. For example, the scientific name of the San Salvador or Central Bahamian

Rock Iguana is *Cyclura rileyi*, in which *Cyclura* refers to the genus of rock iguanas and *rileyi* refers to the species, which was named after the collector, Joseph Harvey Riley. Some widespread species may be further subdivided into subspecies or races that may differ in appearance or behavior from other populations of the same species on other islands (see subspecies examples in the accounts of the various species of rock iguanas).

Species are traditionally defined as groups of populations that interbreed and produce fertile offspring, but which are reproductively isolated from populations of other species (i.e., incapable of producing fertile offspring in crosses between species). Reproductive isolation may be verified when two or more presumed species do not interbreed or produce fertile offspring when they occur together on the same island. When the presumed species have geographic ranges that do not overlap (e.g., each population is isolated on a different island), the taxonomic decision to designate each of the geographically isolated populations as separate species on the respective islands becomes more difficult. The taxonomic decision to divide (split) populations into separate species or to combine (lump) populations into the same species often depends on the degree of morphological, behavioral, and genetic differences among the populations in question. The classification of organisms may change as more is learned about the organisms' biology. This clarification is especially true as molecular genetic techniques are applied to different groups of organisms and uncover differences or similarities in the genetic code (DNA) of various organisms, thus resulting in re-classification (e.g., lumping or splitting) of species or even re-arranging taxonomic groups (e.g., moving genera into new families, or even creating new families). As The Bahamas archipelago and its biota become better known to science, we expect new species will be added to the list of species known from the region, especially for the poorly known taxa from the archipelago, such as fungi and invertebrates. Little is currently known about fungi in the archipelago; at least one of the species of fungi featured in our guide lacks an appropriate scientific name, and several of the fungal species lack basic information on their natural history. Even for the relatively well-studied species in the archipelago, such as the reptiles, we expect that future studies will reveal new species and perhaps species endemic to the archipelago.

Conservation Status
For certain species, the conservation status based on the International Union for Conservation of Nature and Natural Resources (IUCN) is provided in the species accounts. The IUCN compiles and publishes the IUCN Red List of Threatened Species, which assesses the conservation status of species worldwide (www.iucnredlist .org).

Species Distributions
Species are not evenly distributed in the archipelago. We reference distribution by island(s) and emergent banks (see map on inside cover) and, for some taxa, we

describe ranges by the following island groups: northern islands (Grand Bahama, Abaco, the Biminis, Berry Islands, Andros, Cay Sal Bank, New Providence, and Eleuthera); central islands (Exumas, Cat Island, Ragged Island, Long Island, Rum Cay, and San Salvador); and the southern islands (Crooked and Acklins, Samana Cay, Great and Little Inagua, Mayaguana, and the TCI).

Potential Hazards in the Bush

Sun and heat can be intense in the archipelago, so a hat, sunglasses, a liberal application of sun block, and full-length trousers and long-sleeve shirts are all helpful to combat UV radiation. Carry plentiful amounts of water while exploring the bush. Given the uneven terrain, including sharp limestone rocks and sinkholes that may be concealed by fallen leaves, watch where you place your feet (strong footwear is recommended) and use caution while walking in the bush. It is also worth pointing out that it can be surprisingly easy to get disorientated in the bush, especially on the less inhabited islands, so a compass or GPS are recommended for off-road work.

Fortunately, the Bahamian bush is relatively benign in terms of natural threats and hazards to naturalists, as compared with continental subtropical and tropical habitats. Several of the authors of this book have spent years in the bush with no problems; we do not want to over-emphasize the small risk posed by the species we list below. Nonetheless, visitors should acquaint themselves with the appearance of plants and animals that can potentially cause harm.

Several plant species should be avoided. For example, Poison Wood, a small- to large-sized tree, is widespread in pineyards and coppice habitats throughout the archipelago; in the pineyards understory it can be common and low growing and shrubby. Contact with its leaves or bark can cause skin rashes, some severe, especially for those allergic to plants in the Anacardiaceae. Similarly, the less common Manchineel tree, found throughout the archipelago, should be recognized to avoid contact with leaves and bark, which also can cause severe skin rashes. Avoid standing under this tree if it rains as the sap is water soluble, and it also has deadly poisonous fruit. Plants with thorns and spines are common, but in most instances the thorns or spines are readily apparent in the field (e.g., various species of cactus, Chaney Briar, Wild Lime, Haulback or Devil's Claw [*Pisonia aculeata*], and Century Plant [*Agave*]) and can be avoided by staying on cleared trails or roads.

Animal threats to visitors are few and can be avoided if careful. Mosquitos and sand flies can be especially challenging in the wet season or in some coastal habitats, so use of repellents, long trousers, and long-sleeve shirts can be helpful in reducing bites. We have even found it useful to use a screened head net while exploring some habitats during the wet season. Keep an eye out for spiders, such as the Black Widow and Brown Recluse, which can bite causing severe pain and additional medical complications. Be especially careful when turning over rocks, logs, and debris on the ground as spiders and scorpions are common in these sites and all can inflict painful bites. Bees and wasps are common, and as elsewhere in the world, can have painful stings that can be dangerous for those allergic to their stings. If you have such allergies, carry appropriate medication to prevent anaphylactic shock.

The Cuban Tree Frog, found throughout the archipelago, secretes a toxic mucus from its skin when handled, which can cause an allergic reaction, asthma, or an intense burning sensation should it come in contact with the eyes and nose. Similarly, the Cane Toad (currently only on New Providence) secretes a toxic mucus. Contact with these species should be avoided.

The Brown Racer is the only venomous reptile in the archipelago. Fortunately, it is only mildly venomous. Those likely to be bitten are those who try to capture them, which is difficult because as per their name they are fast and flee from people. A Racer bite may cause the affected appendage to swell, which can be treated with ice and by taking antihistamines to reduce swelling.

Caution is warranted while beach combing not to touch the colorful float or long thread-like tentacles (typically at least 10 m [33 ft] long, normally longer) of the Portuguese Man of War (*Physalia physalis*). These jellyfish-like animals wash up on beaches throughout the archipelago and have stingers (pneumatocysts) that contain a neurotoxin; the sting is extremely painful and on rare occasions has been known to be fatal to humans.

Species Accounts

Fungi

Fungi are neither plants nor animals but represent a separate kingdom of organisms that diverged during evolution from a lineage closer to animals. Unlike plants, their cell walls lack cellulose and instead have chitin, which is also found in invertebrate animals, such as insects and crabs. Fungi obtain their energy and nutrients from live or dead plants and animals. Some fungi form beneficial symbioses with tree roots known as mycorrhizae (myco = fungus + rrhiza = root). Mycorrhizal fungi help their host plants obtain nutrients and water from soil, and they protect them from infections in exchange for sugar. Many fungi decompose dead leaves or wood from trees and are important for recycling nutrients.

Fungal fruiting bodies, such as mushrooms and shelf fungi (wood ears) are used to identify macrofungi (visible without magnification). Fungal fruiting bodies make up less than 10% of the fungus, and the vegetative part, called a mycelium, grows mostly within the substrate and often goes unnoticed. The most prominent fungi in the archipelago are basidiomycetes, including mushrooms, boletes, chanterelles, earthballs, stinkhorns, and shelf fungi. Mushrooms, boletes, chanterelles, and wood ears have a cap (pileus) and either gills (lamellae) or pores on the lower surface below the cap (see illustration). Long-distance dispersal by most basidiomycete fungi is by way of airborne spores that are forcibly released from the gills or pores. Color of spore deposits is often useful for identification of the family and genus. Forcible spore release requires that the fungus be hydrated, and many groups have protective veils that retain moisture; veils may also protect against fungus-eating insects. One veil type is known as a universal veil, which covers the entire fruit body when it is young. Universal veils leave an attached ring or a cuplike structure at the base of the stalk, known as a volva; they may also leave warts on the cap surface. Warts are easily removed, whereas scales (squamules) are an integral part of the cap surface and are not removable. Partial veils extend from the stalk to the margin of the cap and cover the gills or pores when young, leaving a ring (annulus) on the stalk or remnants on the margin of the cap. A few basidiomycete fungi, often those growing in dry habitats, lack forcible spore release, for example, earthballs, puffballs, and stinkhorns. Earthballs and puffballs use wind to disperse their spores whereas stinkhorns attract flies to carry sticky spores.

This group is poorly studied; there are no documented surveys of fungi in the archipelago. Visitors to these islands can make significant contributions to the knowledge of macrofungi of the archipelago by taking photographs of the upper and lower sides of fruit bodies and submitting them with sizes, habitat, and possible host plant associations to Mushroom Observer (http://mushroomobserver.org) or iNaturalist (https://www.inaturalist.org/).

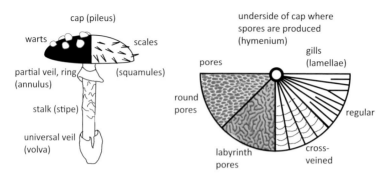

cap (pileus)

warts

scales

partial veil, ring (annulus)

(squamules)

stalk (stipe)

universal veil (volva)

underside of cap where spores are produced (hymenium)

pores

gills (lamellae)

round pores

labyrinth pores

cross-veined

regular

Fungal structure, line illustration. D. Jean Lodge.

Wood Rotters

Fungi that decay wood include a diverse array of basidiomycetes and ascomycetes, but the most noticeable are basidiomycetes that produce visible fruiting bodies in the form of mushrooms (with caps, gills, and usually stalks), shelf fungi or wood ears (tough or hard, usually shell-shaped but sometimes stalked), or spherical lattice balls emitting odors of rotting meat (a type of stinkhorn). The following wood decomposer fungi are ordered by form (stinkhorns, polypores, mushrooms), then by order and genus.

Latticed Stinkhorn   *Clathrus crispus*
Order: Phallales (Stinkhorns)
Family: Clathraceae
Range: Archipelago-wide, the Neotropics, the Gulf Coast.
Description: Globose lattice structure (receptacle) 7–15 cm (2.8–5.9 in) diameter with red upper parts, paler below; lattice edges corrugated and coated with olive-green, sticky spore mass (gleba) with odor of rotting meat. The fruit body expands overnight from a round jelly-filled "egg" (volva), and the remnants of the leathery exterior and gelatinous interior of the egg can often be found at the base. The foul odor attracts flesh flies that mistake it for rotting meat and end up distributing the spores. Fruit bodies normally appear on bare ground, but the rubbery white cords at the base can often be traced to the true source of nutrition—rotting wood. Stinkhorn fungi use their cord systems to locate new food bases and to move mineral nutrients into fresh pieces of wood. **Habitat:** Found in most coppice and pine habitats except those on coastal dunes and mangroves. **Similar species:** *C. ruber* (possibly introduced; native to Europe); lacks corrugated rims and the spore mass is spread on the inside of the lattice arms.

Yellow-edged Shelf Fungus   *Phellinus gilvus*
Order: Hymenochaetales
Family: Hymenochaetaceae
Range: Archipelago-wide; the Americas and Australia in tropical and temperate broad-leaved forests.

Clathrus crispus

Clathrus crispus

Phellinus gilvus

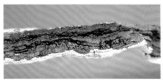

Earliella scabrosa, pileus

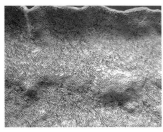

Earliella scabrosa, habitat

Earliella scabrosa, pores

Description: Shelf fungus 5–15 cm (2.0–6.0 in) broad, shell-shaped, leathery to somewhat woody, brown and roughened on upper side with a yellow edge when young, minute pores on lower side, often yellow at first, then brown. **Habitat:** Grows on wood of broad-leaved tree species in a range of coppice habitats.

Corrugated Polypore   *Earliella scabrosa*
Order: Polyporales
Family: Polyporaceae (Polypores)
Range: Archipelago-wide, the Neotropics, the Gulf Coast.
Description: Shelf fungus 10–50 cm (3.9–19.7 in) diameter, tough, leathery, with a two-tone cap, inner part corrugated or roughened, reddish brown and somewhat shiny, contrasting with a pale rim. Can be shell-shaped or more irregular and expansive with the pore surface extending over the wood (effused-reflexed). The pore surface can be off-white to deep cream, and the pores are large and often irregularly elongated, somewhat labyrinth-like on sloping surfaces. **Habitat:** Found in all coppice habitats with large pieces of wood.

Hairy Shelf Fungus   *Hexagonia hydnoides*
Order: Polyporales
Family: Polyporaceae (Polypores)
Range: Archipelago-wide, the Gulf Coast. Pantropical.
Description: Shelf fungus 8–25 cm (3.2–9.8 in) broad, shell-shaped, dark brown, with stiff, coarse dark brown hairs on the upper surface and minute brown pores on the lower surface. **Habitat:** This wood-rotting fungus tolerates hot, dry conditions and is often on standing dead trees. Found in all woodland types.

Caribbean Turkey Tail Fungus   *Trametes ectypa*
Order: Polyporales
Family: Polyporaceae (Polypores)
Range: Archipelago-wide, the Caribbean, Florida, the Gulf Coast.
Description: Large shell-shaped wood ears 5–15 cm (2.0–5.9 in) diameter, which are banded and velvety, growing in overlapping clumps (imbricate); the pores below are white or cream colored and minute. **Habitat:** Grows on wood and stumps in coppice in the archipelago. **Similar species:** Turkey Tail Fungus (*T. versicolor*) is similar in color and structure but is smaller. *Note:* Morphology is not always reliable in separating species in this complex.

Blood-pored Shelf Fungus   *Trametes sanguinea (=Pycnoporus sanguineus)*
Order: Polyporales
Family: Polyporaceae (Polypores)
Range: Archipelago-wide, pantropical, the eastern U.S.
Description: Shelf fungus 5–10 cm (2.0–4.0 in) diameter, shell-shaped, bright orange, stiff, with minute red-orange pores on the lower surface. **Habitat:** Grows in dry woodlands and exposed sites, preferring hot temperatures and moisture levels that are five times drier than those at which a tomato plant would permanently wilt and die; tolerates the drying effects of salt in mangrove forests.

Lentinus berteroi **and** *L. crinitis*
Order: Polyporales
Family: Polyporaceae (Polypores)
Range: *L. berteroi* throughout the archipelago and the Neotropics; *L. crinitus* throughout the archipelago, the Neotropics, the southeast U.S.
Description: Fruit bodies 2–4 cm (0.8–1.6 in) diameter, shell-shaped with short stalks or funnel-shaped with short to long stalks, gills running down the stalk below, leathery, coarsely hairy above, cream to dark brown hairs over lighter ground. The gills of *L. berteroi* are yellow when fresh whereas gills of *L. crinitis* are white, and a honeycomb pattern is more likely present in *L. berteroi* where the lamellae join the stalk. Resembles a true mushroom, but they are more closely related to polypore fungi (wood ears). Unlike fragile mushrooms, such as the similar-looking oyster mushrooms, polypores and their *Lentinus* relatives are difficult to tear. **Habitat:** Wide range of coppice habitats. Can tolerate open, dry conditions.

Hexagonia hydnoides

Hexagonia hydnoides

Trametes cf. *ectypa* (cf. denotes uncertainty in the species identification)

Trametes sanguinea

Trametes sanguinea, cap and pores

Lentinus sp.

Lentinus crinitis and *L. berteroi* compared

Split-gill Wood Ear   *Schizophyllum commune*
Order: Agaricales (Mushrooms)
Family: Schizophyllaceae
Range: Worldwide in tropical and temperate forests.

Description: Small, 0.5–2 cm (0.2–0.8 in) diameter, shell-shaped, pale gray, cream, or brown, leathery fruit bodies on wood with hairs above and white or cream-colored split gills below. The diagnostic feature of the genus is the split gill edges (schizo = split; phyllum = plate or leaf). Although this fungus resembles a true mushroom, it is more closely related to polypore fungi (wood ears). Unlike mushrooms, polypores and their relatives are difficult to tear. **Habitat and use:** Grows on dead wood that is suspended above the ground (e.g., dead in situ trees) in dry or open sites in most habitats and tolerates hot, dry environments. Despite their tough texture, the young fruit bodies are roasted and eaten throughout the tropics.

Litter and Humus Decomposers

These mushrooms decompose twigs usually less than 1 cm (0.4 in) in diameter, fallen leaves, or in some cases, plant litter partly decomposed by other fungi (e.g., humus) or microbial symbionts of termites (termite nest carton). The following litter decomposer fungi are ordered alphabetically.

Neotropical Gymnopus   *Gymnopus neotropicus*
Order: Agaricales (Mushrooms)
Family: Marasmiaceae
Range: Archipelago-wide and the Neotropics.

Description: Mushroom caps 2–6 cm (0.8–2.4 in) diameter, pinkish brown on the low knob in center, fading to beige near the margin, deeply radially grooved, translucent when fresh, becoming opaque when dry; gills 2 mm (0.1 in) broad, widely spaced, dirty white; flesh tough and elastic. **Habitat:** Grows on leafy and woody litter in coppice.

Yellow Parasol Mushroom or Yellow Houseplant Mushroom
Leucocoprinus birnbaumii
Order: Agaricales (Mushrooms)
Family: Agaricaceae
Range: Pantropical and in greenhouses and potted plants throughout the world.

Description: Caps bright or pale yellow with soft scales on the surface, margin pleated, 1.5–5 cm (0.6–2.0 in) diameter, conic at first, expanding to parasol-shaped, thin and readily collapsing; lamellae light yellow, not attached to the stalk; stalk yellow, with a ring formed from a partial veil, thickest at base, often joined at the base. **Habitat:** Typically observed on leaf litter or soil that has abundant organic matter but can also be found on decaying wood.

Gymnopus neotropicus

Schizophyllum commune

Marasmiellus cubensis

Schizophyllum commune

Leucoprinus
birnbaumii

Cuban Marasmiellus   *Marasmiellus cubensis*
Order: Agaricales (Mushrooms)
Family: Marasmiaceae
Range: Archipelago-wide and the Caribbean.
Description: Small (≤ 1 cm [≤ 0.4 in]) diameter, white, fragile mushrooms with a
pleated, translucent cap above and cross-veins between the gills below, and an open
cup (volva) at the base of the white stalk; grows in groups on wood. **Habitat:** Likely

restricted to coppice. **Similar species:** In the Neotropics, three related white species of *Marasmiellus* that have a volva (remnants of a universal veil covering the mushroom in the immature, button stage) occur. Of these, *M. cubensis* has the most restricted range, whereas *M. coilobasis* and *M. volvatus* are widespread. The rim of the volva is free from the stalk in *M. cubensis* but sheaths the stalk like a sock in *M. coilobasis* and *M. volvatus*.

Cross-veined Mycena   *Mycena tesselata*
Order: Agaricales (Mushrooms)
Family: Mycenaceae
Range: Archipelago-wide and the Caribbean.
Description: Small, 1–2 cm (0.4–0.8 in) diameter, white with pink and cream color tints, cap translucent, grooves corresponding to lamellae and cross-veins below; faint radish odor. This species is a tropical member in a group related to the temperate zone *M. pura* (lilac bonnet), and it has a similar radish odor. Resembles the Cuban Marasmiellus (see above) in having a white, translucent cap and cross veins between the lamellae, but differs in having a distinct odor, pink and yellow tints, and lacks a cup at the base of the stalk. **Habitat:** Grows on leaf litter in coppice.

Ectomycorrhizal Symbionts

These fungi form mutually beneficial relationships with plant roots. Fungi provide their plant hosts with mineral nutrients and water from the soil in exchange for sugar. The following ectomycorrhizal symbionts are arranged alphabetically by order and then by genus.

Sand-loving Amanita   *Amanita arenicola*
Order: Agaricales (Mushrooms)
Family: Amanitaceae
Range: Archipelago-wide, neotropical coastal forests.
Description: Mushroom caps 2–7 cm (0.8–2.8 in) diameter, pale gray to white with pleated margin, often bearing a load of sand, lacking a ring (annulus) on the stalk but having remnants of a small, fragmented white cup (volva) at the base. Ephemeral, emerging from several cm below the sand surface around 02:00–04:00 a.m. and often collapsing by mid-morning. Unlike most mushrooms, it often emerges fully expanded, which may help it retain moist sand that protects the cap. **Habitat:** Apparently restricted to coastal coppice under Sea Grape (*Coccoloba uvifera*) with which it forms a beneficial symbiosis.

Pineapple Bolete   *Boletellus ananas*
Order: Boletales
Family: Boletaceae (Boletes)
Range: Pine islands of the archipelago, the Caribbean, the Gulf Coast to Central America.

Mycena tesselata

Amanita arenicola, cap

Amanita arenicola, habitat

Boletellus ananas

Boletellus ananas

Description: Caps 4–8 cm (1.6–3.2 in) diameter, hemispherical coarsely scaly, various shades of pink, often with yellow flesh showing between scales, veil remnants clinging to margin; cap flesh quickly staining blue when cut; pores soft, yellow, quickly bruising blue; stalk white with a pinkish-red band (sometimes with yellow) at top, cylindrical, often with a tuft of white mycelium at base. **Habitat:** Pine habitats; ectomycorrhizal symbiont of pine tree roots. Sometimes observed emerging from sides of sinkholes.

Red-pored Bolete   *Caloboletus firmus* (*= Boletus firmus, = B. piedmontensis*)
Order: Boletales
Family: Boletaceae (Boletes)
Range: Andros (probably archipelago-wide), southeast U.S. to Central America
Description: Cap 6–16 cm (2.4–6.3 in) diameter, pale white, pale pink, or pale gray, contrasting with the red pores below; cap, pores and flesh staining blue; stalk variously pale yellow and pink, staining blue above and reddish purple below, without a honeycomb ornamentation at the top or only faintly netted. Taste bitter (but not edible). This species is one of many red-pored boletes, but the pale, dull cap contrasting with the yellow tubes and red pores, the blue bruising reaction, the weak or absent netting at the top of the stalk and the bitter taste distinguish it from other species. **Habitat:** Pine habitats.

Sculpted Slimy Bolete   *Fistulinella sp.*
Order: Boletales
Family: Boletaceae (Boletes)
Range: Archipelago-wide and the Caribbean.
Description: Cap 3–6 cm (1–2 in) diameter, sculpted, brown with a gelatinous covering that extends to the white stalk, forming a gelatinous ring (annulus) on the stalk when it breaks, thereby exposing the soft white pores beneath the cap that turn brownish pink as the spores mature. Stalk base stains yellow when bruised, and the gelatinous veil becomes slimy when handled. **Habitat:** A range of coppice habitats; is an ectomycorrhizal symbiont of *Coccoloba* spp. roots and roots of Small and Large Leaf Blolly *Guapira* spp. and possibly Rams Horn (*Pithecellobium keyense*).

Sea Grape Chanterelle   *Cantharellus coccolobae*
Order: Cantharellales (Chanterelles)
Family: Cantharellaceae
Range: Archipelago-wide, the Caribbean, the Gulf Coast.
Description: Mushroom caps 3–6 cm (1–2.5 in) diameter, smooth, orange, somewhat indented; gills repeatedly and irregularly forked, narrow, thick, and wrinkle-like; flesh thick with an odor of dried apricots. **Habitat:** Coastal habitats and coastal forest, usually in sand, near Sea Grape (*Coccoloba uvifera*); is a beneficial root symbiont of Sea Grape under which it grows. As are many other Chanterelles, it is edible but sometimes difficult to remove the sand.

Boletus firmus, red-pored side   *Boletus firmus*, pores

Fistulinella sp., showing   *Fistulinella* sp.   *Cantharellus coccolobae*
mature spores and pink pores

Fistulinella sp.

Cantharellus coccolobae

Bermuda Earthball   *Scleroderma bermudense*

Order: Sclerodermatales

Family: Sclerodermataceae (Earthballs)

Range: Archipelago-wide, Bermuda, coastal Central and South America; also reported from Malaysia.

Description: Globose at first, 1.5–3.5 cm (0.6–1.4 in) diameter, pale, off-white, light yellow to pale gray with grayish brown root-like structures on the surface, splitting open at the top with 4–5 rays that are cream color and then pinkish brown or darker with age, surrounding a powdery purple-brown mass of spores and tissue (gleba). This species has a pale outer surface when young, unlike *S. stellatum*, which always has a dark surface. **Habitat:** A range of coppice habitats; is an ectomycorrhizal symbiont of the roots of Sea Grape and of Small Leaf and Big Leaf Blolly.

Tremelloscypha dichroa

Order: Sebacinales

Family: Sebacinaceae (Encrusting or Coralloid Jelly Fungi)

Range: Archipelago-wide and the Caribbean to lowland Central America.

Description: Cap 5–15 cm (2.0–5.9 in) diameter, flat or funnel-shaped, coarsely hairy, often concentrically banded, often surrounding the stem of a live tree; lower surface smooth, gelatinous, cream to tan color. Originally described from The Bahamas as *Stereum dichroum* but now considered to be a type of jelly fungus (Sebacinales). **Habitat:** A range of coppice habitats; is an ectomycorrhizal symbiont of Pigeon Plum (*Coccoloba diversifolia*) and possibly Rams Horn (*Pithecellobium keyense*).

Uncertain Nutrition

How these species obtain their energy is currently unclear.

Colonial Earthballs   *Diplocystis wrightii*

Order: Boletales

Family: Diplocystidiaceae

Range: Probably Archipelago-wide, also the Caribbean.

Description: Typically found as acorn-like spore-bearing sacks nested in cups that are joined to a slightly domed mat 5–10 cm (2–4 in) diameter. Earlier stages can be found below the sand or soil surface with the unopened spore sacks on the inside of a vertically compressed hollow ball. As the spores mature and the humidity changes, the outer part of the ball splits and recurves, pushing the fruiting body out of the soil and exposing the spore sacks. This unique fungus is the only species in its genus. **Habitat:** Coastal and inland coppice, also pine forests. *Diplocystis* is thought to be mycorrhizal with Sea Grape (*Coccoloba uvifera*) and other *Coccoloba* spp., *Neea* spp. (Family Nyctaginaceae), and possibly pine (*Pinus* spp.).

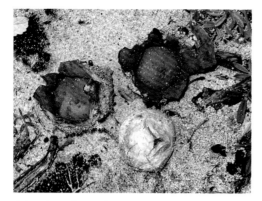

Scleroderma bermudense

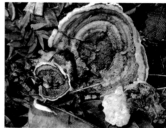

Tremelloscypha dichroa

Diplocystis wrightii

Tremelloscypha dichroa

Tremelloscypha dichroa

Diplocystis wrightii

Plants

C ompared to the flora on many islands, the flora of the archipelago is relatively young because of its low elevation and cyclical inundation during interglacial periods. Many plant species are shared with the Greater Antilles (Cuba especially) and Florida.

The current flora of the archipelago comprises approximately 1400 vascular plant species, approximately 50 lower vascular plants (e.g., spore-producing plants including ferns and mosses), 4 gymnosperms (cone-producing plants), and c. 1350 angiosperms (flowering plants).

The angiosperms include c. 320 monocotyledons (monocots; represented by 21 families)—whose seedlings typically have only one embryonic leaf and mature plants have parallel veins in the leaves and flower parts in multiples of three, for example, the grasses and orchids—and c. 1030 dicotyledons (dicots; represented by 98 families), whose seedlings have two embryonic leaves and mature plants have netted leaf veins (usually with a midvein and branching smaller veins) and floral parts in multiples of four or five.

Unique to the archipelago are 89 endemic species (c. 6.5%): 88 angiosperms and 1 gymnosperm, with more endemicity in the southern part of the archipelago. The most endemic-rich families, archipelago-wide, are the Rubiaceae (14 endemic species), Euphorbiaceae (11 endemic species), and Asteraceae (9 endemic species).

The flora exhibits adaptations to nutrient-poor soils and lack of surface freshwater, especially in more environmentally stressed habitats (e.g., coastal areas). These include species with succulent leaves, which store water and reduce leaf area to limit evapotranspiration (e.g., Sea Purslane, Black Ink Berry); species with waxy, hairy, or spiny leaf surfaces to reflect heat and reduce water loss to the plant (e.g., Bay Lavender, Buttonwood, Turk's Head Cacti); and species, such as orchids, with mycorrhizal associations (through which the plant benefits from the fungi's higher absorptive capacity for water and mineral nutrients). Furthermore, the north–south environmental gradient that defines the island chain (specifically, rainfall and temperature; see introduction) also affects growth forms with vegetation in the southern drier islands being typically shorter and more scrub-like than in the wetter northern archipelago.

Plant identification: Variation can occur in all the major parts of vascular plants, for example, roots, shoots, leaves, sporangia, cones, or flowers, but leaves are most variable in terms of size, shape, margins, and leaf tips, so more than a single leaf should be observed when identifying specimens. By contrast, cones, flowers, and fruits, which are involved with reproductive structures, are typically more uniform within a species and often more reliable as a means of identification. In some cases, identification of a plant can be made only if it is flowering or in fruit. Close examination of flowers (e.g., the species is hermaphroditic if the stamens [male] and pistil [female] parts of a flower are on the same flower, monoecious if the male and female flowers are separate but on the same plant, or dioecious if male and female flowers are separate and on different plants) or fruits may be needed for identification. In addition, different species may be restricted to certain habitats, which can help to distinguish among similar species.

A glossary is provided to describe plant characteristics. Some common terms are illustrated in the figures below. We recommend the Leon Levy Native Plant Preserve website (www.levypreserve.org) as an excellent resource for Bahamian plants and their identification. For a complete guide to the identification of all The Bahamas archipelago plant species, consult publications in the references section.

The plants are ordered as follows: spore-bearing plants (alphabetically by family and then by genus), Gymnosperms (cone-bearing plants), and Angiosperms (details of their arrangement are provided in the relevant sections).

Several plant species found in the archipelago have toxic or poisonous secretions or sap, fruits, and seeds, for example, Poison Wood and the Manchineel tree. A range of medical issues may occur if the foliage, branches, and/or trunks are touched or handled, and/or if the fruits or seeds are eaten. Potentially dangerous species are highlighted in the text.

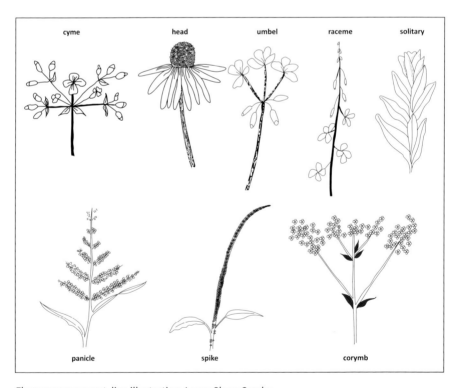

Flower arrangement, line illustration. Laura Sloan Crosby.

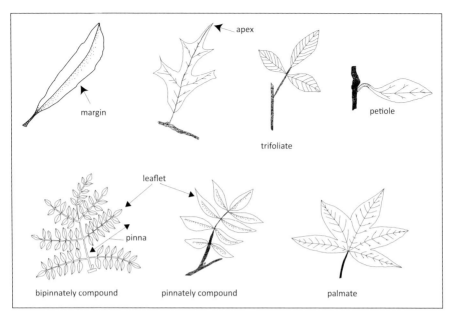

Leaf parts and shape, line illustration. Laura Sloan Crosby.

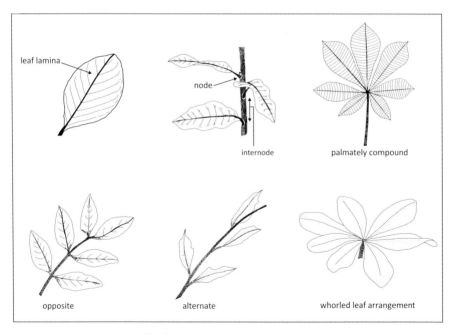

Leaf parts and arrangement, line illustration. Laura Sloan Crosby.

Ferns are in the class Polypodiophyta and constitute only a small proportion of the archipelago's flora. Although ferns can be found in a variety of habitats, they typically occur in cooler, moister locations. Their growth forms include vines, those spreading by rhizomes, or species with a clumping habit, and the leaves vary from pinnate (leaflets arising from a central stem) to multiple levels of division (leaves appearing lacy). Many species have distinctive spore-producing structures (sori) whose variation in shape, size, and location on leaves or leaflets can be used for identification. See figures and glossary for specific terminology.

Bracken Fern   *Pteridium aquilinum*
Family: Dennstaedtiaceae
Range: Archipelago-wide; worldwide in temperate, subtropical, and tropical regions.
Description: Grows from underground stem (rhizome), covered with lance-shaped, brown scales. The alternate stiff fronds are up to 3 m (10 ft) long with leaves arranged on opposite sides of a central stalk. Each frond may be divided pinnately four to six times. The lower frond surface is covered with scales of varying shapes and sizes (triangular, round, shield-shaped, or pointed). Sporangia (spore-bearing structures) are grouped together in marginal sori on the underside of the middle and upper frond lobes. **Habitat:** Coppice, pine woodlands, and disturbed habitats; in drier habitats than most ferns.

Giant Leather Fern   *Acrostichum danaeifolium*
Family: Pteridaceae
Range: North and central archipelago, Florida, Caribbean, Mexico, Central and South America.
Description: Grows with a basal rosette of fronds (a whorl of fronds radiating from a crown or center) up to 4 m (13 ft) long. Both fertile fronds (with spores in sori) and sterile fronds (no spores) occur. The sterile fronds are pinnately compound with pointed pinnae up to 60 cm (24 in) long. The fertile fronds are similar but longer, more erect, and with spore-producing sporangia covering the entire lower surface of the leaflets. **Habitat:** Along edges of brackish and freshwater ponds, sinkholes, and marshes with standing water year-round.

Maidenhair Fern   *Adiantum capillus-veneris*
Family: Pteridaceae
Range: Northern archipelago, also temperate, subtropical, and tropical regions.
Description: Delicate, light green fronds are bipinnate up to 50 cm (20 in) in length. Grows from underground stems (rhizomes), emerging every 20–60 cm (8–24 in). Stem is black and wiry, and the leaflets are notched at the outer edges. Entire frond is lax and hangs down. The sori (spore-producing structures) are on the edges of the leaflets. **Habitat and use:** Moist areas including sinkhole walls in coppice. Used in horticulture.

Pteridium aquilinum

Acrostichum danaeifolium

Adiantum capillus-veneris

TREES AND SHRUBS

Only four species of gymnosperms, non-flowering plants, grow in the archipelago, and they occur in two families: Pinaceae (pines) and Zamiaceae (coonties or cycads). The pines and junipers form medium- to large-sized trees with needles or scales, whereas the coonties are low growing with pinnately compound leaves on the crown of a fleshy thickened root. See figures and glossary for specific terminology.

Caribbean Pine   *Pinus caribaea* var. *bahamensis*
Division: Pinophyta (Conifers and allies)
Family: Pinaceae (Pines)
Range: Endemic subspecies. Northern archipelago (Grand Bahama, Abaco, New Providence, Andros) and southern archipelago (Pine Cay, North Caicos, Middle Caicos) but not in central islands. Much more common and widespread in northern archipelago; species also in Cuba and Central America.
Description: Unbranched, crown-forming tree to 30 m (100 ft) tall and trunk up to 75 cm (30 in), with dark reddish-brown flaking bark. The leaves are needles to 25 cm (10 in) long and in clusters (fascicles) of two or three. Male and female cones occur on the same tree. The hanging (pendulous) male cones develop from the base of new shoots, grow to 4 cm (2 in) long, in groups of one to four. The large female cones are solitary and up to 15 cm (9 in) long, with hardened scales with a central protruding bump and spine. Seeds are winged for wind dispersal. **Habitat and use:** Extensively logged in the 20th century for lumber and cardboard pulp. The national tree of the TCI.

Coontie   *Zamia integrifolia*
Division: Cycadophyta (Cycads)
Family: Zamiaceae (Sago Palms)
Range and status: Northern archipelago, Florida, Cuba; threatened throughout much of its range.
Description: Grows from underground stem or tuber to 75 cm (30 in) high. Pinnately compound leaves to 50 cm (20 in) long with narrow leaflets, each to 20 cm (8 in) long and 0.5 cm (0.25 in) wide, with a slightly toothed tip, an entire margin (occasionally irregular), and covered with rusty brown hairs when young. Separate male and female plants each produce cones covered with reddish hairs. Male cones (8 cm by 2 cm [2 in by 0.5 in]) are narrower and shorter than female cones (13 cm by 4 cm [5 in by 2 in]), the latter having five to eight series of hexagonal structures with seeds. **Habitat and use:** Coastal coppice, shrublands/dwarf shrublands, and in pine woodlands. Host plant of the rare Atala butterfly. It is listed as Near Threatened.

Pinus caribaea var. *bahamensis*

Zamia integrifolia

Angiosperms (Flowering Plants)

Most of the flora of the archipelago are flowering plants, which are found across all habitats. Major families represented are legumes, for example, peas and beans (Family Fabaceae), spurges (Euphorbias; Family Euphorbiaceae), asters and daisies (Family Asteraceae/Compositae), members of the coffee family (Family Rubiaceae), orchids (Family Orchidaceae), and grasses (Family Poaceae).

Vegetative characteristics that aid in their identification include 1) growth form (woody or non-woody; see below for more details); 2) leaf characteristics including arrangement, size, shape, color, surface texture, and level of division; 3) floral characteristics including arrangement of the flowers, the number of parts and color of each level (sepals, petals, stamens, and pistil); and 4) fruit or seed characteristics (size, type, arrangement, and the color of the fruit at maturity). See figures and glossary for specific terminology.

To further facilitate identification, the Angiosperms are grouped into two general categories: trees and shrubs—woody species having single trunk (tree) or multiple stems (shrub); and wildflowers—mostly herbaceous or non-woody plants including grasses, sedges, rushes, cacti, and vines; some may be partially woody, often at the base; many with showy flowers (e.g., orchids, daisies, morning glories, and lilies), but others without showy flowers (e.g., wind-pollinated grasses). Within each of these two categories the species are ordered alphabetically by family, and within a given family they are ordered alphabetically by genus name.

Prickly Bush   *Oplonia spinosa*
Family: Acanthaceae (Acanthuses)
Range: Archipelago-wide, Greater Antilles.
Description: Thin, twining shrub/vine to 3 m (10 ft) long, trailing over other vegetation. The stem is less than 1 cm (0.25 in) thick with backward curving (recurved)
spines and opposite, simple, oblong leaves to 2.5 cm (1 in). Bilateral, five-parted,
purplish-pink, hermaphroditic flowers have petals fused into a tube, with the flat
lobes at the open end being divided into two upper and three lower. Flowers are
solitary in leaf axils. Fruit is a dry capsule that splits to release the seeds. **Habitat:**
Coppice.

Glasswort   *Salicornia virginica*
Family: Amaranthaceae (Goosefoots)
Range: Archipelago-wide, Caribbean, U.S., Europe, North Africa.
Description: Low, mat-forming succulent (fleshy plant, parts of which retain water),
that grows from branching rhizomes, up to 50 cm (20 in) long, with reddish-green
stems that become woody with age. Leaves are reduced to tiny scales and the
green (photosynthetic) stems are segmented. Green, radial flowers are in threes
and embedded in stems, forming a spike at the branch tip that resembles fat green
stems with tiny, yellow projections, which are the stamens. Fruit is an inflated
single-seed bladder (utricle) covered in curved hairs. **Habitat:** Saline areas and
mangrove flats.

Brazilian Pepper   *Schinus terebinthifolius*
Family: Anacardiaceae (Cashews)
Range: Introduced. Invasive from South America now in northern archipelago,
Florida.
Description: Shrub or tree to 8 m (26 ft) tall with peeling, scaly bark. Alternate, pinnately compound leaves to 16 cm (6 in) long, with seven to eleven sessile (stalkless)
leaflets. Elongate/oblong leaflets to 5 cm (2 in) with a notched leaf margin. Radial,
five-parted, white, dioecious flowers are in dense panicles (clusters) in leaf axils.
Mature fruit is a bright red drupe that hangs in large bunches. **Habitat:** Invades
disturbed habitats and freshwater wetlands. Birds consume fruits and disperse the
seeds. **Important:** Contact with leaves or branches can cause dermatitis in some
people.

Poison Wood   *Metopium toxiferum*
Family: Anacardiaceae (Cashews)
Range: Archipelago-wide, Florida, Caribbean.
Description: Shrub to large tree up to 15 m (50 ft) tall with a trunk diameter to 50 cm
(20 in), and orange and brown bark that peels in plates. The trunk and branches
produce a black sap. Alternate, pinnately compound leaves with five to seven ovate

Oplonia spinosa

Schinus terebinthifolius

Metopium toxiferum

Salicornia virginica

Salicornia virginica

leaflets to 10 cm (4 in), which cluster at branch tips, and develop black spots as they age. Radial, five-parted, dioecious, white flowers are in panicles (clusters). Mature fruit is an orange-yellow drupe c. 1 cm (0.5 in) long. **Habitat and use:** Coppice, shrublands, and pine woodlands, where it is often the dominant understory species. Fruits are consumed by the White-crowned Pigeon and the Cuban Parrot. **Important:** Oils from the sap can cause dermatitis ranging in severity from an irritating red rash to intense blistering of the skin. Avoid contact with leaves, branches, and trunk.

Pond Apple   *Annona glabra*
Family: Annonaceae (Custard-apples)
Range: Archipelago-wide, Florida, Caribbean, Central and South America, West Africa.
Description: Tree to 8 m (26 ft) tall with simple, alternate leaves to 20 cm (8 in) long, occurring in one plane. Radial, fleshy, yellowish-white, three-parted flowers are solitary in leaf axils and hang down like Chinese lanterns. Apple-sized, green-yellow fruit is a fleshy aggregate (made up of fruitlets embedded in a fleshy matrix) containing c. 100 seeds. **Habitat and use:** Freshwater wetlands. Fruit is edible, and the wood is used for fishing floats.

Frangipani   *Plumeria obtusa*
Family: Apocynaceae (Dogbanes)
Range: Archipelago-wide, Greater Antilles, Central America.
Description: Shrub up to 8 m (26 ft) tall with a trunk diameter to 20 cm (8 in), and whitish-gray bark. Alternate, oblong leaves to 25 cm (10 in) are clustered at branch tips. All parts exude a milky sap when broken. Radial, five-parted, yellow-centered, white flowers are arranged in umbel-like panicles (umbrella-shaped clusters) in leaf axils. Floral parts are fused forming a tube with the free lobes overlapping to one side to form a pinwheel shape. Flowers fragrant, especially at night. Mature fruit is a brown, dry follicle that opens along one side only to release the seeds. Seeds have tufts of hair at one end that aid in aerial dispersal. **Habitat and use:** Coppice and shrub scrublands. Often planted as a garden ornamental. Flowers pollinated at night by sphinx moths, and the foliage is consumed by their brightly striped caterpillars.

Piss-A-Bed   *Vallesia antillana*
Family: Apocynaceae (Dogbanes)
Range: Archipelago-wide, Florida, Greater Antilles.
Description: Shrub up to 4 m (13 ft) tall with a trunk diameter to 10 cm (4 in), and yellowish-gray bark with fissures. Alternate, simple, lanceolate leaves are up to 10 cm (4 in) long. Radial, five-parted, white, hermaphroditic flowers have the petals fused into a tube with the free lobes overlapping to one side to form a pinwheel shape. Flowers are in dichotomous cymes (two-part clusters). Mature fruit is a white berry. **Habitat and use:** Coppice and shrublands. Used to treat gastrointestinal problems.

Vallesia antillana

Vallesia antillana

Annona glabra

Annona glabra

Plumeria obtusa

Silver Top Palm   *Coccothrinax argentata*
Family: Arecaceae (Palms)

Range: Archipelago-wide, Florida, Caribbean, Mexico.

Description: Medium-sized palm tree up to 10 m (33 ft) tall with a trunk diameter to 15 cm (6 in) and smooth bark. Palmately lobed leaves to 75 cm (30 in) long, are arranged spirally at the top of the trunk. The un-split petiole (leaf stalk) base is expanded into a fibrous sheath surrounding the trunk. The lower leaf surface has silvery hairs (hence the common name). Radial, pale yellow, three-parted, hermaphroditic flowers are arranged along a central stalk in a narrowly branched panicle (cluster). Fruit is a small drupe that turns from green to black at maturity and hangs in large, open, grape-like bunches. **Habitat and use:** Mostly in sandy coastal areas where fruits are consumed by birds. Often the dominant species forming a palm coppice. Used to treat pain.

Thatch Palm   *Leucothrinax morrisii*
Family: Arecaceae (Palms)

Range: Archipelago-wide, Florida, Caribbean, Central America.

Description: Medium-sized, unbranched palm tree up to 5 m (16 ft) tall with a trunk diameter to 20 cm (8 in) and smooth bark. Large, simple, palmately lobed leaves, up to 1.3 m (4.5 ft) wide, are deeply divided and arranged spirally at the top of the trunk. Upper leaf surfaces are dark green while the lower are covered in dense, light-colored pubescence (fuzz). The petioles (leaf stalks) split at their bases. The radial, three-parted, pale yellow, hermaphroditic flowers are arranged in multi-branched panicles (long clusters). Mature fruit is a white to light brown drupe. **Habitat and use:** Coppice. Used in strengthening teas and to treat pain.

Buccaneer Palm   *Pseudophoenix sargentii*
Family: Arecaceae (Palms)

Range: Archipelago-wide, Florida, Caribbean, the Yucatan Peninsula.

Description: Unbranched palm tree to 10 m (33 ft) tall with a trunk diameter to 30 cm (12 in) and with distinctive circular leaf scars. Pinnately compound leaves, to 2 m (6.5 ft) long, are deeply divided and arranged spirally. Radial, three-parted, yellowish-white flowers are in panicles (clusters) up to 1 m (39 in) long. Mature fruit is a red berry, with two or three lobes, that hangs in grape-like bunches. **Habitat and use:** Coastal and interior coppice. Sap used to make alcoholic beverages, the fruits for animal feed, and the trees in horticulture.

Coccothrinax argentata

Coccothrinax argentata

Pseudophoenix sargentii

Leucothrinax morrisii

Pseudophoenix sargentii

Sabal Palm   *Sabal palmetto*
Family: Arecaceae (Palms)
Range: Archipelago-wide, Caribbean, North and Central America.
Description: Palm tree to 15 m (50 ft) tall with an unbranched trunk to 70 cm (28 in) in diameter. Leaves to 3 m (10 ft) are spirally arranged with blades up to 2 m (6.5 ft). Petiole extends into the leaf blade causing it to bend back toward the trunk (recurved). Radial, three-parted, greenish-white, hermaphroditic flowers are in panicles (clusters) to 2 m (7 ft) long. Fruit is an edible berry that turns from green to black at maturity and hangs in large, loose bunches. **Habitat and use:** In or near ephemeral freshwater ponds in areas that are sandy or with exposed limestone rock. Indicator of freshwater. Birds consume its fruits. Used for making baskets, roofing, and brooms.

Broom Bush   *Baccharis dioica*
Family: Asteraceae (Asters, Daisies, and Sunflowers)
Range: Archipelago-wide; also Florida, the Caribbean.
Description: Many-branching shrub to 3 m (10 ft) tall. Simple, alternate leaves to 5 cm (2 in) long, oblong to obovate, and a rounded or notched tip. Radial, five-parted, white flowers are in monoecious heads; male heads have twenty to twenty-five flowers and female heads have fifty to sixty flowers. Flower heads are in panicles (clusters). Fruit is a hard, dry, one-seeded achene, with hairs to aid seed dispersal. **Habitat:** Coastal coppice and coastal scrublands and edges of freshwater wetlands.

Sea Ox-eye   *Borrichia arborescens*
Family: Asteraceae (Asters, Daisies, and Sunflowers)
Range: Archipelago-wide; also Florida, the Caribbean, Mexico.
Description: Semi-succulent perennial shrub to 1.5 m (5 ft) tall. Opposite, simple, lanceolate leaves to 12 cm (5 in) long, glabrous (without hairs) or with silvery hairs, and a toothed or entire leaf edge. Composite, "daisy-like," flower heads consists of central, radial, five-parted, yellow flowers that produce fruit, and peripheral, bilateral, three-parted, yellow flowers with obvious petals. Fruit is a hard, dry, dark-colored, one-seeded, needle-like, small achene. **Habitat:** Dunes and rocky shores.

Yellow Top   *Flaveria linearis*
Family: Asteraceae (Asters, Daisies, and Sunflowers)
Range: Northern archipelago; also Florida, the Caribbean, Mexico.
Description: Woody perennial shrub to 1 m (3 ft) tall. Opposite, simple, sessile (stalk-less), linear leaves to 15 cm (6 in). Flower heads are unlike the typical "daisy" of other Asteraceae. Each small head has two to eight, tiny, yellow flowers with no obvious petals, and sometimes a solitary flower with a noticeable (fused) petal. Flower heads are in flat-topped clusters, which may cover the whole plant. Fruit is a hard, dry, one-seeded achene. **Habitat and use:** Pine woodlands and edges of freshwater wetlands. Used in horticulture.

Baccharis dioica

Sabal palmetto

Borrichia arborescens

Borrichia arborescens

Flaveria linearis

77

Horse Bush *Gundlachia corymbosa*
Family: Asteraceae (Asters, Daisies, and Sunflowers)
Range: Archipelago-wide; endemic to the Caribbean.
Description: Semi-woody herb/shrub to 1.5 m (5 ft) tall. Alternate, simple, linear/ oblanceolate leaves to 9 cm (4 in) long. Flowers are arranged in panicles (clusters) of small heads subtended by bracts (phyllaries) that look like a collar or slight swelling of the stem just below the flower head. Radial, five-parted, white flowers are central, and the bilateral, one-parted, white flowers with obvious petals are peripheral. Fruit is a hard, dry, one-seeded achene, with hairs to aid seed dispersal. Fruiting heads resemble a dandelion but with shorter hairs and a more open cluster. **Habitat and use:** Diverse coppice habitats and disturbed sandy habitats. Used to treat pain, dermatological problems, colds, and flu.

Jack-Ma-Da *Koanophyllon villosum*
Family: Asteraceae (Asters, Daisies, and Sunflowers)
Range: Archipelago-wide; also Florida, the Caribbean.
Description: Small- to medium-sized shrub up to 2 m (7 ft) tall with hairy stems and leaves. Opposite, simple, ovate leaves to 7 cm (3 in) long with a slightly toothed edge. Small, radial, five-parted, white to pink flowers have their petals fused into a tube. Up to 15 flowers are arranged in heads with a green, collar-like swelling beneath. Heads are arranged in a corymbose series of heads (flat-topped clusters). Mature fruit is a hard, black, dry, one-seeded achene, with hairs to aid seed dispersal. **Habitat and use:** Edge of coppice, pine woodlands, and disturbed habitats. Used to treat circulatory problems, colds and fevers, diabetes, gastrointestinal illnesses, and infant issues.

Rong Bush *Wedelia bahamensis*
Family: Asteraceae (Asters, Daisies, and Sunflowers)
Range: Endemic to central and southern archipelago.
Description: Woody perennial shrub to 2 m (7 ft) tall. Opposite, simple, ovate/lanceolate leaves to 15 cm (6 in) long with a toothed leaf edge. Bright yellow "daisy-like" flower heads (2.5 cm [1 in] wide) on long stalks. Each flower head divided into two groups: outer or peripheral bilateral, three-parted, female-only flowers, with showy petals; and central radial, five-parted, male and female flowers. Mature fruit is a wedge-shaped, brown, hard, dry, one-seeded achene, c. 4 mm (0.2 in) long, with a shallow cup. **Habitat:** Coastal coppice and shrublands in sand.

Black Mangrove *Avicennia germinans*
Family: Avicenniaceae (Mangroves)
Range: Archipelago-wide; also Florida, the Caribbean, Central and South America.
Description: Low shrub to tall tree up to 10 m (33 ft) tall with dark trunks up to 75 cm (30 in) in diameter. Opposite, lanceolate, dull-grayish yellow-green leaves to 10 cm (4 in). Leaves exude salt crystals from their lower surface. Roots produce pneumatophores (upward growing roots) protruding to 20 cm (8 in) above ground.

Koanophyllon villosum

Wedelia bahamensis

Avicennia germinans

Avicennia germinans

Gundlachia corymbosa

Bilateral, yellowish-white, hermaphroditic flowers are arranged in terminal and axillary spikes up to 7.5 cm (3 in) long. Fruit is a dry capsule c. 2 cm (1 in) long. Seeds germinate while attached to the parent plant and then drop into the water. **Habitat and use:** Saline and brackish environments, often intermixed with Red Mangroves. Used for honey production, woodcarving, and medicinal treatments for gastrointestinal and dermatological problems and hemorrhoids.

Five Finger   *Tabebuia bahamensis*
Family: Bignoniaceae (Bignonias)
Range: Archipelago-wide, Cuba.
Description: Shrub to medium-sized tree up to 10 m (32 ft) tall with a trunk diameter to 30 cm (12 in) and furrowed bark in older trees. Opposite, palmately compound leaves have three or five leaflets all arising from one point like the fingers of a hand. Large c. 6 cm (2.5 in), showy, bilateral, five-parted, funnel-shaped, white to pink flowers have two lips and are in panicles (clusters). Fruit is a cylindrical pod-like capsule c. 12 cm (5 in) long, containing winged seeds. **Habitat and use:** Coppice on both limestone and sand with good drainage and in pine woodlands. Used in strengthening and aphrodisiac teas, for pain, gastrointestinal issues, circulatory, and respiratory problems.

Yellow Elder   *Tecoma stans*
Family: Bignoniaceae (Bignonias)
Range: Archipelago-wide, New and Old World tropics and subtropics.
Description: Large shrub to small tree to 5 m (16 ft) tall with a trunk diameter to 20 cm (8 in). Opposite, pinnately compound leaves have five to thirteen ovate to lanceolate leaflets with a toothed margin. Vegetation may have fuzz (pubescence) or fuzz may be absent (glabrous). Showy, yellow, bilateral, five-parted, tubular, yellow flowers have two lips, and are in terminal racemes (clusters). Mature fruit is a green-brown, pod-like capsule, 10–30 cm (4–12 in) long, containing papery winged seeds. **Habitat and use:** Disturbed habitats and edges of coppice and scrublands. It is the national flower of The Bahamas and is often planted as a garden ornamental.

Strong Back   *Bourreria succulenta*
Family: Boraginaceae (Borages)
Range: Archipelago-wide, Florida, Cuba.
Description: Shrub to small tree to 10 m (33 ft) tall with a trunk diameter to 20 cm (5 in). Alternate, simple leaves are obovate and up to 13 cm (5 in). Small, radial, white, five-parted flowers are tubular with the petals spreading into five flat lobes at the open end. They are arranged in cymes (flat clusters). Mature fruit is a round, red-orange berry in small clusters. **Habitat and use:** Coppice or shrublands. Used to treat diarrhea, fevers, nerves, pain, skin infections and inflammations, weak bladders, and as a component of strengthening and aphrodisiac teas. Fruits are edible to humans and birds. Flowers attract butterflies and other insects.

Tabebuia bahamensis

Tecoma stans

Bourreria succulenta

Cordia sebestena

Cordia sebestena

Geiger Tree   *Cordia sebestena*

Family: Boraginaceae (Borages)

Range: Archipelago-wide, New World tropical and subtropical regions.

Description: Medium-sized tree to 10 m (32 ft) tall. Alternate, ovate leaves up to 25 cm (10 in) are covered with numerous short, very stiff hairs. Showy, bilateral, orange, five-parted flowers have petals fused into a tube that opens out into five to seven, flat, crinkly lobes. They are in terminal corymbs or cymes (flat clusters) at branch ends. Pear-shaped fruit, c. 4 cm (1.5 in) long, is a nut enclosed in the fleshy floral base and it hangs down in clusters, turning from green to white at maturity. **Habitat and uses:** Coastal coppice forest to shrublands. Used to treat gastrointestinal issues and in horticulture.

Bay Lavender   *Tournefortia gnaphalodes*
Family: Boraginaceae (Borages)
Range: Archipelago-wide; also Florida, the Caribbean, Central America.
Description: Woody perennial to 2 m (7 ft) tall. Simple, alternate leaves to 15 cm (6 in) in whorls at branch ends. Stems and leaves are covered with a dense gray pubescence (fuzz). Small, bilateral, white (with a pinkish-yellow center), fragrant, five-parted flowers are fused to form a tube, which opens out into flat lobes, one of which is slightly longer than the others. They are held in two rows along curved stems at twig ends. Fruit is a capsule that hangs down from curved stems. **Habitat:** Dunes and other coastal habitats.

Granny Bush   *Varronia bahamensis*
Family: Boraginaceae (Borages)
Range: Archipelago-wide; also Florida, Cuba.
Description: Shrub to small tree up to 4 m (13 ft) tall. Bark is black with whitish raised spots (lenticels or breathing pores). Simple, alternate, linear/elliptic to 10 cm (4 in) long leaves have a slightly toothed margin and are covered with scabrous (rough) scales. Small, radial, five-parted, white flowers, with both sepals and petals fused into a cup, are arranged in heads. Mature fruit is a red berry, which retains the cup-like remnants of the flower. **Habitat and use:** Coppice, primarily in shrublands and pine woodlands. Flowers attract insects and birds. In the northern Bahamas it attracts the rare Atala Butterfly.

Gammalamme   *Bursera simaruba*
Family: Burseraceae (Torchwoods)
Range: Archipelago-wide; also Florida, the Caribbean, Central and South America.
Description: Low bush to large tree to 15 m (50 ft) tall with distinctive peeling reddish-brown bark. Alternate, compound leaves have three to nine leaflets. Male and female flowers occur on the same plant (monoecious). Radial, greenish-white flowers, five-parted in male and three-parted in female, are in panicles (clusters). Fruit, produced in March–April, is a large green (unripe) to purplish (ripe) berry with three seeds. Fruits are conspicuous after leaves have been dropped; plant is often leafless in March–April. **Habitat and use:** Coppice and pine woodlands. Fruits are consumed by a variety of birds. Used to treat circulatory problems in strengthening and aphrodisiac teas and in horticulture.

Tournefortia gnaphalodes

Tournefortia gnaphalodes

Varronia bahamensis

Bursera simaruba

Varronia bahamensis

Bursera simaruba

Cinnamon Bark   *Canella winterana*
Family: Canellaceae (Canellas)
Range: Archipelago-wide; also Florida, the Caribbean.
Description: Medium- to large-sized tree up to 10 m (36 ft) tall with a light gray trunk to 30 cm (12 in) in diameter. Alternate leaves to 10 cm (4 in) long, cluster toward branch tips. Leaves and bark are aromatic. Radial, red, yellow-centered, five-parted flowers are in panicles (clusters) at branch ends and may be seen in the same cluster as developing fruit. Mature fruit is a soft, purplish berry with black seeds. **Habitat and use:** Grows in coppice. Fruits are consumed by birds. Has been used to treat toothaches and gastrointestinal issues. Fruits and leaves can be used as a spice.

Australian Pine   *Casuarina equisetifolia*
Family: Casuarinaceae (Beefwoods)
Range: Introduced. Archipelago-wide; also Florida, the Caribbean, tropics worldwide. Native to Australia.
Description: Large pine-like tree up to 25 m (82 ft) tall with a trunk diameter to 1 m (3.3 ft) and peeling, gray-brown bark. Leaves are reduced to triangular scales, and the long, hanging, needle-like branches, up to 38 cm (15 in), are photosynthetic. Radial, two-parted, highly reduced flowers are arranged in terminal male spikes and female "cones." Round fruit is an aggregate of woody capsules c. 1.5 cm (0.5 in) forming a dry, warty, brown "cone." Seeds are single-winged samaras. **Habitat:** Disturbed habitats including yards, roadsides, and cleared fields; along shorelines, dunes, rocky shores, and in coastal coppice–palm woodlands. Highly invasive in coastal zones and contributes to loss of beach dunes and native plants.

Pain-in-Back   *Trema lamarckianum*
Family: Celtidaceae (Hackberries)
Range: Archipelago-wide; also Florida, the Caribbean.
Description: Large shrub to small tree up to 5 m (16 ft) tall with a trunk diameter to 20 cm (5 in). Simple, alternate, ovate/oblong, rough leaves to 7 cm (3 in), with a serrate (toothed) margin and three prominent midveins. Radial, monoecious, five-parted, greenish-white flowers are in clusters at leaf axils. Mature fruit is a pink drupe. **Habitat and use:** Disturbed habitats, edges and occasionally the interior of coppice. Used in strengthening teas.

Coco Plum   *Chrysobalanus icaco*
Family: Chrysobalanaceae (Coco Plums)
Range: Archipelago-wide; also Florida, the Caribbean, Mexico south to northern South America.
Description: Shrub to 1.5 m (5 ft) tall with brown bark. Simple, glabrous (hairless), alternate, elliptic/ovate/obovate, two-ranked leaves 10 cm (4 in) long, with entire leaf margin and obtuse/acute leaf tip. Young stems and petioles (leaf stalks) are reddish brown and lightly pubescent (fuzzy). Radial, white, five-parted flowers are

Canella winterana

Chrysobalanus icaco

Canella winterana

Casuarina equisetifolia

Trema lamarckianum

in axillary or terminal few-flowered cymes. Round to oval fruit is a white or purple edible drupe c. 2.5–4 cm (1–1.5 in) in diameter. **Habitat and use:** Coastal coppice and shrublands and around freshwater and saline wetlands. Fruits consumed by birds. Popular garden plant often used as a hedge.

Autograph Tree   *Clusia rosea*
Family: Clusiaceae (Garcinias)
Range: Archipelago-wide; also Florida, the Caribbean, Central and South America.
Description: Medium to large tree up to 15 m (50 ft) tall and with a trunk diameter to 75 cm (30 in), often with adventitious and prop (modified support) roots present. Simple, opposite, stiff, obovate leaves are to 20 cm (8 in) long. Radial, white to pink, six-parted flowers are solitary at either branch ends or in leaf axils. Mature fruit is a light green, slightly fleshy capsule that splits along six suture lines. All portions of the plant exude yellowish latex. **Habitat and use:** Grows in and around sinkholes and wet areas in coppice where its fruits are consumed by birds. Distinctive rose-like flowers are popular in horticulture. Writing on the leaves creates a mark that remains until the leaf dies.

Buttonwood   *Conocarpus erectus*
Family: Combretaceae (White Mangroves)
Range: Archipelago-wide; also Florida, the Caribbean, subtropical and tropical Central and South America, Africa.
Description: Small shrub to large tree up to 20 m (66 ft) tall. Simple, alternate leaves to 15 cm (6 in) can be covered with a fine gray fuzz (pubescence) or can be hairless (glabrous) and green. The petiole (leaf stalk) has two glands at the leaf base. Inconspicuous, tiny (less than 2 mm [0.1 in]), fragrant, greenish flowers are crowded in a ball less than 6 mm (0.25 in) in diameter. Purplish-brown, ball-shaped, hard, cone-like fruit heads c. 1 cm (0.5 in) in diameter are composed of many one-seeded fruits each c. 3 mm (0.15 in) long. **Habitat and use:** Near saline environments along shorelines and interior ponds, but typically only in locations that have occasional flooding. Fruits float, which aids dispersal. Used to treat sores, induce vomiting, for charcoal production, smoking fish, barbecuing, tanning leather, and in horticulture.

White Mangrove   *Laguncularia racemosa*
Family: Combretaceae (White Mangroves)
Range: Archipelago-wide; also Florida, the Caribbean, Central America, Western Africa.
Description: Medium- to large-sized trees to 6 m (20 ft) tall with trunk diameter to 50 cm (20 in). Opposite, oval, hairless (glabrous) leaves 3–8 cm (1–3 in) long, and the petioles have two glands at the base of the leaf blade. Small, radial, white, five-parted, hermaphroditic, bell-shaped, stalkless flowers are arranged in terminal or axillary spikes 5–10 cm (2–4 in) long. Mature fruit is an edible, elongate, reddish-brown drupe, 12–20 mm (0.5–0.8 in) long, with longitudinal ridges. **Habitat and use:** Edges of saline environments both coastal and along interior ponds and lakes. Has been used for honey production, tanning leather, and medicinally as an antiseptic.

Laguncularia racemosa

Conocarpus erectus

Terminalia catappa

Clusia rosea

Clusia rosea

West Indian Almond   *Terminalia catappa*
Family: Combretaceae (White Mangroves)

Range: Archipelago-wide, introduced from Old World tropics and now in Western Hemisphere tropics/subtropics.

Description: Tree to 25 m (82 ft) tall with a trunk diameter to 1.5 m (5 ft). Simple, ovate/oblong alternate leaves to 25 cm (10 in) long, often wider near the tip with an entire leaf margin. Leaves cluster at the ends of short branches. Radial, five-parted, greenish-white, monoecious flowers are in spikes. Large fruit is an edible drupe,

5–7 cm (2–3 in) long and 3–5.5 cm (1–2 in) wide, green at first, then yellow, and finally red when mature, containing a single seed. **Habitat and use:** Disturbed habitats and along edges of beaches and freshwater wetlands. Salt tolerant. Fruit and seeds spread by floating away from parent. Has been used to treat eye problems.

Stiff Cock   *Diospyros crassinervis*
Family: Ebenaceae (Persimmons/Ebonies)
Range: Archipelago-wide; also Greater Antilles.
Description: Shrub to small tree to 8 m (26 ft) tall. Stiff, simple, alternate, ovate leaves are to 8 cm (3 in), dark green above and light green to yellow below. Male and female flowers are on separate trees (dioecious). Radial, three-parted, bell-shaped, small, hairy, yellow-tan flowers emerge in groups (male flowers), or solitarily (female flowers), from leaf axils. Fruit is a yellow-green berry that turns black at maturity. **Habitat and use:** Coppice shrublands. Used in strengthening teas and for gastrointestinal issues.

False Cocaine   *Erythroxylum areolatum*
Family: Erythroxylaceae (Cocas)
Range: Archipelago-wide; also the Caribbean, Central America.
Description: Shrub or small tree to 10 m (33 ft) tall with a trunk diameter to 25 cm (10 in). Simple, alternate, elliptical (pointed) leaves are to 15 cm (6 in) long. The leaf underside shows two faint lines running parallel to the prominent midvein. Small, radial, five-parted, monoecious white flowers arise from the axils above leaf scars. Fruit is a red drupe. **Habitat and use:** Coppice, pine woodlands, and edges of freshwater wetlands. Burning the wood has been used to repel mosquitoes.

Cascarilla   *Croton eluteria*
Family: Euphorbiaceae (Spurges)
Range: Archipelago-wide; also the Caribbean, Mexico, South America.
Description: Shrub to small tree to 4 m (13 ft) tall with the young stems and leaves having silvery-brown scales. Alternate, simple, ovate leaves are to 8 cm (3 in) long. Radial, five-parted, dioecious, white flowers are in axillary or terminal racemes (clusters). Mature fruit is a pea-sized, green, three-parted capsule. **Habitat and use:** Coppice shrublands. Used medicinally to treat appetite, coughs, diarrhea, flu, indigestion, stomach pain, and to prevent vomiting. The bark is used as a flavoring of Campari (an alcoholic liqueur).

Granny Bush   *Croton linearis*
Family: Euphorbiaceae (Spurges)
Range: Archipelago-wide; also Florida, the Caribbean.
Deccription: Small- to medium-sized shrub to 2 m (7 ft) tall. Simple, alternate, linear-elongate leaves with an entire margin have glands at leaf base. The upper surface is glabrous (smooth, hairless), the lower surface has yellowish hairs. Radial, five-parted, white, dioecious flowers are in axillary or terminal racemes (clusters).

Diospyros crassinervis

Croton linearis

Croton eluteria

Gymnanthes lucidus

Erythroxylum areolatum

Erythroxylum areolatum

Inconspicuous mature fruit is a yellow-green, three-parted capsule. **Habitat and uses:** Coppice shrublands and dunes. Used to treat obstetric and gynecological issues, pain, gastrointestinal issues, and circulatory problems.

Crab Wood   *Gymnanthes lucidus*
Family: Euphorbiaceae (Spurges)
Range: Archipelago-wide; also Florida, the Caribbean, Mexico, Central America.
Description: Shrub to small tree up to 10 m (33 ft) tall. Simple, alternate, elliptic/ oblanceolate leaves to 12 cm (5 in) long are glandular toothed along the upper

leaf margin with small emergent lobe to either side of the leaf base. New leaves are reddish. Inconspicuous, bilateral, yellow-green, fragrant, dioecious flowers are situated above a bract and are arranged in terminal and axillary racemes (clusters). Fruit is a pea-like, three-lobed capsule to 1 cm (0.4 in) that turns from green to reddish-brown to black as it ripens and hangs down on a long stalk. **Habitat and use:** Range of coppice habitats. Used to treat gastrointestinal problems, pain, hemorrhoids, and for wood carving. The sap can be a skin irritant.

Manchineel   *Hippomane mancinella*
Family: Euphorbiaceae (Spurges)
Range: Archipelago-wide; also Florida, the Caribbean, Central and South America.
Description: Tree to 20 m (66 ft) tall with a trunk diameter to 50 cm (30 in). Simple, oval, alternate leaves to 12 cm (7 in) with a toothed (crenate) leaf margin, and a circular gland on the petiole. Leaves and stems produce milky latex. Flowers inconspicuous, greenish, with eight to fifteen blossoms in terminal spikes; several male flowers at apex of the spike and one or two female flowers at the base. Mature fruit is a green berry containing flattened seeds. **Habitat and use:** Coastlines in scrubland coppice or behind dunes. **Important:** One of the most poisonous trees in the world. Avoid contact with foliage; sap causes extreme dermatitis. The sap is water soluble and standing under the tree in the rain can also cause dermatitis. The small, green, apple-like fruit may be fatal to humans if consumed.

Rock Bush   *Phyllanthus epiphyllanthus*
Family: Euphorbiaceae (Spurges)
Range: Archipelago-wide; also Florida.
Description: Small- to medium-sized shrub to 2 m (7 ft) tall. There are no leaves; instead the branches are modified photosynthetic leaf-like structures (phyllodia). Alternate, curved phyllodia are narrow and taper toward the tip (lanceolate) with a serrate (toothed) margin. Clusters of tiny, reduced, radial, five- or six-parted, monoecious flowers varying from yellow to red arise from the notches of the phyllodia. Mature fruit is a three-lobed, brown capsule. **Habits and use:** Open rocky areas, edges of coppice and scrublands. Used to treat colds and fevers, coughing, dermatological issues, headaches, hemorrhaging, stomach pains, sore throats, mouth sores, toothaches, and as an anti-emetic and as strengthening teas.

Maiden Bush   *Savia bahamensis*
Family: Euphorbiaceae (Spurges)
Range: Archipelago-wide; also Florida, the Caribbean.
Description: Shrub to small tree to 5 m (16 ft) tall. Simple, alternate, elliptic to oblong/obovate leaves to 7 cm (3 in) long, with rounded leaf tip and entire margin. Paired stipules (small leaf-like appendages) grow at the petiole base. Radial, five-parted, green, dioecious flowers are in leaf axils. Mature fruit is a three-lobed, small, brown capsule that may be slightly pubescent (fuzzy). **Habitat:** Diverse coppice habitats from dwarf shrubland, scrubland to forest and woodland.

Savia bahamensis

Hippomane mancinella

Hippomane mancinella

Phyllanthus epiphyllanthus

Acacia choriophylla

Cinnecord   *Acacia choriophylla*

Family: Fabaceae (Legumes [Peas and Beans])

Range: Archipelago-wide; also Florida, Cuba.

Description: Medium-sized tree up to 10 m (33 ft) tall with a trunk diameter to 40 cm (16 in) that is brown with whitish fissures in the outer layer of the bark. Bipinnately compound leaves are alternate. One to three pairs of pinnae, each with four to ten pairs of dark green, oblong leaflets to 1 cm (0.5 in) in length. Radial, five-parted, hermaphroditic flowers each have numerous, bright orange-yellow stamens that are twice as long as the petals. Flowers are arranged in tight heads that look like small, orange balls that occur terminally or in leaf axils. Mature fruit is a dry, brown pod up to 6 cm (2 in) in length. Seeds are embedded in a white fleshy pulp. **Habitat and use:** Coppice, pine woodlands, sand dunes, and disturbed habitats. Pulp surrounding the seeds is edible.

Brasilleto   *Caesalpinia vesicaria*
Family: Fabaceae (Legumes [Peas and Beans])
Range: Mostly in northern and central archipelago; also the Caribbean, Mexico.
Description: Shrub or small tree to 8 m (26 ft) tall. Alternate, bipinnately compound leaves to 30 cm (12 in) long, with three to six pairs of pinnae. Obovate leaflets are in one to three pairs, to 5 cm (2 in). Spines may be present at the leaf axils. Bilateral, five-parted, yellow, hermaphroditic flowers have the lower petal bent to form a keel and are in long, open racemes (clusters), often at branch tips. Fruit is a pod, up to 10 cm (4 in) long, that turns bright red when mature. **Habitat and use:** Coppice and scrublands. Used medicinally to treat low blood pressure, and the wood produces a red dye. Eleutheran adventurers sent ten tons of wood to Massachusetts in the 16th century to repay for supplies. Money from the sale of the wood was used to purchase the first building at Harvard University.

Coastal Senna   *Chamaecrista lineata*
Family: Fabaceae (Legumes [Peas and Beans])
Range: Archipelago-wide; also Greater Antilles.
Description: Shrub to 2 m (7 ft) tall. Pinnately compound, alternate leaves to 5 cm (2 in) with a round and depressed gland between the lower leaflets. Leaflets are in three to six pairs, gray green, oval, and sessile (stalkless and attached directly to the petiole). Lanceolate stipules occur at the petiole base. Distinctive, bilateral, five-parted, yellow, hermaphroditic flowers are in axillary groups of one to five. Fruit is a pod 3–5 cm (1.3–1.8 in) long that twists and splits at maturity. **Habitat:** Dunes and coastal coppice, shrublands, and palm woodlands.

Royal Poinciana   *Delonix regia*
Family: Fabaceae (Legumes [Peas and Beans])
Range: Introduced. Archipelago-wide; also the Caribbean, Florida, Central America. Native to Madagascar.
Description: Tree up to 15 m (50 ft) with a trunk diameter to 1 m (3 ft), often with buttresses. Bark is gray with prominent lenticels. Bipinnately compound, alternate leaves to 50 cm (20 in), with ten to twenty-five pinnae, each with ten to forty oblong leaflets to 1 cm (0.5 in). Showy, bilateral, five-parted, bright red-orange flowers are in panicles (clusters) and may cover the whole tree. Mature fruit is a long, flat pod up to 50 cm (20 in), containing up to ten seeds. **Habitat and use:** Disturbed habitats. Used in landscaping because of the large, showy inflorescences.

Caesalpinia vesicaria

Chamaecrista lineata

Haematoxylum campechianum

Delonix regia

Delonix regia

Logwood   *Haematoxylum campechianum*

Family: Fabaceae (Legumes [Peas and Beans])

Range: Introduced from Central America. Archipelago-wide; also Florida, the Caribbean.

Description: Tree or shrub to 10 m (33 ft) tall with trunk diameter to 50 cm (20 in), branches often spiny. Evenly pinnately compound, alternate leaves to 7 cm (2.8 in), with the two to four oval leaflets each to 2 cm (1 in), with rounded leaf tips, which may have a small central notch (and appear heart-shaped) and an entire (untoothed) margin. Slightly bilateral, five-parted, reddish-purple and yellow flowers are in dense, axillary racemes (lateral clusters). Fruit is a brown pod, 3–5 cm (1–2 in) long, and containing two or three seeds. **Habitat and uses:** Disturbed habitats. Used medicinally to treat circulatory issues and in strengthening teas and has been traditionally used to make dyes for textiles and microscopy. A blue dye can be extracted from the heartwood.

Jumbay   *Leucaena leucocephala*
Family: Fabaceae (Legumes [Peas and Beans])
Range: Introduced. Archipelago-wide, tropics and subtropics worldwide.
Description: Shrub to medium-sized tree up to 10 m (33 ft) tall. Alternate, evenly bipinnately compound leaves, with four to nine pinnae to 10 cm (4 in), each with eleven to seventeen pairs of leaflets. Between the lower pinnae there may be a small, circular depressed gland. Base of the petiole and leaflets are swollen. Radial, five-parted, hermaphroditic flowers each have ten stamens that are five times as long as the petals. Flowers are arranged in tight heads that resemble powder-puff balls and occur in axillary racemes (clusters). Mature fruit is a brown pod, 14–26 cm (5–10 in) long, containing eight to eighteen flat, hard, shiny seeds. **Habitat and use:** Disturbed habitats and edges of coppice. Used to treat colds, fevers, and flu, in strengthening teas, circulatory problems, to calm nerves, to treat tuberculosis, to reduce back pain and menstrual cramps. Leaves and seeds are edible if cooked. It was introduced for livestock feed, but it causes horses to lose their hair. Plants can be invasive on disturbed sites.

Wild Tamarind   *Lysiloma latisiliquum*
Family: Fabaceae (Legumes [Peas and Beans])
Range: Archipelago-wide; also Florida, the Caribbean, Central America.
Description: Medium- to large-sized tree up to 20 m (66 ft) tall with a trunk diameter to 1 m (4 ft) and smooth, grayish-white bark. Alternate, bipinnately compound leaves have large ovate appendages at the leaf node (stipules). There are two to five pairs of pinnae, each with ten to thirty-five pairs of leaflets. Radial, five-parted, hermaphroditic flowers each have numerous, showy, white stamens that are eight to ten times longer than the petals. Flowers are arranged in heads that resemble delicate powder-puff balls. Heads occur in axillary racemes (clusters). Fruit is a pod, 13–20 cm (5–8 in) long, that becomes slightly twisted at maturity, with the outer coating peeling away so that the fruit appears white and brown; contains five to ten dark brown seeds. **Habitat and use:** Coppice, pine woodlands, and disturbed areas. Used for wood carving and boat building.

Horseflesh   *Lysiloma sabicu*
Family: Fabaceae (Legumes [Peas and Beans])
Range: Northern and central islands of archipelago; also Florida, Greater Antilles.
Description: Medium- to large-sized tree up to 12 m (40 ft) tall with a trunk diameter to 1 m (4 ft). Bark is gray-brown and peels in strips. Sapwood has a distinctive dark red color when cut. Alternate, bipinnately compound leaves to 15 cm (6 in) have

Lysiloma latisiliquum

Leucaena leucocephala

Lysiloma latisiliquum

Lysiloma sabicu

Lysiloma sabicu

two to four pinnae, each with three to seven pairs of leaflets. Radial, five-parted, hermaphroditic flowers have numerous greenish-white stamens that are three to four times longer than the petals. Flowers are arranged in heads that resemble powder-puff balls. Heads occur in axillary racemes (clusters). Fruit is a pod, 7–15 cm (3–6 in) long, that becomes slightly twisted at maturity. Over time as the fruit ages the outer coating peels away so that the fruit appears white and brown-black. **Habitat and use:** Coppice. Used for wood carving, furniture, house trim, and boat building.

Dogwood   *Piscidia piscipula*
Family: Fabaceae (Legumes [Peas and Beans])
Range: Archipelago-wide; also Florida, the Caribbean, Central America.

Description: Medium- to large-sized tree to 15 m (50 ft) tall with gray bark, having prominent lenticels (breathing pores). Alternate, pinnately compound leaves have five to nine leaflets. Leaflets are oblong to obovate, to 12 cm (5 in). The lower leaflet surface is covered with short hairs. Bilateral, five-parted, hermaphroditic flowers have white or light pink petals, tinged with red or pink on the keel. They are arranged in panicles (clusters). The upper petal of each is enlarged to form a standard or banner (petal), and two lower petals are fused to form a keel giving a "pea type" flower. Fruit is a winged pod, 7.5–10 cm (3–4 in) long, which is brown at maturity and contains black seeds. **Habitat and use:** Edges of coppice. Used to treat chiggers and rheumatism as well as in charcoal production and for fence posts, boat building, and in horticulture. The name derives from its use in ship-building where it was carved to make the central axis or "dog" on a boat.

Ram's Horn   *Pithecellobium keyense*
Family: Fabaceae (Legumes [Peas and Beans])
Range: Archipelago-wide, Florida, Greater Antilles, Mexico.

Description: Shrub to small tree to 7 m (23 ft) tall. Typically, the trunk is narrow, less than 15 cm (6 in) diameter, multi-branched and can be rough, with shallow fissures with occasional prickles on the stems. Alternate, bipinnately compound leaves have two to four obovate, sessile leaflets. Tiny, radial, five-parted, hermaphroditic flowers are in heads 2–3 cm (0.8–1.2 in) wide, and they vary in color from white to pink to reddish. The numerous stamens of each flower are much longer than the petals, so mature heads resemble powder-puff balls. Flower heads are arranged in panicles (clusters). Fruit is a pod, 10–15 cm (4–6 in) long, that coils as it matures (hence the name Ram's Horn) and contains black seeds surrounded by a red fleshy tissue (aril). **Habitat and use:** Coppice, pine woodlands, and disturbed habitats. The red arils are edible and consumed by birds. Used in horticulture because of its flowers and distinctive coiled fruits.

Necklace Pod   *Sophora tomentosa*
Family: Fabaceae (Legumes [Peas and Beans])
Range: Archipelago-wide; also the Caribbean, New World tropics and subtropics.

Description: Shrub to 3 m (10 ft) tall with yellowish-brown bark. Alternate, pinnately compound leaves to 35 cm (14 in) with eleven to nineteen oval leaflets to 5 cm (2 in) that hang down either side of the stem. Young stems and leaves are covered with fine, grayish pubescence (fuzz). Bilateral, five-parted, yellow flowers are in racemes (clusters) to 50 cm (20 in) in length, from twig ends. Mature fruit is a brown pod, up to 20 cm (8 in) long, and very constricted between the two to ten brown seeds, resembling a necklace; hanging in dense clumps. **Habitat and use:** Coppice, shrublands, dunes, and rocky shores. Flowers attract insects and birds, and it is used in horticulture.

Piscidia piscipula

Piscidia piscipula

Sophora tomentosa

Piscidia piscipula

Pithecellobium keyense

Pithecellobium keyense

Pithecellobium keyense

White Zapoteca *Zapoteca formosa*
Family: Fabaceae (Legumes [Peas and Beans])
Range: Archipelago-wide; also Cuba.

Description: Shrub to 3 m (10 ft) tall. Bipinnately compound, alternate leaves to 15 cm (6 in) with pinnae in two to three pairs, each with eleven pairs of oblong/ obovate leaflets, with rounded leaf tips and a smooth margin. Radial, five-parted, hermaphroditic flowers are arranged in round powder-puff heads with numerous showy, white stamens seven to ten times the length of the petals. Mature fruit is a brown pod, up to 15 cm (6 in) long, that splits open to shed seeds that are brown or black with a mottled surface. **Habitat:** Coppice, scrublands, pine woodlands, and edges of disturbed habitats.

Black Ink Berry *Scaevola plumieri*
Family: Goodeniaceae
Range: Archipelago-wide; also Florida, the Caribbean, tropical and subtropical Western Hemisphere.

Description: Low shrub to 1.5 m (5 ft) tall, rooting at the nodes and forming mounds. Simple, alternate leaves are semi-succulent, glabrous (smooth), ovate, obovate, or oblanceolate to 15 cm (6 in) and orange along the margins. No hairs in the axils. Bilateral, five-parted, pinkish-white flowers are in cymes (small clusters) in leaf axils. Lobes of the flowers are vertically split and spread like a fan, giving a half-flower appearance. Mature fruit is a glossy, dark blue-black drupe. **Habitat:** Sand dunes.

White Ink Berry *Scaevola taccada*
Family: Goodeniaceae
Range: Introduced. Invasive. Archipelago-wide; also Florida, the Caribbean. Native to South Pacific and Hawaii.

Description: Shrub to 4 m (13 ft) tall occasionally rooting at the nodes. Simple, alternate, oval leaves to 20 cm (8 in), with a notched leaf tip, are thick and fleshy. White hairs in the axils and branch tips. Small, bilateral, five-parted, pinkish-white, hermaphroditic flowers are in cymes (small clusters) arising in leaf axils. Floral parts split along one line forming a half-flower appearance that has five lobes with wavy edges. Mature fruit is a white drupe to 2 cm (1 in). **Habitat:** Sand dunes, rocky shores, and disturbed habitats. Also known as beach cabbage or sea lettuce.

Bastard Stopper *Petitia domingensis*
Family: Lamiaceae (Mints)
Range: Northern and central archipelago; also Florida, the Caribbean.

Description: Shrub to tree to 15 m (50 ft) tall. Simple, opposite, ovate leaves to 20 cm (8 in) with an entire leaf margin and a pointed leaf tip. Lower surface is lighter colored with prominent yellow veins. Leaf has a rough appearance and is stiff. Small,

Scaevola plumieri

Zapoteca formosa

Scaevola plumieri

Scaevola taccada

Petitia domingensis

Petitia domingensis

radial, four-parted, white, monoecious flowers look like small, green bells, with flat white petals spreading at the open end, and are in panicles (clusters) in leaf axils. Mature fruit is a red-black berry occurring in dense grape-like bunches. **Habitat and use:** Coppice and pine woodlands. Used to treat respiratory problems.

Lancewood   *Ocotea coriacea*

Family: Lauraceae (Laurels)

Range: Archipelago-wide; also Florida, the Caribbean, Central America.

Description: Tree up to 12 m (50 ft) tall with a trunk diameter to 50 cm (20 in). Alternate, simple, lanceolate leaves to 15 cm (6 in) with petioles in a distinct hook shape. Radial, white, three-parted flowers are arranged in tight panicles (clusters). Mature fruit is a black drupe set in a cup-shaped base. **Habitat and use:** Coppice. Birds consume its fruits. The wood contains a large volume of oil and branches can be burned as a light source, thus another common name is Torchwood.

Guana Berry   *Byrsonima lucida*

Family: Malpighiaceae

Range: Archipelago-wide; Florida, the Caribbean.

Description: Large shrub to small tree to 5 m (16 ft) tall. Simple, opposite, obovate/spathulate (spatula-shaped) leaves 6 cm (2.5 in) with leaf tip rounded with central indentation (retuse) and fine hairs (trichomes) throughout. Small, radial, five-parted, hermaphroditic flowers, with spoon-shaped petals arranged like spokes on a wheel, are initially white but turn red and occur in racemes or panicles (clusters) at the end of branches. A series of paired sepal glands occur at the base of the flower, which remain on the fruit. Mature fruit is an edible fleshy, orange-brown drupe. **Habitat and uses:** Diverse coppice habitats, pine woodlands, and near freshwater wetlands. Used for obstetric and gynecological issues, gastrointestinal problems, in strengthening teas, and in horticulture.

Sea-side Mahoe   *Thespesia populnea*

Family: Malvaceae (Hibiscus)

Range: Archipelago-wide, pantropical.

Description: Shrub to tree to 8 m (26 ft) tall with peeling, scaly bark. Simple, alternate, ovate to round leaves to 13 cm (5 in) long, with a smooth margin, sharp pointed tip, and heart-shaped base with glands. Large, radial, five-parted, yellow with a red-purple center, hermaphroditic flowers are bell-shaped and occur in the leaf axils. Petals darken to almost purple with age. Mature fruit is a brown woody capsule that breaks apart, containing hairy seeds. **Habitat and use:** Near fresh and saline wetlands, dunes, and coastal disturbed habitat. Used medicinally to treat pain (headaches).

Wild Guava   *Tetrazygia bicolor*

Family: Melastomataceae (Melastomes)

Range: Northern and central archipelago; also Florida, Cuba.

Description: Shrub to small tree to 5 m (16 ft) tall. Simple, opposite, lanceolate leaves to 20 cm (8 in) long with three primary parallel veins that have lateral veins at a 90-degree angle. Bilateral, white, five-parted, hermaphroditic flower with unfused petals, and distinctive hook-shaped stamens with yellow anthers are in panicles

Byrsonima lucida

Ocotea coriacea

Tetrazygia bicolor

Thespesia populnea

Tetrazygia bicolor

(clusters). Mature fruit is a purple berry 1.3 cm (0.5 in) in diameter, held erect in small clusters. **Habitat and use:** Coppice, scrubland, and pine woodlands where a variety of birds feed on its fruits. Used to treat colds, coughs, and night sweats and used in horticulture.

Mahogany or Madiera *Swietenia mahagoni*
Family: Meliaceae (Mahoganies)
Range: Archipelago-wide; also Florida, the Caribbean, Central and northern South America.
Description: Medium- to large-sized tree up to 15 m (50 ft) tall with a trunk diameter to 1 m (4 ft). Bark is smooth in young specimens, darkening and becoming furrowed with age. Alternate, pinnately compound leaves have leaflets to 8 cm (3 in), with an unequal leaflet base, with one side being narrower than the other, resulting in a curved mid-rib. Radial, five-parted, greenish-white, hermaphroditic flowers are in inconspicuous, small panicles (clusters). Mature fruit is a brown, oval, fleshy capsule the size of a baseball that when mature breaks apart to release flat, long-winged seeds. **Habitat and use:** Coppice and pine woodlands. A valuable timber tree. Used in strengthening and aphrodisiac teas, to treat dermatological issues, for pain, and to help with obstetric and gynecological problems.

Golden Wild Fig *Ficus aurea*
Family: Moraceae (Figs or Mulberries)
Range: Archipelago-wide (except Inagua and the TCI); also Florida, the Caribbean.
Description: Tree to 20 m (66 ft) tall with a trunk diameter to 1.25 m (4 ft) and branches producing aerial roots that can become secondary trunks. Simple, oblong, alternate leaves to 25 cm (10 in) long. A prominent ring occurs on the stem where the petiole attaches. Leaves and branches exude a milky latex when broken. Highly reduced, radial, monoecious flowers are borne entirely within each fig (known as a synconium) and are fertilized by wasps. The sessile (stalkless), berry-like fig is yellowish-red to purple and contains tiny fruits with seeds. **Habitat and use:** Coppice woodlands or shrublands around sinkholes. Occasionally found in Sabal Palm woodlands. Used to treat cancer, gastrointestinal problems, circulatory issues, dermatological matters, and for pain. Fruits consumed by bats and birds.

Myrtle of the River *Calyptranthes zuzygium*
Family: Myrtaceae (Myrtles)
Range: Northern archipelago; also Florida, Greater Antilles.
Description: Large shrub to small tree up to 11 m (36 ft) tall. Simple, opposite, elliptic to ovate leaves to 8 cm (3 in) long, with translucent dots on the lower surface. Leaves, when crushed, have a slight odor. Radial flowers have no petals but have numerous white stamens and are in short terminal or lateral panicles (clusters). Mature fruit is an edible, dark blue berry. **Habitat:** Coppice and scrublands.

White Stopper *Eugenia axillaris*
Family: Myrtaceae (Myrtles)
Range: Archipelago-wide; also Florida, the Caribbean, Central and South America.
Description: Shrub to small tree to 9 m (30 ft) tall with a trunk diameter to 30 cm (12 in). Simple, opposite, ovate to elliptical leaves to 8 cm (3 in) long. Petioles

Swietenia mahagoni

Calyptranthes zuzygium

Ficus aurea

Eugenia axillaris

Eugenia axillaris

and young developing leaves are reddish and release a menthol-like odor when crushed. Radial, four-parted, white, hermaphroditic flowers have numerous showy stamens and are arranged in short racemes (clusters) in leaf axils. Mature fruit is an edible, reddish-black berry c. 10 mm (0.4 in) in diameter. **Habitat and use:** Wide range of coppice. Fruits are consumed by birds. Used to treat gastrointestinal problems, respiratory issues, pain, and for strengthening teas, a bathing solution for women, and in horticulture.

Spanish Stopper   *Eugenia foetida*
Family: Myrtaceae (Myrtles)
Range: Archipelago-wide; also Florida, the Caribbean, Mexico, Central America.
Description: Large shrub to small tree to 7 m (23 ft) tall with a trunk diameter to 30 cm (12 in). Simple, opposite, obovate/oblanceolate leaves to 5 cm (2 in) long; when crushed have a pungent odor. Lower leaf surface has small black dots. Radial, four-parted, white, hermaphroditic flowers have numerous showy stamens and are arranged in short terminal or axillary racemes (clusters). Mature fruit is an orange-yellow to brownish-black berry, 8 mm (0.3 in) in diameter. **Habitat and use:** Mature and shrubland coppice. Fruits are consumed by birds. Used in horticulture.

Sweet Margaret   *Mosiera longipes*
Family: Myrtaceae (Myrtles)
Range: Archipelago-wide; also Florida, the Caribbean, Mexico.
Description: Low shrub to small tree to 5 m (16 ft) tall with a trunk diameter to 30 cm (12 in). Simple, opposite, lanceolate leaves to 6 cm (2.5 in) long. Young stems and petioles often a reddish color and have a menthol-like odor when crushed. Radial, four-parted, white to pink, fragrant, hermaphroditic flowers have numerous showy stamens and occur in groups of one to four in leaf axils. Mature fruit is an edible, dark red berry that retains the four sepals at the top like a crown. **Habitat and uses:** Coppice, pine woodlands, rocky shores, and saline-mangrove flats above high tide line. Used in general soothing teas, to treat colds, diarrhea, stomachaches, and as an aphrodisiac tea for men.

Paper Bark Tree   *Melaleuca quinquenervia*
Family: Myrtaceae (Myrtles)
Range: Introduced. Archipelago-wide; also Florida, the Caribbean. Native to Australia, Papua New Guinea, New Caledonia.
Description: Shrub to small tree to 12 m (39 ft) tall with a trunk diameter to 30 cm (12 in). Bark is brownish-white, corky, and peels in layers. Stiff, simple, alternate, elliptic to lanceolate, gray leaves to 10 cm (4 in) long with prominent veins. Petioles and leaf edges are a reddish color and will release a menthol-like (camphor) odor when crushed. Sessile (stalkless), radial, five-parted, white, hermaphroditic flowers have numerous showy stamens arranged in short spikes up to 16 cm (6 in) long and resemble bottle brushes. Woody dry fruit is a cylindrical capsule, 2.5–4 mm (0.1–0.2 in) long, and clustered, spike-like along the branches. Each holds 200–300 tiny, orange-brown seeds. **Habitat:** An aggressive nonnative invasive of disturbed habitats, freshwater wetlands, and pine woodlands.

Small Leaved Blolly   *Guapira discolor*
Family: Nyctaginaceae (Four O'Clocks)
Range: Archipelago-wide. also Florida, the Caribbean.
Description: Small- to medium-sized tree to 7 m (23 ft) tall with a trunk diameter to 25 cm (10 in). Simple, opposite, oblong (sometimes fleshy) leaves to 12 cm (5 in)

Eugenia foetida

Eugenia foetida

Melaleuca quinquenervia

Guapira discolor

Guapira discolor

Mosiera longipes   *Mosiera longipes*

long have a translucent midvein. Leaves can be variable in shape within the same plant. Branches hang or droop. Very small, five-parted, greenish-yellow, monoecious flowers are borne on loose, long-stalked clusters from ends of twigs. Mature fruit is a bright red drupe that is ribbed when it dries. **Habitat and use:** Coppice and pine woodlands. Used in horticulture, and birds consume its fruits.

Big Leaf Blolly   *Guapira obtusata*
Family: Nyctaginaceae (Four O'Clocks)

Range: Archipelago-wide; also Florida, Greater Antilles.

Description: Large shrub to medium-sized tree to 8 m (26 ft) tall with a trunk diameter to 35 cm (14 in). Dark green, simple, opposite, elongate, stiff leaves to 15 cm (6 in), with a white midvein and recurved (rolled inward) leaf margins. Radial, funnel-shaped, small, dioecious, yellowish-green flowers, with red pubescence (fuzz), are arranged on branched stalks in dense, small panicles (clusters). Mature fruit is a bright red drupe, which develops ribs as it dries. **Habitat and use:** Coppice. Fruits are consumed by birds.

Alvaradoa   *Alvaradoa amorphoides*
Family: Picramniaceae

Range: North and central archipelago; also Florida, the Caribbean, Central and South America.

Description: Large shrub to small tree to 10 m (33 ft) tall with a trunk diameter to 25 cm (10 in) and pubescent (fuzzy) branches. Pinnately compound, alternate leaves have nineteen or more leaflets. The leaflets are light green, oval-oblong, and the lower surface is pubescent. Radial, five-parted, yellow, dioecious flowers are in long, thin, hanging racemes (clusters). Fruit is a reddish samara covered with pubescence that hangs in long clusters. **Habitat and use:** Coppice. Used in the horticultural industry.

Pigeon Plum   *Coccoloba diversifolia*
Family: Polygonaceae (Buckwheats)

Range: Archipelago-wide; also Florida, the Caribbean, Central and South America.

Description: Medium shrub to large tree up to 10 m (33 ft) tall with a gray-brown (slightly orange) trunk to 60 cm (24 in) in diameter, with the bark flaking off in jigsaw puzzle–like pieces. Simple, alternate leaves are highly variable in shape: shade and adventitious shoot leaves are large, elongate to lanceolate to 35 cm (14 in) long, and sun leaves are small, less than 10 cm (4 in) long, and elliptical. A ring of green tissue surrounds the stem above the petiole. Dioecious, five-parted, whitish-green flowers, with protruding stamens in male flowers, are arranged in tall racemes (clusters). Mature fruit is an edible, dark red-purple drupe c. 1.5 cm (0.5 in), in grape-like bunches. **Habitat and use:** Common in diverse mature and early succession coppice, shrublands, pine woodlands, rocky shores, and dunes. Fruits are consumed by birds. Used to treat gastrointestinal problems and in strengthening teas and for carving and horticulture.

Guapira obtusata

Coccoloba diversifolia

Alvaradoa amorphoides

Coccoloba uvifera

Sea Grape   *Coccoloba uvifera*

Family: Polygonaceae (Buckwheats)

Range: Archipelago-wide; also Florida, the Caribbean, Central and South America.

Description: Low, small shrub to large trees over 17 m (56 ft) tall with alternate, large, rounded leathery leaves up to 25 cm (10 in) long. A sheath of green tissue surrounds the stem above the petiole base. Dioecious, radial, five-parted, white flowers with protruding stamens in male flowers are arranged in tall racemes (clusters). Mature fruit is an edible, reddish-purple drupe, which grows in large, long, grape-like bunches. **Habitat and use:** Shorelines in coastal coppice, rocky shores, and dunes. Used to treat gastrointestinal problems, boils, and headaches. Fruits consumed by birds, especially White-crowned Pigeons.

Marlberry *Ardisia escallonioides*
Family: Primulaceae (Primroses)
Range: Northern archipelago; also Florida, the Caribbean, Central America.
Description: Shrub to small tree up to 8 m (26 ft) with a trunk diameter to 20 cm (8 in). Simple, alternate, lanceolate leaves are to 20 cm (8 in) and may be slightly fleshy. Radial, white to pink, five-parted, hermaphroditic flowers with bright yellow stamens occur in slightly hanging panicles (clusters). Mature fruit is a black berry that hangs in open bunches. **Habitat:** Coppice habitats.

Ironwood *Krugiodendron ferreum*
Family: Rhamnaceae (Buckthorns)
Range: Archipelago-wide; also Florida, the Caribbean, Central America.
Description: Large shrub to small tree to 10 m (33 ft) tall with a trunk diameter to 35 cm (14 in). Simple, opposite, oval/ovate leaves to 7 cm (3 in) long with a notched leaf tip. Radial, five-parted, yellowish-green, star-shaped, hermaphroditic flowers are in axillary clusters. Mature fruit is an edible, purple-black drupe 5–7 mm (0.2–0.3 in) in diameter. **Habitat and use:** Coppice. Used in strengthening teas, to treat pain and mouth infections, and for wood carvings.

Darling Plum *Reynosia septentrionalis*
Family: Rhamnaceae (Buckthorns)
Range: Archipelago-wide; also Florida, northern Caribbean.
Description: Large shrub to medium-sized tree to 10 m (33 ft) tall with a trunk diameter to 20 cm (5 in) that has red-brown peeling bark. Simple, opposite, oblong leaves to 6 cm (3 in) long with a notched leaf tip. Small, radial, five-parted, star-shaped, hermaphroditic flowers have yellowish-green sepals, no petals, and stamens with black anthers. They are in axillary or terminal umbels (umbrella-shaped clusters). Mature fruit is an edible, dark purple drupe up to 2 cm (0.8 in). **Habitat and use:** Coppice. Used in strengthening teas and the fruit is edible.

Red Mangrove *Rhizophora mangle*
Family: Rhizophoraceae (Red Mangroves)
Range: Archipelago-wide, pantropical and subtropical.
Description: Low shrub to trees over 10 m (33 ft) tall producing arching, prop roots. Simple, opposite, lanceolate leaves cluster at the ends of branches to 15 cm (6 in) long. Radial, four-parted, yellow flowers are in panicles (clusters). Fruit is a leathery drupe with one seed that germinates while attached to the parent plant. The pendulant (hanging) seedling's root emerges while attached to the tree, and the seedling will embed itself in the substrate when it falls. **Habitat and use:** Saline and brackish environments along coastlines, estuaries, and inland lakes and ponds near shoreline. Protects shorelines and provides habitat for juvenile reef organisms.

Rhizophora mangle

Rhizophora mangle

Ardisia escallonioides

Reynosia septentrionalis

Ardisia escallonioides

Krugiodendron ferreum

Snow Berry *Chiococca alba*
Family: Rubiaceae (Coffees)
Range: Archipelago-wide; also the Caribbean, southern U.S., Central America.

Description: Shrub to 3 m (10 ft) tall with several trunks with sprawling habit, may climb vine-like on other shrubs and trees. Simple, opposite, ovate/lanceolate leaves to 12 cm (5 in) with small lance-shaped stipules at leaf base, typical of this family. Small, radial, five-parted, yellow-white, bell-shaped flowers are in hanging panicles (clusters). Mature fruit is a small, white berry to 0.5 cm (0.2 in) and can occur in dense bunches. Often fruits in late fall or early winter. **Habitat and use:** Range of coppice habitats. Fruit consumed by birds. Used in strengthening teas, to stop bed-wetting, and to treat tuberculosis.

Black Torch *Erithalis fruticosa*
Family: Rubiaceae (Coffees)
Range: Archipelago-wide; Florida, the Caribbean, Central and South America.

Description: Medium shrub to small tree up to 4 m (13 ft) tall, but usually less than 2 m (6 ft) tall, and a trunk diameter to 15 cm (6 in). Simple, opposite, obovate/oblanceolate leaves to 7 cm (3 in) long. Petioles have an appendage (stipule), typical of the family, that is triangular with an abrupt tip. Radial, five-parted, white, star-shaped, hermaphroditic flowers are arranged in panicles (clusters). Fruit is a berry up to 0.5 cm (0.25 in) in diameter that turns from white to dark purple or black at maturity. Fruits often seen hanging in clusters. **Habitat and use:** Diverse natural and disturbed coppice habitats especially coastal coppice. The wood is burned as a torch for light, especially for gathering land crabs at night. Used to treat hemorrhoids, measles, and as a diuretic. Flowers visited by bees, and fruits consumed by birds.

Golden Creeper *Ernodea littoralis*
Family: Rubiaceae (Coffees)
Range: Archipelago-wide; also Florida, the Caribbean, Central and South America.

Description: Shrub to 1 m (3 ft) tall with simple, opposite, linear to lanceolate leaves up to 5 cm (2 in) long, without stalks, and with up to seven longitudinal veins running the length of the leaf. Radial, four-parted, red (sometimes white) flowers are sessile (stalkless) and solitarily in leaf axils. Petals are partly fused, forming a long tube with their free ends curved back and the stamens and style protruding. Mature fruit is a one- or two-seeded golden-yellow berry with a crown of remaining sepals. **Habitat:** Grows on sandy soils in the understory of coppice, pine woodlands, and disturbed areas.

Prince Wood *Exostema caribaeum*
Family: Rubiaceae (Coffees)
Range: Archipelago-wide; also Florida, the Caribbean, Mexico, Central America.

Description: Shrub to small tree to 12 m (39 ft) tall with a grayish trunk to 15 cm (6 in) in diameter with horizontal fissures that develop with age. Simple, opposite, lance-

Ernodea littoralis

Exostema caribaeum

Erithalis fruticosa

Chiococca alba

Chiococca alba

olate leaves to 12 cm (5 in) long with slightly undulating margin and pointed stipules at the petiole base are typical of this family. Delicate, radial, five-parted, white or pinkish, hermaphroditic flowers to 5 cm (2 in) are solitary in leaf axils. Petals are partially fused to form a tube and have elongated lobes that hang down around it, with the stamens protruding. Mature fruit is a brown capsule to 1 cm (0.5 in) containing many winged seeds. **Habitat and use:** Coppice and pine woodlands. Used to treat anemia, diarrhea, hemorrhoids, low blood pressure, stomachaches, and ringworm, to increase appetite, as strengthening teas, and in horticulture.

Seven-year Apple   *Genipa clusiifolia*
Family: Rubiaceae (Coffees)
Range: Archipelago-wide; also Florida, the Caribbean, Bermuda.
Description: Medium shrub to 4 m (13 ft) tall. Opposite, simple, obovate leaves 20 cm (8 in) long. Radial, five-parted, monoecious, fragrant flowers are white with a yellow center and arranged in dense panicles. Petals are partially fused to form a tube with their free lobes spreading like a delicate five-pointed star. Fruit is a large, 5–7 cm (2–3 in), hard, green berry that turns yellow then dark brown-black at maturity. **Habitat and use:** Coastal coppice, shrublands, rocky shores, and behind dunes. Used in horticulture.

Smooth Wild Coffee   *Psychotria ligustrifolia*
Family: Rubiaceae (Coffees)
Range: Archipelago-wide; also Florida, Greater Antilles.
Description: Shrub to 2 m (7 ft) tall. Simple, opposite, lance-shaped (lanceolate) leaves to 16 cm (6 in) long with large papery, extended stipules at the petiole base, typical of this family. Radial, five-parted, white flowers are in panicles (clusters). Petals are partially fused to form a tube with their free ends curled back to about half the length of the tube, and there are protruding stamens. Mature fruit is a small, few-seeded, red berry. **Habitat and use:** Understory shrub in coppice and pine woodlands. Used in horticulture. Caffeine content is insufficient to be used as a coffee substitute. Flowers attract butterflies, and fruits are consumed by birds and iguanas.

Box Briar   *Randia aculeata*
Family: Rubiaceae (Coffees)
Range: Archipelago-wide; also Florida, the Caribbean, Central and South America.
Description: Spiny or unspiny shrub to 5 m (16 ft) tall. Simple, opposite, ovate to obovate leaves to 6 cm (3 in) long and clustered at the ends of reduced branches. Paired stipules are located at the petiole base, typical of the family. Radial, five-parted, white, dioecious flowers have a short, tubular base with dense, white hairs at the throat and spreading blunt-tipped petals. They are clustered in leaf axils or solitary at branch tips. Mature fruit is a small, white berry up to 1 cm (0.5 in). **Habitat and use:** Range of coppice habitats and pine woodlands. Used in horticulture. Elsewhere in the Caribbean, the berries are used in dyes and inks and for various medicinal uses.

Wild Thyme   *Rhachicallis americana*
Family: Rubiaceae (Coffees)
Range: Archipelago-wide; also Florida, the Caribbean, Mexico.
Description: Shrub up to 1.5 m (5 ft) (typically > 1 m [4 ft]) tall. Tiny, fleshy, linear, sessile (stalkless), opposite leaves to 1 cm (0.5 in) long have white pubescence (fuzz) on the lower surface and are clustered at branch ends. Triangular stipules are located at the base of the petiole, typical of the family. Radial, orange-yellow,

Genipa clusiifolia

Randia aculeata

Psychotria ligustrifolia

Psychotria ligustrifolia

Rhachicallis americana

Rhachicallis americana

four-parted sessile (stalkless), solitary flowers are tubular with four lobes and occur in the leaf axils. A stipular sheath is situated below the flowers. Mature fruit is a small capsule. **Habitat and use:** Coastal dunes, rocky shores, and inland rock flats. The wood is burned to keep away mosquitos and sand flies.

Mosquito Bush   *Strumpfia maritima*
Family: Rubiaceae (Coffees)
Range: Archipelago-wide; also Florida, the Caribbean, Mexico.
Description: Shrub to 2 m (7 ft) tall. Simple, opposite, linear, needle-like, sessile (stalkless) leaves to 3 cm (1 in) are grouped in threes at joints (nodes) and clumped near branch tips. Lower leaf surface is pubescent (fuzzy) on either side of the midvein forming two parallel white lines. Radial, five-parted, pinkish-white, star-shaped flowers, with a yellow center, are in few flowered axillary racemes (clusters). Mature fruit is a round, white drupe c. 1 cm (0.5 in) in diameter. **Habitat and use:** Dunes, rocky shores, and inland rock flats. Used to treat cuts and sores, and when burned its smoke helps keep away mosquitos. **Important:** Can cause miscarriages and sterility in women. Should be used sparingly and weakly.

White Torch   *Amyris elemifera*
Family: Rutaceae (Citruses)
Range: Archipelago-wide; also Florida, the Caribbean, Central America.
Description: Shrub to small tree up to 6 m (20 ft) tall. Compound, opposite, leaves have three to five, ovate/rhombic leaflets each to 7 cm (3 in), with a leaf margin with blunt-rounded teeth (crenate) and clear dots throughout leaf surface. Leaves are aromatic when crushed. Radial, four-parted, white, hermaphroditic flowers are in terminal cymes (clusters). Mature fruit is an edible, dark blue-black drupe to 0.5 cm (0.2 in). **Habitat and use:** Mature to scrubby coppice. Used to treat colds and flu, burned as a light source (torch), and fed upon by butterfly caterpillars of the Bahamian Swallowtail.

Wild Lime   *Zanthoxylum fagara*
Family: Rutaceae (Citruses)
Range: Archipelago-wide; also the Caribbean, southern U.S., Mexico, Central America.
Description: Large shrub to small tree to 10 m (33 ft) tall, a tan trunk to 25 cm (10 in) in diameter, and branches with hooked spines below the leaves. Alternate, oddly pinnately compound leaves have wings along the stalk. Five to nine, ovate/elliptical sessile (stalkless) leaflets to 2 cm (1 in), have leaf margins with blunt-rounded teeth (crenate). Leaflets produce a citrus odor when crushed. Radial, four-parted, yellowish, dioecious flowers are in panicles of racemes (bunches of small clusters). Mature fruit is a dry capsule that splits to release the single, shiny black seed and occurs in bunches. **Habitat and use:** Coppice and shrublands particularly in disturbed habitats. Used medicinally for general strengthening teas and in horticulture and cabinetry.

Butterbough   *Exothea paniculata*
Family: Sapindaceae (Soapberries)
Range: North and central archipelago; also Florida, the Caribbean, Central America.
Description: Medium to large tree to 20 m (66 ft) tall with a trunk diameter to 50 cm (20 in). Alternate compound, evenly bipinnate leaves have four or six oblong/

Exothea paniculata

Strumpfia maritima

Amyris elemifera

Strumpfia maritima

Zanthoxylum fagara

lanceolate leaflets, to 14 cm (5.5 in) long. Small, radial, five-parted, monoecious or hermaphroditic, white flowers are arranged in loose panicles (clusters). Each flower has unfused, white petals with a bright, orange-yellow center. Mature fruit is a dark purple berry. **Habitat and uses:** Diverse coppice types and shrublands. Has been used for cabinetry, making tool handles, toy boats, and other toys; and in horticulture.

Quick Silver Bush   *Thounia discolor*
Family: Sapindaceae (Soapberries)
Range: Endemic, archipelago-wide.
Description: Large shrub to small tree to 6 m (20 ft) tall with a trunk diameter to 20 cm (8 in), with orange and gray mottled bark. The trifoliate, alternate leaves with leaflets to 8 cm (3 in) long each have a blunt tip. Silvery, lower leaf surface is covered with densely matted woolly hairs. Bilateral, four- or five-parted, white monoecious flowers are in racemes (clusters). Fruit is a dry one-seeded samara. **Habitat and use:** Coppice and pine woodlands. Used to treat dermatological problems, in strengthening teas, in obstetrics, for colds and flu (fevers), and pain.

Satin Leaf   *Chrysophyllum oliviforme*
Family: Sapotaceae (Sapodillas)
Range: Archipelago-wide; also Florida, Greater Antilles, Central America.
Description: Medium to large tree up to 17 m (56 ft) tall with a trunk diameter to 50 cm (20 in). Alternate, simple, ovate leaves to 10 cm (4 in) long. Upper surface is dark green and smooth (glabrous), and the lower surface is covered with short, brown fuzz (pubescence) that also occurs on the sepals. Radial, five-parted, hermaphroditic, white flowers grow in groups of two to nine in leaf axils. Mature fruit is a small, stalked, edible, olive-sized, purple-black berry. **Habitat and use:** Diverse coppice woodlands and shrublands. Used to treat circulatory problems and in horticulture.

Wild Dilly   *Manilkara bahamensis*
Family: Sapotaceae (Sapodillas)
Range: Archipelago-wide; also Florida, Greater Antilles.
Description: Medium- to large-sized tree up to 13 m (43 ft) tall with a trunk diameter to 50 cm (20 in). Simple, alternate, lanceolate/oblong leaves to 10 cm (4 in) long, with an acute leaf tip and smooth leaf margin, are clustered at branch tips. Young leaves and branch tips are covered with fine reddish-brown hairs. All parts of the plant produce a milky sap. Radial, six-parted green, hermaphroditic flowers are covered in reddish-brown hairs and are in leaf axils. Petals are fused to form a shallow cup. Mature fruit is a light brown, scaly berry. **Habitat and use:** Coastal coppice and shrublands. Used in soothing and aphrodisiac teas, wood carving, and furniture. It has an edible, but milky, gummy sap.

Sideroxylon americanum

Manilkara bahamensis

Chrysophyllum oliviforme

Thounia discolor

Thounia discolor

Milk Berry *Sideroxylon americanum*

Family: Sapotaceae (Sapodillas)

Range: Archipelago-wide; also Greater Antilles.

Description: Large shrub to small tree up to 6 m (20 ft) tall. Short branches often occur as spines. Simple, alternate, obovate to elliptical leaves, to 10 cm (4 in). The upper glabrous (smooth) surface dark green, and the lower often with reddish pubescence (fuzz). Radial, five-parted, white, hermaphroditic flowers are in groups of two to nine in leaf axils. Mature fruit is an edible, purple-black drupe with milky latex. **Habitat and use:** Coppice. Used in obstetrics.

Mastic *Sideroxylon foetidissimum*

Family: Sapotaceae (Sapodillas)

Range: Archipelago-wide; also Florida, the Caribbean, Mexico, Central America.

Description: Large tree up to 20 m (66 ft) tall with a trunk diameter to 50 cm (20 in). Simple, alternate leaves to 20 cm (8 in) long; lanceolate, elliptic, or obovate with an undulating leaf margin. Numerous small, radial, five-parted, yellow-green, hermaphroditic flowers are in dense clusters in leaf axils. Mature fruit is an oval, orange drupe, c. 3 cm (c. 1 in) long. **Habitat and use:** Coppice, shrubland, and pine woodlands. Used in horticulture, woodworking, and the long straight trunks were once used for sailboat and ship masts.

Willow Bustic *Sideroxylon salicifolium*

Family: Sapotaceae (Sapodillas)

Range: Archipelago-wide; also Florida, the Caribbean, Mexico, Central America.

Description: Large shrub to small tree up to 15 m (50 ft) tall with a trunk diameter to 50 cm (20 in), but the latter is usually smaller. Alternate, simple leaves to 10 cm (4 in) long with an entire margin. Young stems and new leaves have reddish-brown hairs. Radial, five-parted, whitish-yellow, hermaphroditic flowers are in clusters along the stem in leaf axils. Fruit is a small berry, 6–10 mm (0.25–0.4 in) in diameter, initially red and then becoming black at maturity, in small clusters. **Habitat:** Coppice and pine woodlands.

Snake Root *Picramnia pentandra*

Family: Simaroubaceae

Range: Archipelago-wide, Florida, the Caribbean, South America.

Description: Large shrub to small tree up to 7 m (23 ft) tall with a trunk diameter to 15 cm (6 in). Oddly pinnately compound leaves have five to nine alternate ovate/lanceloate leaflets each to 12 cm (5 in) long. Leaflets are reddish during development. Radial, five-parted, green, dioecious flowers are in hanging panicles of racemes (small bunches of many flowers arranged in a long spike-like cluster). Mature fruit is a small, red berry hanging in long bunches. **Habitat and use:** Understory shrub in coppice and around sinkholes and depressions. Used to treat colds and flu, dermatological problems, gastrointestinal issues, in obstetrics and gynecological care, strengthening teas, respiratory maladies, and in horticulture.

Paradise Tree *Simarouba glauca*

Family: Simaroubaceae

Range: Northern archipelago; also Florida, Cuba, Mexico south through Central America.

Description: Tree to 10 m (33 ft) tall with trunk diameter to 30 cm (12 in). Unevenly pinnately compound, alternate leaves to 40 cm (16 in) long, with ten to twenty leaflets. Leaflets to 10 cm (4 in) long, dark green above, lighter below, with an entire margin and rounded leaf tip. Radial, five-parted, yellowish-white, dioecious flowers are in few-flowered, terminal and axillary panicles (clusters). Fruit is a

Simarouba glauca

Picramnia pentandra

Sideroxylon foetidissimum

Sideroxylon salicifolium

Sideroxylon foetidissimum

drupe to 2 cm (1 in) long, turning from pinkish-red to purplish-black at maturity, occurring in lax bunches. **Habitat and use:** Coppice. Used to treat fevers and colds, dermatological problems, and gastrointestinal issues and has been harvested for resin, an edible oil, and other pharmaceutically useful products.

Canker Berry   *Solanum bahamense*

Family: Solanaceae (Nightshades)

Range: Archipelago-wide; also Florida.

Description: Shrub up to 3 m (10 ft) tall with small star-shaped hairs and may also have prickles covering all surfaces. Simple, alternate, oblong or lanceolate leaves to 15 cm (6 in) with an acute leaf tip, entire to undulate leaf margin, and a rough (scabrous) leaf surface texture. Five-parted, hermaphroditic flowers have purple-blue, backward-pointing petals and bright yellow, central, protruding anthers and are arranged in racemes (clusters). Mature fruit is a bright red berry. **Habitat and use:** Coppice edges. Used to treat respiratory problems, sore throats, and thrush.

Bay Cedar   *Suriana maritima*

Family: Surianiaceae

Range: Archipelago-wide; also Florida, the Caribbean, Central and South America.

Description: Medium-sized shrub up to 3 m (10 ft) tall with alternate, sessile, yellowish-green leaves to 5 cm (2 in) long, clustered at the ends of branches and with an odor similar to cedar when crushed, hence the common name. Leaves slightly succulent and covered with a dense pubescence (fuzz). Radial, five-parted, yellow flowers are in few-flowered panicles (clusters) or are solitary among leaves. Mature fruit is brown and hairy, splitting into five single-seeded portions when mature. **Habitat and use:** Coastal dunes and rocky shores. Used for dermatological and gastrointestinal problems, to ease sore throats and toothaches, and strengthening teas.

Joe Wood   *Jacquinia keyensis*

Family: Theophrastaceae

Range: Archipelago-wide; also Florida, Greater Antilles.

Description: Large shrub to small tree up to 7 m (23 ft) tall with trunk diameter to 30 cm (12 in) and smooth gray bark. Simple, alternate, stiff, leathery, oblong to spatula-shaped leaves are to 5 cm (2 in) long, and have edges rolled toward the underside. Fragrant, radial, five-parted, white flowers have petals that are fused into a tube with the upper lobes spread out flat. They are in terminal racemes (clusters). Fruit is a berry that turns whitish-yellow to orange at maturity. **Habitat and use:** Shorelines in coastal coppice, scrublands, and rocky shores. Used in horticulture because of salt-tolerance, bonsai-like appearance, and showy, fragrant flower.

Wooly Booger   *Corchorus hirsutus*

Family: Tiliaceae (Lindens)

Range: Archipelago-wide; also tropical and subtropical New World and East Africa.

Description: Small- to medium-sized shrub up to 2 m (7 ft) tall. Simple, alternate, ovate/oblong leaves 8 cm (3 in) long have a round-toothed leaf margin. Radial, five-parted, yellow flowers have numerous yellow stamens and are in umbels (umbrella-shaped clusters) arising from nodes. Vegetative and floral structures (excluding petals and stamens) are covered with wooly hairs. Fruit is a wooly,

Suriana maritima

Solanum bahamense

Solanum bahamense

Suriana maritima

Corchorus hirsutus

Jacquinia keyensis

twisted capsule that splits to release the seeds. **Habitat and use:** Disturbed habitats and edges of coppice and shrublands near coast. Used to treat colds and flu. Leaves are edible after cooking.

Golden Dew Drop   *Duranta erecta*

Family: Verbenaceae (Verbenas)

Range: Archipelago-wide; also the Caribbean south to South America tropical and Mexico.

Description: Shrub up to 7 m (23 ft) tall. Opposite, simple, ovate, obovate, or elliptic leaves to 8 cm (3.2 in) long, with either a smooth or slightly serrate (toothed) leaf margin. Bilateral, five-parted, purple-pink (sometimes with white edge or striping) flowers with fused petals are in terminal and axillary pendulant racemes (clusters). Mature fruit is a golden-orange, few-seeded berry, which hangs down in open clusters. **Habitat and use:** Coppice and shrublands, pine woodlands, and disturbed habitats. Flowers attract butterflies, and fruits attract birds.

Bahama Sage   *Lantana bahamensis*

Family: Verbenaceae (Verbenas)

Range: Archipelago-wide; also Florida, Cuba.

Description: Shrub 2 m (7 ft) tall with woody stems, but no large trunk. Opposite, simple, ovate leaves to 2 cm (1 in) with leaf margin with blunt-rounded teeth. Bilateral, five-parted, orange-yellow-red flowers are clustered into flat-topped, multi-colored heads at the ends of flower stalks. Younger, lighter-colored flowers are in the center, and the older, darker ones are on the outer edges of the clusters. Mature fruit is a shiny black drupe. **Habitat and use:** Coppice edges, especially coastal and disturbed habitats. Used to treat dermatological problems. Flowers are attractive to butterflies, and fruits attract birds.

White Sage   *Lantana involucrata*

Family: Verbenaceae (Verbenas)

Range: Archipelago-wide; also Florida, the Caribbean, Mexico to South America.

Description: Small- to medium-sized shrub to 3 m (10 ft) tall, with woody stems, but no large trunk. Aromatic, simple, opposite, ovate leaves to 2 cm (1 in) long with blunt-toothed margins. Bilateral, four-parted, white flowers (with a yellow center) are clustered into heads at the end of flower stalks. Flowers are pink in southern archipelago. Mature fruit is a purplish-pink, sometimes white, drupe 3–4 mm (0.1–0.2 in) in diameter. **Habitat and use:** Coppice, shrublands, sandy coastal sites, and disturbed habitats. Used to treat dermatological issues, respiratory problems, lung congestion, and high blood pressure and is used in horticulture. Fruits consumed by birds.

Lignum Vitae   *Guaiacum sanctum*

Family: Zygophyllaceae (Caltrops)

Range: Archipelago-wide; also Florida, the Caribbean, Central and South America.

Description: Small shrub to medium-sized tree up to 10 m (33 ft) tall. Compound, opposite leaves have six to ten pointed-tipped, opposite leaflets each to 2 cm (1 in) long, with stipules at their bases. Radial, five-parted bluish-purple flowers, with ten stamens and pale colored anthers, are solitary in leaf axils. Mature fruit is an

Lantana involucrata

Lantana bahamensis

Duranta erecta

Guaiacum sanctum

orange, winged capsule that splits into five parts to expose the dark-colored seeds surrounded by a red fleshy tissue (aril). **Habitat and use:** Coastal and interior coppice. Used to treat syphilis, fevers, general pain, gout, arthritis, rheumatism, tonsillitis, skin ailments, constipation, fish poisoning, and in strengthening teas and in horticulture. The wood is extremely dense and hard and has been used to produce ball bearings, bowling balls, fish bats, and ornamental wood carvings. It is the national tree of The Bahamas and is protected.

Sea Purslane   *Sesuvium portulacastrum*
Family: Aizoaceae (Stone Plants)
Range: Archipelago-wide, pantropical and subtropical.
Description: Low succulent (having fleshy leaves and stems), grows horizontally forming large mats from runners that branch frequently, to 2 m (7 ft) long, but usually shorter. Photosynthetic stems become reddish with age. Fleshy, opposite, reddish-green leaves to 7 cm (3 in) long. Radial, five-parted, pink, hermaphroditic flowers have sepals but no petals, and occur solitarily in leaf axils. Mature fruit is an inconspicuous capsule, c. 1 cm (0.5 in) long, which splits to release the black, smooth seeds. **Habitat and use:** Dunes, rocky shores, and the edges of mangrove flats. The entire plant is edible but generally just the leaves are eaten as a supplement to salads.

Spider Lily   *Hymenocallis arenicola*
Family: Amaryllidaceae (Amaryllises)
Range: Northern and central archipelago; also Florida, the Caribbean.
Description: Herbaceous. Produces basal rosette (a whorl of leaves radiating from a crown or center) of semi-succulent or fleshy linear leaves from a fleshy underground bulb. Leaves are to 8 cm (3 in) wide and 70 cm (28 in) long with a rounded leaf tip and have parallel veins. Large, radial, three-parted, white, hermaphroditic flowers are arranged in umbels (umbrella-shaped clusters) bearing up to fourteen flowers. Each flower resembles a white, shallow cup with six, long, slim, white, curving appendages making it look spider-like. Mature fruit is a white berry to 4 cm (2 in). **Habitat and use:** Beach dunes and sandy soils. Used to treat respiratory illnesses and in horticulture.

Marsh Pennywort   *Centella asiatica*
Family: Apiaceae (Celeries, Carrots, and Parsleys)
Range: Archipelago-wide, temperate, subtropical and tropical regions.
Description: Low-lying herb with creeping rootstalks, to 50 cm (20 in) tall. Simple, alternate, ovate to rounded leaves are in basal cluster with a smooth or dentate (toothed) margin and rounded leaf tip. Tiny, radial, five-parted, reddish-white flowers are in a few flowered umbels (umbrella-shaped clusters). Mature fruit is a dry, flattened, winged, brown capsule with ridges. **Habitat:** Freshwater wetlands.

Devil's Potato   *Echites umbellata*
Family: Apocynaceae (Dogbanes)
Range: Archipelago-wide; also Florida, the Caribbean, Mexico, Central America.
Description: Shrubby vine climbing over other vegetation with simple, opposite, ovate leaves to 20 cm (8 in). Radial, white, five-parted flowers are arranged in cymes (clusters). Petals are fused forming a tube up to 10 cm (4 in) long with

Sesuvium portulacastrum

Hymenocallis arenicola

Echites umbellata

Centella asiatica

the free lobes at the end overlapping to one side, which creates a pinwheel shape. Fruits are a pair of dry follicles up to 20 cm (8 in), that split along the upper suture, to which the seeds are attached. Seeds have tufts of hair that aid in aerial dispersal. **Habitat and use:** Disturbed habitats including coppice shrublands and scrublands and pine woodlands. As a climber with large, attractive flowers, it is a useful ornamental on trellises and walls.

Agave   *Agave bahamana*
Family: Asparagaceae (Agaves)
Range: Endemic. Northern and central archipelago.
Description: Semi-succulent (fleshy) shrub with sharp-pointed leaves that are arranged in a basal rosette producing new leaves from the center every year. Leaves are grayish-green, up to 2 m (7 ft) and 20 cm (8 in) wide, and leaves curl inward with age. Leaf tip is modified into a spine, and the leaf margin is prickled or smooth along the upper edge. Radial, three-parted, hermaphroditic yellow flowers are arranged in terminal panicles (clusters) to 12 m (39 ft) that develops quickly. Fruit is a light brown, oblong, capsule up to 5 cm (2 in) long that splits at maturity with numerous seeds. The growing tip is used during flowering, and the entire plant dies after the fruits mature. **Habitat and use:** Diverse coppice habitats on sandy substrates, beach dunes, and on limestone. Flowers attract a variety of nectarivores including warblers, vireos, mockingbirds, and thrushes as well as hummingbirds and bananaquits.

Bay Gerana   *Ambrosia hispida*
Family: Asteraceae (Asters, Daisies, and Sunflowers)
Range: Archipelago-wide, New World tropics and subtropics.
Description: Herbaceous creeping perennial to 20 cm (8 in) tall, forming mats. Opposite, pinnately compound leaves to 9 cm (4 in) long are fern-like. Leaves and stems are covered with pubescence (fuzz) that gives the plant a light green to gray appearance. Radial, five-parted, monoecious flowers are arranged in round clusters along tall, erect stems above the leaves. Mature fruit is a hard, dry, one-seeded achene covered with pubescence and tubercles (bumps). **Habitat and use:** Sand dunes. Used to treat colds, flu, gastrointestinal issues (increasing appetite, controlling diarrhea, indigestion, and vomiting), blisters, and general skin irritation. Used to treat worms and high blood pressure in the Caribbean. Wind pollinated and some people may be allergic to its pollen.

Shepherd's Needle   *Bidens pilosa*
Family: Asteraceae (Asters, Daisies, and Sunflowers)
Range: Archipelago-wide; also the Caribbean, tropical and subtropical Central and North America. Introduced in Old World tropics and subtropics.
Description: Herbaceous annual, to 1 m (3 ft) tall. Opposite leaves are three-lobed to pinnately compound with three to five pointed leaflets with serrated edges. Vegetative structures can be smooth or pubescent (fuzzy). The "daisy-like" flower heads are arranged in panicles of heads (clusters); each head is subtended by an involucre. Along the edges of the flower head there are five or six bilateral flowers with one long fused white petal up to 1 cm (0.5 in); while in the center are radial, five-parted, yellow flowers. Seeds are in bunches and resemble short stiff hairs that have two to four barbed spines at their tips, which can be dispersed by adhering to animal fur and clothing. **Habitat and use:** Disturbed habitats such as yards, roadsides, and abandoned fields. Used to treat circulatory problems, colds, fevers, and flu, dermatological issues, intestinal problems, and urinary

Iva imbricata

Agave bahamana, inflorescence

Bidens pilosa

Ambrosia hispida

tract complications. In southern Caribbean it is used to treat eye infections. Vegetative parts are edible.

Beach Iva　　*Iva imbricata*

Family: Asteraceae (Asters, Daisies, and Sunflowers)

Range: Archipelago-wide; also Florida, Cuba.

Description: Fleshy (succulent) perennial herb to 50 cm (20 in) tall, typically forming mats and mounds. Simple, alternate, fleshy leaves are to 6 cm (3 in) long with a smooth margin (occasionally toothed). Flowers are arranged in heads wrapped in a whorl of leaf-like bracts. Radial male flowers are on the outside, but lack the conspicuous, long petals found in other species in this family, whereas the bilateral female flowers are in the center of the head. Heads are borne in elongated arrays and look like stalks of very small, equal-sized brussels sprouts. Fruit is a hard, dry, one-seeded achene covered with pubescence (fuzz) and tubercles (bumps). **Habitat:** Sand dunes along shorelines.

Shanks *Salmea petrobioides*
Family: Asteraceae (Asters, Daisies, and Sunflowers)
Range: Archipelago-wide; also Florida, the Caribbean.
Description: Semi-succulent (fleshy leaves and stems) perennial shrub to 2.5 m (8 ft) tall. Opposite, smooth (glabrous), rounded-obovate leaves to 12 cm (5 in) long with a smooth margin and sharply pointed leaf tip. Flower heads are in terminal, flat-topped clusters. There are eight to ten, white to pinkish, tiny, flowers in each head, borne in the axils of three leaf-like bracts. Outer flowers lack the long "daisy-like" petals of other species in this family and are similar to the inner flowers that produce fruit. Mature fruit is a dry, hard, one-seeded achene with small projections. **Habitat:** Sand dunes, rocky shores, and coastal palm woodlands.

Sea Rocket *Cakile lanceolata*
Family: Brassicaceae (Mustards and Cabbages)
Range: Archipelago-wide; also Florida, the Caribbean, Mexico, Central America, northern South America.
Description: Spreading semi-succulent (fleshy) annual herb to 80 cm (3 ft) tall. Simple, alternate, oblong/lanceolate leaves are to 12 cm (5 in) long with a toothed margin and rounded leaf tip. Radial, four-parted, white flowers are in open racemes (clusters) up to 40 cm (16 in) long. Mature fruit is a cylindrical silique (a long, dry, bi-partitioned pod). **Habitat:** Dunes, coastal coppice, palm woodlands, and edge of saline wetlands.

Wild Pine *Tillandsia utriculata*
Family: Bromeliaceae (Pineapples)
Range: Archipelago-wide; also Florida, the Caribbean, Central and South America.
Description: No discernible stems. Long, narrow, parallel-veined leaves, to 2 m (6 ft) are in a rosette (a whorl of leaves radiating from a crown or center), forming a series of cups at their base that hold water and detritus. Radial, greenish-white, three-parted flowers, each with a subtending bract, are in a panicle of spikes that is two to three times the length of the leaves; this appears as a long central stalk with spikes near the top. Mature fruit is a dry capsule that splits to release numerous seeds, which have hair tufts that aid in aerial dispersal. The entire plant dies after flowering. **Habitat and use:** Epiphyte on plants or on rocks in coppice. Used to treat thrush and sexual impotence in men.

Salmea petrobioides

Cakile lanceolata

Melocactus intortus

Tillandsia utriculata

Turk's Head Cactus   *Melocactus intortus*
Family: Cactaceae (Cacti)
Range: Archipelago-wide; also Lesser Antilles.
Description: Unbranched, leafless, almost cylindrical cactus to 75 cm (30 in) tall
and trunk up to 35 cm (14 in) in diameter, with up to twenty ridges with areoles
(small areas with spines and hairs), 1–2 cm (0.5 in) apart, each having up to fifteen
radiating spines. A distinctive cephalium (elongated stem), up to 15 cm (6 in) tall,
is produced when the cactus reaches maturity. This continues to grow until the
plant dies and is composed of white wool and soft, reddish-brown bristles and
is said to resemble a fez (hence its common name). Small, radial, many-parted,
pink flowers, up to 1 cm (0.5 in) wide, are arranged solitarily on areoles on the
cephalium. Mature fruit is a bright pink-red berry with black seeds. **Habitat and
use:** Exposed locations, such as hilltops in coppice and dwarf shrublands, and
rocky shores. The national flower of the TCI, where it is abundant. Birds consume
the berries. **Conservation status:** All cacti are protected internationally (under
CITES) and should not be collected from the wild.

Prickly Pear   *Opuntia stricta*
Family: Cactaceae (Cacti)
Range: Archipelago-wide; also Florida, the Caribbean, Mexico.
Description: Succulent, multi-branched cactus up to 2 m (6.5 ft) tall with a trunk to
   10 cm (4 in) in diameter. From the trunk a series of photosynthetic joints (modified branches) arise. The elliptic joints are 15–35 cm (6–14 in), with a crenate
   margin (round-toothed or scalloped), and areoles (small areas with spines and
   hairs) in three to four rows. Each areole is up to 4 cm (1.5 in) apart, slightly raised
   with one to four yellow spines and numerous hair-like spines (glochids). Radial,
   multi-parted, red flowers are solitary, developing from areoles along the upper
   edges of the joints. Mature fruit is a red berry with numerous seeds. **Habitat and
   use:** Dunes and palm woodlands. Used for circulatory problems, dermatological
   and gastrointestinal issues, and pain. The fruit is edible. **Conservation status:**
   All cacti are protected internationally (under CITES) and should not be collected
   from the wild.

Dildo Cactus   *Pilocereus polygonus*
Family: Cactaceae (Cacti)
Range: Archipelago-wide; also Hispaniola, Cuba, Florida.
Description: Multi-branched cactus shrub or tree to 7 m (23 ft) tall; each branch to
   15 cm (6 in) wide. Each branch has up to thirteen ridges that have small areas that
   bear spines or hairs (areoles), 1–2 cm (0.5–1 in) apart, with twenty to twenty-five
   radiating spines each. Slightly bilateral, large, showy, bell-shaped, many-parted,
   white flowers are solitary on areoles, at the end of branches, typically on the lee-
   ward side. Mature fruit is a bright red berry. On the stem at the base of the fruit,
   there may be a grouping of hairs. **Habitat and use:** Exposed locations such as
   hilltops in coppice and scrublands. Flowers visited by bats, birds, and insects, and
   fruit consumed by birds. **Conservation status:** All cacti are protected internationally (under CITES) and should not be collected from the wild.

Morning Glory   *Ipomoea indica*
Family: Convolvulaceae (Morning Glories)
Range: Northern and central archipelago, pantropical and subtropical.
Description: Herbaceous perennial vine becoming woody basally. Simple, alternate,
   ovate (sometimes three-lobed) leaves to 10 cm (4 in) long with a heart-shaped
   (cordate) leaf base and pointed tip. Large, radial, five-parted, funnel-like, purple-
   blue, hermaphroditic flowers are in cymes (small clusters) or solitary. Mature fruit
   is a dry capsule, which splits to release the seeds. **Habitat:** Disturbed habitats.

Rail Road Vine   *Ipomoea pes-caprae*
Family: Convolvulaceae (Morning Glories)
Range: Archipelago-wide, coastal pantropical and subtropical.
Description: Spreading perennial vine to 12 m (40 ft), rooting at the nodes. Simple,
   alternate, pleated, ovate, obovate, orbicular, or kidney-shaped leaves to 20 cm (8 in)

Opuntia stricta

Opuntia stricta

Pilocereus polygonus

Pilocereus polygonus

Ipomoea pes-caprae

Ipomoea indica

with an entire margin. Leaf tip broadly notched, folded along midvein, and leaf base is heart-shaped. Radial, five-parted, pink-purple, funnel-shaped flowers, tending to be darker in the center, are solitary in leaf axils. Mature fruit is a round, dry capsule that splits to release hairy black seeds. **Habitat:** Beach side of dunes.

Havana Clustervine   *Jacquemontia havanensis*
Family: Convolvulaceae (Morning Glories)
Range: Archipelago-wide; also Florida, the Caribbean, Central America.
Description: Vine becoming woody with age, often seen growing over neighboring plants. Simple, oblong, alternate leaves are to 9 cm (3.5 in). Vegetation is smooth or with tiny star-shaped hairs. Radial, white, five-parted flowers have fused petals with free, elongated lobes, which are star-shaped when viewed from above; they are borne in cymes, axillary clusters, or are solitary. Mature fruit is a dry, brown capsule, which splits to release slightly winged seeds. **Habitat:** Diverse coppice habitats from forest to dwarf shrublands and pine woodlands, and dunes.

Saw Grass   *Cladium jamaicense*
Family: Cyperaceae (Rushes)
Range: Archipelago-wide; also the Caribbean, North America.
Description: Clumping, rush-like herb to 3 m (10 ft) producing rhizomes. Leaves are arranged alternately, parallel veined, to 1.5 m (5 ft) in length, and have sharp saw-toothed margins (hence the common name). Small, inconspicuous, radial flowers are arranged in a panicle of spikelets (clusters of stalkless flowers) each subtended by two to three scales, and in groups of two to six at branch ends. No sepals or petals and only two stamens; the lower flowers in each spikelet are infertile. Mature fruit is a dry, hard, brown achene. **Habitat:** Fresh and brackish wetlands and in Sabal Palm woodlands.

Spike Rush   *Eleocharis cellulosa*
Family: Cyperaceae (Rushes)
Range: Archipelago-wide; also the Caribbean, eastern North America.
Description: Clumping herb that produces underground stems (rhizomes) with round stems to 75 cm (2.5 ft) and to 4 mm (0.2 in) wide. Leaves are reduced to a reddish sheath at the base of the stem. Small, radial flowers occur in spikes (clusters of stalkless flowers) at the end of each stem. A bract is at the base of each flower. Sepals and petals are reduced to six bristles. Mature fruit is a hard, dry, one-seeded achene with a conical tubercle (bump). **Habitat:** Fresh and brackish wetlands and Sabal Palm woodlands.

White-head Rush   *Rhynchospora floridensis*
Family: Cyperaceae (Rushes)
Range: Archipelago-wide; also Florida, the Caribbean, Mexico, Central America.
Description: Clumping grass-like herb to 40 cm (16 in) tall. Simple, alternate, parallel-veined leaves to 15 cm (6 in) long on a stem with triangular cross section. Tiny radial flowers are arranged in spikelets (clusters of stalkless flowers). Each spikelet has a large, showy white bract at the base that looks like a petal, and these are grouped together so they look like a flower head; they are attached on leafless stalks that arise from the root, which are taller than the leaves. Mature fruit is a hard, dry, one-seeded achene, with hairs to aid seed dispersal. **Habitat:** Near freshwater wetlands, in Sabal Palm woodlands, pine woodlands, and brackish mangrove systems.

Eleocharis cellulosa

Eleocharis cellulosa

Rhynchospora floridensis

Rhynchospora floridensis

Jacquemontia havanensis

Cladium jamaicense

Cladium jamaicense

Slender Nut-rush   *Scleria lithosperma*
Family: Cyperaceae (Rushes)
Range: Archipelago-wide, pantropical.
Description: Perennial, clumping, grass-like herb to 40 cm (16 in) tall. Reddish, alternate, parallel-veined leaves are on a triangular stem. At the base of the leaf blade is a triangular hairy extension. Bilateral, highly reduced flowers are terminal and axillary, each subtended by one scale. Mature fruit is a white, hard, dry, one-seeded achene. **Habitat:** Understory of diverse coppice habitats.

Milky Weed   *Euphorbia mesembryanthemifolia*
Family: Euphorbiaceae (Spurges)
Range: Archipelago-wide; also Florida, the Caribbean, Central America.
Description: Erect or sprawling semi-succulent herb, stem and leaf surfaces lack hairs or down. Up to 1 m (4 ft) long (tall). Simple, opposite, folded, ovate to elliptical leaves to 1 cm (0.5 in) with acute leaf tip and entire leaf margin and may have a reddish color. At the base of the petiole are two triangular or lobed and hairy appendages (stipules). Vegetation produces a thick, milky sap. Radial, white, highly reduced flowers are in tight clusters of cyathia (a characteristic group of flowers found only in the genus *Euphorbia* that resembles a single cup-shaped flower enclosing greatly reduced male flowers or a single female flower) in leaf axils. Fruit is a three-lobed capsule. **Habitat and use:** Dunes and sandy coastal areas. Used to treat dermatological issues and gastrointestinal problems. **Important:** All parts of the plant are extremely toxic, and the sap should not be ingested!

Grey Knickers   *Caesalpinia bonduc*
Family: Fabaceae (Legumes [Peas and Beans])
Range: Archipelago-wide, pantropical.
Description: Large climbing vine up to 7 m (23 ft) long (tall) forming large impenetrable clumps. Stems and leaves are covered with recurved prickles and spines. Bipinnately compound, alternate leaves are up to 45 cm (18 in) long with four to five pinnae each with four to eight leaflets. The leaflets are ovate with a long, attenuated leaf tip and smooth margin. Persistent stipules to 3 cm (1 in). Bilateral, five-parted, greenish-yellow, hermaphroditic flowers are in racemes (clusters). Mature fruit is a brown spiny pod, 5–8 cm (2–3 in) long, containing two or three hard, shiny, gray seeds. **Habitat:** Disturbed habitats and coastal areas near dunes.

Bay Bean   *Canavalia rosea*
Family: Fabaceae (Legumes [Peas and Beans])
Range: Archipelago-wide; also Florida, the Caribbean, Mexico.
Decription: Long-lived perennial vine to 15 m (49 ft) long (tall). Trifoliate, alternate leaves with ovate, oblong, or lanceolate leaflets to 8 cm (3 in) with a rounded or slightly notched leaf tip and smooth margin. Bilateral, five-parted, pink-purple, "pea-like" flowers to 5 cm (2 in) long (with a large upper, reflexed petal) are in axillary and terminal racemes (clusters) on long stalks. Mature fruit is a flat, brown, woody, ridged pod, 10–15 cm (4–6 in) long, and slightly compressed between the

Scleria lithosperma

Euphorbia mesembryanthemifolia

Canavalia rosea

Caesalpinia bonduc

Caesalpinia bonduc

seeds. **Habitat and use:** Coastal areas, primarily dunes. The vines are good for trellises and grow quickly near the coast. Beans when well roasted are edible and can be added to coffee. **Important:** Raw beans are poisonous.

Butterfly Pea   *Centrosema virginianum*
Family: Fabaceae (Legumes [Peas and Beans])
Range: Archipelago-wide; also the Caribbean, southern U.S., Central America, northern South America.
Description: Trailing or climbing perennial vine to 2 m (7 ft) long. Trifoliate (three-parted), alternate leaves to 12 cm (5 in) long with linear (slender) leaflets, to 8 cm (3 in) long and with a pointed leaf tip and smooth margin. Bilateral, five-parted, pink-purple, hermaphroditic flowers are "pea-like" but are inverted with the larger standard petal below the upper keel petals. They are solitary in leaf axils and subtended by two bracts at the flower stalk base. Mature fruit is a pod, 7.5–15 cm (3–6 in) long, that turns brown and splits in half, with both halves remaining on the vine. **Habitat:** Disturbed habitats and coppice edges.

Love Vine   *Cassytha filiformis*
Family: Lauraceae (Laurels)
Range: Archipelago-wide, pantropical.
Description: Parasitic vine with yellowish-orange, thread-like stems up to 7 m (23 ft) long, which may form dense tangles. Highly reduced, scale-like leaves are arranged spirally. Vines attach to stems and branches of host plants by means of a haustorium, a modified structure to absorb nutrients and water. Radial, three-parted, white flowers are arranged in lateral spikes. Mature fruit is a white drupe. **Habitat and use:** Grows on other plants in coppice, pine woodlands, dunes, rocky shores, wetlands, and disturbed habitats. Used to treat pain, dermatological issues, obstetric and gynecological concerns, and in strengthening and aphrodisiac teas.

Pineyard Pink   *Bletia purpurea*
Family: Orchidaceae (Orchids)
Range: Northern archipelago; also Florida, the Caribbean, Central and northern South America.
Description: Semi-epiphytic but primarily among leaf detritus and soil. Round, underground pseudobulbs up to 10 cm (4 in) from which grow the thin, linear, parallel-veined leaves to 50 cm (20 in) and 3 cm (1 in) wide. Bilateral, three-parted, pink-purple, hermaphroditic flowers are arranged in terminal racemes (clusters) up to 75 cm (30 in) in height and contain ten to fifteen flowers. Each flower has three purple-pink, petal-like sepals and three petals of the same color that are differentiated to form a hood over the fused sex organs and a three-lobed, frilly lip below. The hanging fruit is a dark capsule to 4 cm (2 in), which splits along six suture lines at maturity. **Habitat:** Pine woodlands and ephemeral freshwater wetlands. **Conservation status:** All orchids are protected internationally (under CITES) and should not be collected from the wild.

Christmas Orchid   *Encyclia altissima*
Family: Orchidaceae (Orchids)
Range: Archipelago-wide; also Florida, Cuba.

Encyclia altissima

Encyclia altissima

Cassytha filiformis

Centrosema virginianum

Bletia purpurea

Bletia purpurea

Description: Epiphyte (grows on other plants without taking nourishment from them) or grows among leaf detritus, debris, and rocks with elongate pseudobulbs to 25 cm (10 in) from which simple, flattened, linear leaves grow to 1 m (3 ft) long. Bilateral flowers are in terminal panicles (clusters) to 1.3 m (4 ft) tall; each panicle with thirty to forty-five flowers. The upper floral parts are greenish-yellow, but the lower three-lobed labellum starts yellowish-white and develops reddish radiating lines with age. Two outer lobes wrap around the fused sex organs. Flowers November to February. Fruit is a smooth, hanging, purple capsule to 7 cm (3 in) that splits along six suture lines at maturity. **Habitat:** Variety of coppice habitats. **Conservation status:** All orchids are protected internationally (under CITES) and should not be collected from the wild.

Rufous Orchid   *Encyclia rufa*
Family: Orchidaceae (Orchids)
Range: Archipelago-wide; also Florida, Cuba.
Description: Epiphyte (grows on other plants without taking nourishment from them) or among leaf detritus, debris, and rocks. The ovate pseudobulbs are up to 10 cm (4 in) long from which grow the simple, linear, green leaves to 40 cm (16 in) in length. Bilateral, greenish-yellow flowers are in terminal panicles (clusters) up to 90 cm (36 in) long with fifty to sixty flowers in each. The lower petal or labellum may develop reddish lines as it ages. Flowers April to June. Fruit is a hanging capsule to 4 cm (2 in) that splits along six suture lines at maturity. **Habitat:** Diverse coppice habitats. **Conservation status:** All orchids are protected internationally (under CITES) and should not be collected from the wild.

Butterfly Orchid   *Oncidium ensatum*
Family: Orchidaceae (Orchids)
Range: Archipelago-wide; also Florida, Cuba.
Description: Herb growing on rock and in leaf litter. Oblong pseudobulbs up to 16 cm (6 in) and enclosed by leaf bases arranged alternately in two opposite vertical rows. Simple, bright green, linear leaves to 1.2 m (4 ft) long and 3 cm (1 in) wide. Bilateral, yellow and brown, hermaphroditic flowers are in terminal panicles (clusters) to 1.2 m (4 ft) with a bract subtending each flower. The large, lower central petal or labellum of each flower is yellow with reddish-brown spots and may have a notched or scalloped margin and two sets of tubercles. Fruit is a green capsule that splits at maturity. **Habitat:** Grows on rocks in coppice, pine woodlands, and mangrove wetlands. **Conservation status:** All orchids are protected internationally (under CITES) and should not be collected from the wild.

Bahamian Passion Flower   *Passiflora bahamensis*
Family: Passifloraceae (Passion flowers)
Range: Endemic to Andros, Abaco, New Providence, Grand Bahama.
Description: Vine with smooth surface (glabrous) with older portions becoming corky with age. Simple, alternate, three-lobed leaves to 8 cm (3 in) with tendrils present at the joint of the stem and leaf (nodes). Large, radial, showy flowers with five blue-white sepals and petals, five green stamens, and three stigmatic lobes. Also has a crown-like corona of filaments: the outer filaments are longer than the petals and purple with a white base, while the inner are short and white. Three bracts occur below each flower. Mature fruit is an edible, red berry. **Habitat and use:** Pine woodlands and Sabal Palm woodlands. Although the fruit is edible, it is not very tasty.

Encyclia rufa

Encyclia rufa

Passiflora bahamensis

Oncidium ensatum

Oncidium ensatum

Passiflora bahamensis

Wild Watermelon   *Passiflora cupraea*
Family: Passifloraceae (Passion flowers)
Range: Archipelago-wide; also Greater Antilles.
Description: Vine with older portions becoming corky. Simple, alternate leaves to
   15 cm (6 in) long with tendrils at the joints with the stem (nodes). Leaves variably shaped from lanceolate to two- or three-lobed. Large, radial flowers have five
   green sepals, no petals, five stamens, and a superior ovary with three stigmatic
   lobes, with both of the latter protruding from the flower. The outer ring of the
   crown-like corona is dark red and recurved and looks like petals, and the inner is
   formed of dark red filaments. The edible fruit is a dark purple berry. **Habitat:** All
   terrestrial habitats excluding rocky shores.

Sandbur   *Cenchrus incertus*
Family: Poaceae (Grasses)
Range: Archipelago-wide; also Florida, the Caribbean, Central and South America.
Description: Annual or biannual plant that creeps along ground to 90 cm (35 in) long
   forming large, dense clumps. Simple, parallel-veined, alternate leaves to 20 cm
   (8 in) long and edges rough to the touch. Bilateral flowers are spikelets, composed
   of incomplete flowering structures and enclosing bracts, which are in terminal
   panicles (clusters). Spikelets are surrounded by bracts that are sharpened into
   spines that attach to fur and clothing for dispersal. Mature fruit is a dry, black,
   hard, single-seeded caryopsis, typical of grasses and cereals. The entire spikelet,
   or burr, is the dispersal unit. **Habitat:** Dunes and sand substrate palm woodlands.

Bamboo Grass   *Lasiacis divaricata*
Family: Poaceae (Grasses)
Range: Archipelago-wide; also Florida, the Caribbean, Central and South America.
Description: Climbing or trailing perennial bamboo-like grass, becoming woody
   at the base, to 4 m (13 ft) growing laxly over other vegetation. Simple, alternate,
   parallel-veined leaves have a sheath that is pubescent (fuzzy) along the margin.
   Bilateral, highly modified flowers are in spikelets grouped in terminal panicles
   (clusters). Mature fruit is small c. 2 mm (< 0.1 in), black, single-seeded caryopsis,
   typical of grasses and cereals. **Habitat and use:** Understory of coppice. Used to
   stop water retention.

Sea Oats   *Uniola paniculata*
Family: Poaceae (Grasses)
Range: Archipelago-wide; also the Caribbean, North, Central, and South America.
Description: Perennial grass forming large colonies or clumps connected by rhizomes; up to 1.5 m (5 ft) tall. Simple, alternate, very narrow, parallel-veined leaves
   to 50 cm (20 in) long with a pubescent (fuzzy) sheath along the stem. Bilateral
   flowers are in flattened spikelets in terminal panicles (clusters) to 2 m (7 ft) tall.
   Fruit is a caryopsis (a dry one-seeded fruit, typical of grasses and cereals) contained in golden-brown seed heads like oats (hence the common name). **Habitat
   and use:** Sand dunes. Often used in dune stabilization and restoration efforts.

Stemodia maritima

Lasiacis divaricata

Uniola paniculata

Passiflora cupraea

Cenchrus incertus

Obeah Bush *Stemodia maritima*

Family: Scrophulariaceae (Snapdragons)

Range: The Bahamas; also the Caribbean, South America.

Description: Perennial herb to 1 m (3 ft) tall, becoming woody at its base and covered with glands and pubescence (fuzz). Simple, opposite, sessile (stalkless), lanceolate leaves to 3 cm (1 in) with a serrate (toothed) margin. Small, bilateral, five-parted, purple-blue, hermaphroditic flowers are solitary in leaf axils. Mature fruit is a tiny capsule. **Habitat and use:** Freshwater wetlands, Sabal Palm woodlands, coastal zones, and disturbed habitats. Used to treat colds and flu, dermatological issues, gastrointestinal problems, pain, and obstetric and gynecological issues.

Chaney Briar   *Smilax havanensis*
Family: Smilacaceae (Greenbriers)
Range: Archipelago-wide; also Florida, the Caribbean.
Description: Semi-woody vine of indeterminate length that grows over other vegetation and has stems with short prickles. Simple, alternate, ovate leaves to 15 cm (6 in) long with a pointed leaf tip and stipular tendrils. Stem and leaf margins may have prickles or spines. Radial, three-parted, greenish-white, monoecious flowers are in umbels (umbrella-shaped clusters). Mature fruit is a bluish-black berry that hangs in round clusters. **Habitat and use:** Coppice, pine woodlands, and disturbed habitats. Used in strengthening teas, berries have been fermented to make alcohol, roots are a starch source, and young tips are eaten like asparagus.

Buttercup   *Turnera ulmifolia*
Family: Turneraceae
Range: Archipelago-wide, pantropical.
Description: Woody herbaceous shrub to 1 m (3 ft) tall. Simple, alternate, lanceolate leaves to 15 cm (6 in) long with a dentate (toothed) margin, paired glands at the leaf base, and can be glabrous (smooth) or with hairs. Radial, five-parted, yellow, hermaphroditic flowers are solitary in leaf axils. Mature fruit is a dry capsule. **Habitat and use:** Dunes, disturbed habitats, and coppice edge. Used medicinally to treat gastrointestinal problems, colds and flu, circulatory difficulties, obstetric and gynecological issues, and dermatological concerns. Also used in horticulture.

Capeweed   *Phyla nodiflora*
Family: Verbenaceae (Verbenas)
Range: Archipelago-wide, tropical and subtropical regions.
Description: Creeping herb up to 10 m (33 ft) long rooting at its nodes. Simple, opposite leaves to 8 cm (3 in) have serrate (toothed) leaf margins. Bilateral, four-parted flowers have purple or white petals forming a short tube with five free lobes and a yellow center. Flowers are in compact heads that elongate to form a spike on long stalks, up to 15 cm (6 in) tall, arising from leaf axils. Fruit is a dry capsule that splits at maturity to release the seeds. **Habitat:** Edges of ponds, ephemeral freshwater wetlands, Sabal Palm woodlands and savannas, and disturbed habitats.

Blue Flower   *Stachytarpheta jamaicensis*
Family: Verbenaceae (Verbenas)
Range: Archipelago-wide, New World tropics and subtropics.
Description: Woody herb up to 1 m (3 ft) long (tall), typically creeping along ground. Ovate to elliptical, opposite, pubescent (fuzzy) leaves to 7 cm (3 in) with a toothed

Phyla nodiflora

Stachytarpheta jamaicensis

Turnera ulmifolia

Smilax havanensis

leaf margin. Small, bilateral, five-parted, purple-blue flowers have the petals fused into a tube and spreading at the open end. They are in terminal spikes up to 50 cm (20 in) tall, usually shorter. Flowers open continuously for months, with two to three new ones every few days, so there may be only a small number open on the spike at any time. The dry fruit is a capsule that splits at maturity to release the seeds. **Habitat and use:** Salt flats, ephemeral freshwater wetlands, savannas, coppice edges, and disturbed habitats. Used to treat circulatory problems, gastro-intestinal issues, respiratory conditions, blisters and boils, chills, fevers, and colds and flu. Flowers attract insects and hummingbirds and bananaquits. Used in horticulture.

Invertebrates

Invertebrates, animals that lack a vertebral column, comprise the vast majority of animal species globally (97%) and include marine, freshwater, and terrestrial species (e.g., insects, worms, clams, crabs, octopus, snails, and starfish). The invertebrate fauna of the archipelago is closely related to the mainland fauna, particularly that of southern Florida, as well as to the Caribbean. Many orders are poorly known and new species are likely to be discovered.

Described invertebrates are grouped in two phyla: 1) Mollusca (the mollusks, ordered alphabetically by family) and the 2) Arthropoda. The Arthopods are listed in the following order: crustaceans (crabs and lobsters); arachnids (spiders, scorpions, and myriapods [centipedes and millipedes]); and insects (dragonflies and damselflies, mantis, cockroaches and termites, hymenopterans [ants, bees, and wasps], butterflies and moths, flies, beetles, bugs, grasshoppers and crickets, stick insects, and ant lions).

Care should be taken with some invertebrates found in the archipelago as several species can give painful stings (bees, wasps, and scorpions) or bites (e.g., spiders and centipedes). Normally, unless the victim has pre-existing medical conditions, the sting or bite may be unpleasant but additional complications are unlikely. If medical attention is required, however, collect the offending invertebrate for identification, whenever possible. Potentially poisonous or otherwise dangerous species are highlighted in the text.

Phylum Mollusca (Mollusks)

Large, diverse invertebrate phylum of eight extant families and c. 85,000 species inhabiting a wide range of marine, freshwater, and terrestrial habitats worldwide, including familiar sea shells and snails as well as sea slugs and cephalopods (squids and octopus).

CLASS GASTROPODA (Snails and Slugs)

Gastropods are the most numerous mollusks comprising 80% of species (c. 60,000). Those with shells are referred to as snails while those lacking shells are known as slugs. Most gastropods are marine and either herbivores (plant eaters and algae grazers) or predators (carnivores). Several terrestrial genera of air-breathing snails and slugs are also found in the archipelago.

◼ FAMILY ACHATINIDAE

Family of medium to large terrestrial snails originally from sub-Saharan Africa. Several species have been introduced intentionally (for food) and/or accidentally throughout the subtropics and tropics, and in some cases, they have become serious agricultural pests and vectors of disease. At least one species is recorded in the archipelago.

Giant African Land Snail *Achitina* spp.

Range and status: Introduced. *A. fulica* recorded in northern and central Bahamas and perhaps elsewhere on the archipelago. Native to East Africa, now recorded throughout the tropics including the Americas, Asia, the Caribbean.

Description: Shell length 70–200 mm (2.8–7.9 in). Very large terrestrial snail with a brown conical shell with irregular darker streaking. Care is needed to distinguish between the smaller juveniles and similar native snail species, for example, *Bulimulus* sp. (see below). **Habits and habitat:** Potentially long lived and highly fecund species. Introduction was accidental through the horticultural trade. Worldwide is now considered one of the most invasive invertebrate species. Causes damage to crops, competes with native snail species, and is a potential vector of animal and plant diseases. As in other parts of the world, the predatory Rosy Wolf Snail (*Euglandia rosea*) was introduced to the archipelago as a biological control, but it has also been damaging to native snail species.

◼ FAMILY CEPOLIDAE

Family of terrestrial snails that occur throughout North America and the Caribbean.

Sea Grape or Variegated Bush shells *Hemitrochus* spp.

Range and status: Archipelago-wide, Greater Antilles.

Description: Shell length c. 15 mm (0.6 in). Commonly seen, variably colored tree-dwelling snails found in most terrestrial habitats in the archipelago, often in groups. At least fifteen species are described; however, identification is difficult as inter-individual variation within a species can be extreme and differences between species can be subtle. In general, all species have approximately five convex whorls. Color is variable ranging from creamy white to dark brown either with or without bands. Banding patterns are also variable and include brown banding on a creamy background, pale banding on a brown background or with less distinct and/or interrupted bands on either a dark or light background. Some species are endemic to a single island whereas others are widely distributed throughout the archipelago.

Plagioptycha spp.

Range and status: Archipelago-wide; at least ten species are described; also the Caribbean.

Description: Shell length c. 10 mm (0.4 in). Small- to medium-sized, typically dull-colored, dorsal-ventrally flattened snails. The small tree-dwelling species *P. duclosiana salvatoris* (from San Salvador, Cat Island, and Long Island) is shown. *Plagioptycha duclosiana utowana* is recorded from Eleuthera.

Plagioptycha duclosiana salvatoris

Achitina sp.

Hemitrochus sp.

Hemitrochus sp.

Hemitrochus sp.

■ FAMILY CERIONIDAE

Family of small- to medium-sized terrestrial snails, with the extant species native to the Caribbean.

Peanut (or Beehive) Snails   *Cerion* spp.

Range and status: Archipelago-wide. Genus is endemic to the Caribbean.

Description: Shell length c. 20 mm (0.8 in). Small- to medium-sized tropical terrestrial land snails, which derive their common name from their distinctive elongated shape resembling a peanut. **Habits and habitat:** They often hang from plants in coastal areas but can also be found on the ground inland. The most common land snail in the archipelago and can occur at very high densities (up to 100 shells/m²). High densities of empty shells (up to 200 shells/m²) are found scattered throughout the undergrowth. They are primarily nocturnal feeders on plant material but are also active during the day after rain. During stressful conditions, e.g., during high summer temperatures, these snails can become temporarily inactive (dormant) by sealing their shell opening to prevent desiccation. Individuals can remain inactive for most of their lives, and even when "active" they exhibit limited mobility, frequently spending several years at a time on the same plant. This group exhibits great variation in shell morphology. Many species are restricted to one island. Local island populations may exhibit a diagnostic size, shape, and color, and more than 300 species have been described in the archipelago. Specimens from Eleuthera, San Salvador, Acklins Island, and East Plana Cay are shown.

■ FAMILY CHITONIDAE (Chitons)

Marine species with a dorsal shell that is composed of eight flexible plates surrounded by a fleshy ridge known as a girdle. Often attached to rocks along the shoreline. If dislodged, it curls into a ball.

West Indian Fuzzy Chiton   *Acanthopleura granulata*

Range and status: Archipelago-wide; also Florida, the Caribbean.

Description: Shell length up to 70 mm (3 in). Medium-sized, dull-colored chiton in which the surface of the eight plates is typically eroded and the girdle is spiky and has irregular black bands. **Habits and habitat:** Found on rocky shores, often seen near high-water mark.

■ FAMILY LITTORINIDAE (Periwinkles)

Small, herbivorous marine snails with a worldwide distribution, usually found on rocks, stones, or pilings between high- and low-tide marks. Shells may have sharp or round protuberances.

Beaded Periwinkle   *Cenchritis muricatus*

Range and status: Archipelago-wide; also south Florida, the Caribbean.

Cerion sp., Eleuthera

Cerion sp., East Plana Cay

Cerion sp., Acklins Island

Cerion sp., San Salvador

Cenchritis muricatus

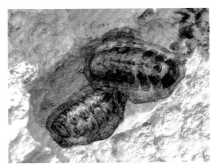

Acanthopleura granulata

Description: Shell length 13–30 mm (0.5–1.2 in). Common pale-colored snail. **Habits and habitat:** Observed from the waterline to well above the highest reaches of the splash zone, often seen attached to vegetation. They can remain inactive for long periods by sealing their shell.

■ FAMILY NERITIDAE (Nerites)

Family of small- to medium-sized typically marine species (but also include fresh and brackish water species) with a worldwide distribution; most species are in the Southern Hemisphere.

Nerita spp.

Range and status: Archipelago-wide. Four species recorded.

Description: Shell length c. 20 mm (0.8 in). Small- to medium-sized rounded or elongate marine species sometimes with small teeth-like projections at the shell opening. **Habits and habitat:** Common on rocky shores. Four-Toothed Nerite (*N. versicolor*) is shown. Variably colored; smaller individuals are white with sporadic black markings (often with flecks of red), and larger individuals often have algae growing on the shell and thus a green appearance.

■ FAMILIES ORTHALICIDAE AND BULIMULIDAE

Medium- to large-sized terrestrial snails found throughout the Caribbean and subtropical and tropical Americas, often commonly termed tree snails.

Range and status: Archipelago-wide; more common in north and central archipelago.

Description: Shell length 20–70 mm (0.8–2.8 in). Locally common, medium- to large-sized terrestrial snails with oval or elongate, typically smooth, glossy shells. Variable and/or distinctive coloration ranging from ivory (white) to tan with or without darker markings. **Habits and habitat:** Seen hanging on vegetation in many natural and disturbed habitats. *Orthalicus undatus* (6.5 cm [2.5 in]); unidentified *Bulimulus* sp.; and Lined Tree Snail (*Drymaeus multilineatus*, c. 22 mm [0.9 in]) are shown.

■ FAMILY STROMBIDAE (True Conchs)

Medium- to large-sized marine snails with worldwide distribution, typically in tropical and subtropical seas. They often have thick-walled decorative shells and a long, narrow aperture or opening. Several species are economically important.

Queen or Pink Conch   *Lobatus gigas*

Range and status: Archipelago-wide; also the tropical western Atlantic from Bermuda south through the Caribbean.

Description: Shell length up to 352 mm (13.9 in). Large, culturally and commercially important marine mollusk with a horny chitinous outer shell and a distinctive glossy pink inner shell. Individuals reach sexual maturity at approximately four years (with a shell length c. 180 mm [7 in]); adults have a wide flaring lip, whereas the lip of the juveniles is sharp and unflared. Long-lived, up to thirty years. **Habits and habitat:** Grazes on detritus, macroalgae, and epiphytes. Typically found in sea grass beds in shallow water (1–30 m [3–100 ft]), sometimes at high densities

Orthalicus undatus

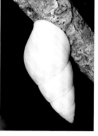

Bulimulus sp.

Drymaeus multilineatus

Nerita versicolor

Lobatus gigas

Lobatus gigas, juvenile shell above, mature shell below

Lobatus gigas, pile of empty shells

of 1–2 per m². **Conservation status:** Conch were used by the native peoples for food and tools. Present-day conch fishing is important both economically and culturally throughout the archipelago; it is a significant component of local diet and is also a multimillion dollar industry through their export to overseas markets, primarily the U.S., with an annual harvest in The Bahamas (at the time of publication) of c. 600–700 MT (590–689 t). Conch fishery is important for supplementing income during the closed season for Caribbean Spiny Lobster (see below), the largest marine commercial industry in the archipelago. Currently, conch is one of the largest and most commercially important mollusks in the Caribbean; however, because of overexploitation populations are declining range-wide. The importance of the harvest can be seen in the large piles of empty shells in many coastal areas of the region. **Important:** CITES listed and a permit is required (including for tourists) to export conch shells from either The Bahamas or TCI.

■ FAMILY TEGULIDAE

Family of marine gastropods (in the superfamily Trochoidea) with a worldwide distribution typical of the inter- and subtidal zones. Shells typically have only a few whorls and all have a pearl-like inner shell.

West Indian Top Shell   *Cittarium pica*

Range and status: Archipelago-wide; also the Caribbean.

Description: Shell length 90–130 mm (3.5–5 in). Large grazing edible marine snail with a pale-colored heavy shell and black-brown zigzag spots on each whorl. **Habits and habitat:** Commonly found on rocky shores. Third most economically important invertebrate species in the Caribbean, after the Caribbean Spiny Lobster (crawfish) and the Queen Conch, and has become locally extinct in the region from overexploitation. Empty shells are used by the Caribbean Hermit Crab (see below), which carry the shells to terrestrial habitats; empty shells can be found some distance from marine habitats.

■ FAMILY VERONICELLIDAE (Leather Leaf Slugs)

Family of terrestrial, herbivorous slugs with a pantropical distribution.

Caribbean Leatherleaf Slug   *Sarasinula plebeia*

Range and status: Possibly introduced. Distribution poorly known; currently only recorded from northern islands of the archipelago.

Description: Shell length up to 70 mm (2.8 in). Terrestrial gray-brown slug with small, fine black markings. Resembles rubbery or leathery leaf when inactive, hence its common name. **Habits and habitat:** Recorded from both suburban and natural habitats. Potentially a serious agricultural pest of beans, sweet potatoes, and tomatoes.

Sarasinula plebeia

Cittarium pica, middle and right

Coenobita clypeatus

Phylum Arthropoda (Crustaceans, Arachnids, Myriapods, and Insects)

Arthropods are invertebrates that have an external (exo)skeleton, a segmented body typically in three sections (head, thorax, and abdomen), and jointed limbs or appendages.

ORDER DECAPODA (Crabs and Lobsters)

Land crabs (subphylum Crustacea) are commonly seen throughout the archipelago and will eat just about anything. Most species are nocturnal and require damp conditions and are therefore seen less often during the dry season. All species return to the sea to breed during the wet season. There are two groups of land crabs: the hermit crabs (infraorder Anomura), which have an asymmetrical abdomen and are concealed in an empty mollusk shell that is carried around by the crab; and the true crabs (infraorder Brachyura) in which the abdomen is reduced or completely concealed under the crab's shell (thorax). Several Brachyuran species are both economically and culturally important in the archipelago and are collected for food.

■ **FAMILY COENOBITIDAE** (Terrestrial Hermit Crabs)

Hermit crabs are so named because as they age and grow, they must find a new larger shell to shelter their vulnerable abdomen and therefore abandon the previous one.

Caribbean Hermit Crab   *Coenobita clypeatus*

Range and status: Archipelago-wide; also southern Florida, the Caribbean, South America.

Description: Length up to 70 mm (2.8 in). Only terrestrial hermit crab in the archipelago. Frequently occupies the West Indian Top Shell (see above). **Habits and**

habitat: A slow-moving species found in most terrestrial habitats, sometimes surprisingly far inland. Their distinctive large purple claw, typically the left, can be used to cover the shell opening for protection against predators. Even when inland, they still require damp humid conditions and are frequently nocturnal, often burying themselves or remaining under stones during the day to avoid desiccation. Although they return to the sea to breed, complete submergence is fatal. Land hermit crabs are herbivorous and scavengers; diet includes plant and animal remains, overripe fruit, and even feces. Local name is Soldier Crab.

■ **FAMILY GECARCINIDAE (Land Crabs)**

Land crabs are typically more restricted in their distribution than are hermit crabs, and they are usually associated with terrestrial habitats close to saltwater, such as whiteland or coastal coppice. Despite their terrestrial lifestyle, they return to the sea to breed. They excavate burrows, which can be up to 1.5 m (5 ft) deep, that lead to a living chamber used for food storage and as a retreat during dry conditions. Mostly nocturnal, or at least remaining in dense shady places close to their burrows during the day. Land crabs are more likely to be encountered during the wet season. Their varied diet includes leaves, fruits, grasses, insects, and carrion, often collected near the burrow.

Black Crab   *Gecarcinus ruricola*

Range and status: Archipelago-wide; also the Caribbean.

Description: Carapace length to 70 mm (2.8 in). Medium- to large-sized terrestrial crab. Carapace variably colored, frequently black or purple but can also be red, yellow, or green. Limbs usually match carapace color with paler joints. Typically has a pair of pale bluish-gray spots at the rear end of the carapace and a bright red or orange spot beneath each eye. Eyestalks are purple and the cornea is black. **Habits and habitat:** The most terrestrial of the Brachyuran land crabs in the region. Can be found far inland; elsewhere in the Caribbean it can be found at elevations up to 1000 m (3000 ft) asl. An important food resource in The Bahamas. Other common names include Purple Land Crab and Black Land Crab. **Similar species:** Red morph can be confused with Black-Back Land Crab (*G. lateralis*, see below).

Black-back Land Crab   *Gecarcinus lateralis*

Range and status: Archipelago-wide; also southeast U.S., the Caribbean, Central America.

Description: Carapace length to 70 mm (2.8 in). Medium- to large-sized, distinctively colored terrestrial crab, with a large central dark or black patch on the carapace; the dorsal (top) edge of carapace and limbs is scarlet-orange. Eyestalks are reddish purple with a black cornea. **Habits and habitat:** Often seen in low-lying coastal coppice habitats with moist soils, but not in standing water. Also called Bermuda or Red Land Crab. A food resource in the archipelago.

Gecarcinus ruricola

Gecarcinus lateralis

Cardisoma guanhumi

White Crab   *Cardisoma guanhumi*

Range and status: Archipelago-wide; also southeast U.S., the Caribbean, Central and South America.

Description: Carapace length to 110 mm (4.3 in). Largest terrestrial crab in the archipelago, weighing up to 500 g (1.1 lb). Fully grown individuals are typically pale gray, tan, or a pale blue-lavender color. Smaller individuals are more variable in color, including pink or orange. Males have one claw that is larger than the other; the larger claw can grow to approximately 150 mm (5.9 in) in length. **Habits and habitat:** Typically found in low-lying coastal areas and wetlands. An important food resource in the archipelago.

■ **FAMILY GRAPSIDAE** (Shore Crabs and allies)

Diverse family of crabs occupying a range of aquatic, typically marine habitats including rocky shores, estuaries, marshes, and even pelagic environments, for example, among drifting seaweed and flotsam.

East Atlantic Sally Lightfoot Crab   *Grapsus adscensionis*

Range and status: Archipelago-wide; also Florida, the Caribbean, Central and
South America.

Description: Carapace length 50–80 mm (1.9–3.2 in). Agile, active, dorsally-ventrally
flattened marine species with typically a dark-colored carapace ranging from brown
to black with numerous irregular light blue-gray spots and/or splotches. Eyestalks
are bluish cream below with maroon lines and spots above, while the corneas are
green to black. **Habits and habitat:** Frequently seen in the splash zone on rocky
shores where they dash quickly between incoming waves to feed on algae.

■ FAMILY OCYPODIDAE (Ghost and Fiddler Crabs)

Semi-terrestrial crabs. Ghost crabs (subfamily Ocypodinae) are common throughout
the tropics and subtropics on sandy beaches, occupying burrows in the intertidal zone.
Their common name derives from their nocturnal behavior and their generally pale
coloration. Both sexes have one of the paired claws (chelipeds) slightly larger than the
other, thick and elongated eyestalks, and a box-like body. One species is recorded in
the archipelago.

Fiddler crabs (subfamily Ucinae) are generally small and common throughout the
tropics and subtropics and occupy burrows in a range of marine habitats, such as intertidal mudflats, lagoons, and mangroves. Males have a significantly more pronounced
asymmetry in claw size than do ghost crabs; in females the paired functional feeding
claws are the same size. Male fiddler crabs use their larger, more colorful claw in visual
displays, typically a series of wave gestures to communicate ownership of territory
and during courtship. At least five species are recorded in the archipelago.

Ghost Crab   *Ocypode quadrata*

Range and status: Archipelago-wide; also the western Atlantic from eastern U.S. to
South America.

Description: Carapace length to 50 mm (1.9 in). Medium-sized intertidal species with
a box-shaped body, distinctive thick and characteristic elongated eyestalks, and
one claw larger than the other. Two color phases: an off-white phase found on light-
colored sand and a brown phase on darker-colored beaches. The light phase varies
from bluish white to yellowish cream, and the dark phase has a brown carapace
with cream to pale tan stripes and spots. **Habits and habitat:** Typically found on
sandy beaches between the high- and low-tide lines. Broad diet includes carrion.
Frequently nocturnal, remaining in their burrow during the heat of the day.

Thin-fingered Fiddler Crab   *Uca (Leptuca) leptodactyla*

Range and status: North and central archipelago (Bimini, New Providence, San
Salvador), probably archipelago-wide; also the Caribbean, Central and northern
South America.

Description: Carapace length 11 mm (0.4 in). Small pale-colored species. Large claw
of the male is up to 40 mm (1.5 in). **Habits and habitat:** Fiddler crabs make burrows in the intertidal zone and can occur at surprisingly high densities (29–42 m^2
[2.5–3.7 ft^2]), sometimes considerably higher (depending on species). Females

Grapsus adscensionis

Ocypode quadrata, off-white phase

Uca leptodactyla

Aratus pisonii

select males on basis of claw size and the "quality" of their waving display. A more vigorous waving display with their enlarged claw indicates male quality, while claw size indicates burrow size; females lay their egg in the burrow of the male and can therefore select their preferred size of burrow based on the size of the male's claw.

■ FAMILY SESARMIDAE (Terrestrial Crabs)

Family of (semi-)terrestrial crabs that include some species that do not need to return to the sea to breed.

Mangrove Tree Crab   *Aratus pisonii*

Range and status: Probably archipelago-wide; also eastern Florida, throughout the Caribbean to northern South America, also on Pacific coast from Nicaragua to Peru.

Description: Carapace length 18–27 mm (0.7–1.1 in). Small, dull-colored species with brown or olive-green carapace and brown mottled legs with pointed tips (to facilitate climbing). Pincers have tufts of black hair-like projections. **Habits and habitat:** Salt marshes and mangroves. Frequents the canopy foliage during high tide and the exposed sediments during low tide. **Similar species:** Can coexist with Mangrove Root Crab (*Goniopsis cruentata*), which is larger (63 mm [2.5 in]) and has a dark brown carapace, red legs patterned with yellow spots, and pale inner-colored and hairless pincers.

■ **FAMILY PALINURIDAE (Spiny Lobsters)**

Family of commercially important crustaceans (in the infraorder Achelata) found worldwide in warm temperate, tropical, and subtropical marine habitats. Similar in appearance to true lobsters (Family Nephropidae) but lack the large claws and have a characteristic pair of thick, long, forward-pointing, spine-covered antennae (typically longer than the body).

Caribbean Spiny Lobster   *Panulirus argus*

Range and status: Archipelago-wide; also the Atlantic coasts of the southern U.S., Central and South America, the Caribbean.

Description: Length 200–600 mm (7.9–24 in). Culturally and commercially important large, olive-greenish or brown marine species covered with spines with a scattering of yellowish to cream-colored spots on the carapace and larger yellow to cream-colored spots on the abdomen (usually four to six). Long, spiny antennae are green or brown; legs are usually striped longitudinally with blue and yellow ending in a single spine-like point; and lobes of the tail are yellow and black. **Habits and habitat:** Found 2–100 m (6.5–328 ft) deep in coral, artificial reefs, mangroves, and other habitats. Diet includes mollusks, plants, and carrion. Nocturnal; during the day are usually hidden in crevices or under rocks or corals. Large-scale mass migrations occur, typically in the fall, to deeper water. The lobster is the most important food export of The Bahamas, worth (at time of publication) c. 50–70 million dollars per year. Local name is crawfish.

CLASS ARACHNIDA (Spiders, Scorpions, and Whip Spiders)

This diverse group of primarily terrestrial carnivorous invertebrates are in the subphylum Chelicerata and number more than 100,000 species. Arachnids have two body segments: the cephalothorax (combined head and thorax) and the abdomen. Eight legs are attached to the cephalothorax. Arachnids also have two additional pairs of appendages: 1) the chelicerae, which are specialized mouthparts used for grabbing prey, and in poisonous species modified to inject venom (fangs); and 2) the pedipalps, which are used for feeding and during mating. In addition to their eyes (most have eight), arachnids sense their environment by hairs located in sensory pits that cover the body. Females are usually bigger than males. We provide body length (comprising the cephalothorax and abdomen) as a measure of size.

Panulirus argus

ORDER ARANEAE (Spiders)

Spiders are the most common and familiar arachnids with c. 40,000 described species worldwide. The group is divided into two suborders: Mesothelae, which includes the more primitive spider families; and Opisthothelae, which is split into two infraorders: Mygalomorphae (Tarantulas and allies) and Araneomorphae (sometimes referred to as "true spiders"). The Mygalomorphs are typically large, thickset, hairy spiders in which the fangs (chelicerae) point straight down, whereas in the Araneomorphs the fangs oppose each other and cross in a pinching action. All spiders can produce silk (from specialized structures called spinnerets located on the underside of the abdomen), which has a variety of functions. Many species use silk to build complex webs to trap their insect prey, whereas other species actively hunt their prey. Spiders are an important natural control of harmful insect pests, and although most use venom to kill their prey, the majority pose no threat to humans. The arachnid fauna of the archipelago is poorly described. To date, approximately fifty species are documented for The Bahamas and at least sixty species for the TCI; most are orb-weaving species (Family Araneidae).

■ FAMILY THERAPHOSIDAE (Tarantulas)

Large, hairy, venomous, typically terrestrial spiders with powerful mouthparts comprising a pair of large downward-pointing parallel fangs. They actively hunt and feed on small vertebrates including frogs, lizards, and even birds. Despite their threatening appearance and reputation, they are typically shy and avoid human contact. Bites are painful but are not normally serious. Before biting, New World tarantulas typically signal their intention to attack by rearing up on their hind legs in a "threat posture." Their next step, short of biting, may be to slap their raised front legs down onto the intruder. If that response fails to deter a potential attacker, an individual may next turn away and flick out special fine, barbed (urticating) hairs from their abdomen, which irritate the skin and other soft tissues of the recipient. Exposure of the eyes and respiratory system to these hairs can have serious health effects and should be avoided.

Bahamian Tarantula *Cyrtopholis bonhotei*

Range and status: Endemic. Throughout the Little and Great Bahama Banks. Limited data on distribution within southern Bahamas and unknown from the TCI.

Description: Body length 40–70 mm (1.6–2.8 in). Largest spider in the archipelago with up to c. 150 mm (5.9 in) leg span. Dark-colored hairy species varying from light brown to dark brown to almost black. Primarily nocturnal and rarely seen during the day when it lives in burrows or under rocks or vegetation. Typically terrestrial but can also be a surprisingly fast and capable climber. Diet largely unknown, but similar-sized tarantulas eat insects and other arthropods, lizards, and occasionally larger prey. Usually solitary. Difficult to distinguish sexes (without handling), but males tend to be smaller with narrower and shorter abdomens than the females.

■ **FAMILY ARANEIDAE** (Orb-weaver Spiders)

Members of this family are the most common group of web-building spiders, which construct spiral wheel-shaped webs often found in gardens, fields, and forests. Their common name is derived from the spherical-shaped web, and the group includes many well-known large or brightly colored garden spiders.

Silver Garden Spider or Silver Argiope *Argiope argentata*

Range and status: Archipelago-wide; also southern U.S., Central and South America.

Description: Body length 20–35 mm (0.8–1.4 in). Distinctively silver-colored spider with brown and orange markings at the end of the abdomen and orange, black, and silver banded legs. Web has a distinctive crisscross band of silk through the center that may decrease the visibility of the remainder of the web to prey species. The bite can be painful, but the effects are typically short-lived, although it may have health repercussions for children, seniors, and physically weak people.

Crab Spider *Gasteracantha cancriformis*

Range and status: Archipelago-wide; also southern U.S., Central and South America.

Description: Body length: males 2–3 mm (0.1 in), females 5–9 mm (0.2–0.4 in). Small spider with six characteristic abdominal spine-like projections on the carapace and distinctively short legs; at first glance can resemble plant seeds or thorns hanging in the web. Dorsal color of carapace in females include black spots on a

Gasteracantha cancriformis, female dark morph

Cyrtopholis bonhotei

Gasteracantha cancriformis, female

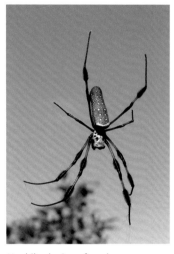

Argiope argentata

Nephila clavipes, female

white or yellow background, and six white (sometimes red, yellow, or black) distinctive spines, while the legs and underside are black with white spots under the abdomen. Males are smaller and have a gray abdomen with white spots and the spines are reduced to four or five stubby projections. **Habits and habitat:** Common along woodland edges and in shrubby gardens. Short-lived, surviving until reproduction; males die soon after mating and females die after egg laying.

Golden Orb or Santa Claus Spider   *Nephila clavipes*

Range and status: Archipelago-wide; also the Caribbean, the Americas.

Description: Body length: males c. 6 mm (0.2 in), females 25–51 mm (1–2 in). Commonly seen sexually dimorphic species. Females are large and distinctively colored with a silvery carapace, yellow spots on a dull orange to tan cylindrical body, and brown and orange banded legs. The latter are specialized for weaving (where their tips point inward, rather than outward as is the case with many wandering spiders) and have black hair-like tufts on the joints. The duller brown-colored, smaller male (three to four times smaller) is often seen on the margins of the female's web. The contrast of dark brown-black and green-yellow serves as a warning to potential predators. The larger female has a venomous bite but will do so only if provoked, and the bite is usually relatively harmless, leading to slight redness and localized pain. The large, yellow, often irregularly threaded sticky webs can be up to 2 m (7 ft) across.

■ **FAMILY SICARIIDAE (Recluse, Fiddle-back, or Violin Spiders)**

Family of spiders with six eyes (other spiders have eight eyes) and are renowned for their poisonous bites that can result in serious tissue damage.

Loxosceles spp.

Range and status: Archipelago-wide; also the Caribbean, the Americas including southeast U.S.

Description: Body length 7–12 mm (0.3–0.5 in). Small, dull-colored spiders usually with a distinctive dark violin-shaped pattern on the thorax. They have six eyes and the leg joints may appear to be a slightly lighter color. The prominent dark pedipalps are normally held horizontally forward. **Habits and habitat:** Found in many natural and urban habitats. Recluse spiders are known for their venomous bite, which contains tissue-destroying toxins capable of causing long-term wounds and may require extensive medical attention. The chelicerae (fangs) are small and unable to penetrate most fabrics, so bites generally occur when the spider unintentionally comes into direct contact with the skin. Seek medical attention if bitten, and if possible, collect the spider for identification. At least two species are found in the archipelago: *L. taino* and *L. cubana*.

■ **FAMILY SPARASSIDAE (Huntsman Spiders)**

These spiders are an agile, fast-moving, hairy nocturnal species. Eight eyes appear in two largely forward-facing rows of four. Many species grow very large and can be identified by their legs, which, rather than being jointed vertically relative to the body, are twisted in such a way that in some instances the legs extend forward in a crab-like manner. Individuals do not spin webs; they hunt by waiting and attacking prey that comes within range.

Huntsman   *Heteropoda venatoria*

Range and status: Archipelago-wide. Pantropical distribution.

Description: Body length 22–28 mm (0.9–1.1 in). Large red or gray-brown, short-haired, nocturnal spider with a leg span of 70–100 mm (2.8–3.9 in). **Habits and habitat:** Despite their formidable appearance, they are shy and flee at the approach of danger. They wait for their prey to come within striking distance. A flattened body allows them to utilize narrow cracks. Often seen in and around human habitation where they feed on insect pest species including cockroaches. Female has

Latrodectus mactans, female

Latrodectus mactans, female, ventral view

Loxosceles sp.

Heteropoda venatoria

a stouter body and shorter, thicker legs than the male has, and when breeding she carries a pillow-like egg case (c. 15 mm [0.6 in] in diameter) under her body. Also known as Banana, Giant Crab, or Cane Spider.

■ **FAMILY THERIDIIDAE** (Tangle-web, Cobweb, and Comb-footed Spiders)

Diverse family of spiders that contains the genus *Latrodectus*, also referred to as the widow spiders because the larger female sometimes consumes the male after mating. Black widow spiders are perhaps the best-known members of the genus.

Black Widow or Southern Black Widow   *Latrodectus mactans*

Range and status: Archipelago-wide; also the Caribbean, the Americas.

Description: Body length: males 3 mm (0.12 in), females 13 mm (0.51 in). Small- to medium-sized, dark-colored, long-legged spider. Large, shiny, black females usually have a characteristic red hourglass mark on the underside of the large abdomen, and sometimes an orange or red patch just above the spinnerets on the top of the abdomen. Adult male black widows are a quarter the size of the female and can be gray or brown rather than black. Juveniles are strikingly different with white-striped gray or black abdomen spotted with yellow and orange. **Habits and habitat:** Web consists of irregular, tangled, sticky silken fibers often under overhanging rocks or vegetation. The bite from a male is not considered dangerous, but the larger adult female has a very toxic bite. This species is listed as the most venomous spider in North America. **Similar species:** Brown Widow Spider (*L. geometricus*) recorded from the southern Bahamas also has a red hourglass shape, but differs from *L. mactans* in being smaller, typically lighter-colored, with a black-and-white geometric pattern on the dorsal side of its abdomen, and striped legs.

■ **FAMILY TETRAGNATHIDAE** (Long-jawed Orb Weavers)

Typically long-legged, thin-bodied spiders; the first pair of legs is usually significantly longer than other legs and extend forward. Their common name is derived from the longer chelicerae (jaws) compared to those of other orb weavers (family Araneidae; see above). The structure of the wheel-like web is similar in both families.

Leucauge argyra

Range and status: Archipelago-wide; also the Caribbean, the Americas.

Description: Body length 19–36 mm (0.8–1.4 in). Medium-sized slender species. Upper surface of the silvery-white elongated abdomen has three black lines that run parallel to about halfway along the back, where the outer two bend inward before flowing parallel once again to the end of the abdomen. Dark underside has two parallel pale lines ending in yellow spots. **Habit and habitats:** Often in loose colonies. Horizontal webs are spun on vegetation or man-made structures; these spiders hang upside-down beneath their web.

ORDER AMBLYPYGI (Amblypygids: Tailless Whip Scorpions)

Tailless whip scorpions, also known as cave spiders, whip spiders, or whip scorpions, are found throughout the tropics including the Caribbean. They have a characteristic flattened spider-shaped body with long, thin, modified front legs referred to as whips (hence their common name) that can extend several times their body length. Pedipalps are modified into spinous raptorial pincers (each spine is termed a patella), and prey are located with the "whips," captured with the pincers, and then dismembered with the chelicerae. Unlike spiders, they lack silk glands and despite their fierce appearance they are not venomous. Bites from larger species can be painful but are not serious. Usually nocturnal, they are found in humid environments (under logs, bark, stones, leaves, and in caves). Two species from one family (two genera) are recorded from the archipelago.

■ **FAMILY PHRYNIDAE**

Family of whip scorpions found throughout the tropical and subtropical Americas.

Phrynus marginemaculatus

Range and status: Archipelago-wide (especially widespread in the northern archipelago); also the Caribbean.

Description: Body length 6–18 mm (0.2–0.7 in). Front legs can reach 100 mm (3.9 in). Variably colored: yellowish-reddish-brown to black.

Phrynus marginemaculatus

Paraphrynus viridiceps

Leucauge argyra

Paraphrynus viridiceps

Range and status: Andros, Eleuthera, New Providence; few records from elsewhere in the archipelago. Also Cuba.

Description: Body length 6–15 mm (0.2–0.6 in). Front legs can reach 100 mm (3.9 in). Similarly colored to *P. marginemaculatus* (see above) and both species can occur together. Can be distinguished from each other by a subtle difference in the structure of the pincer-like pedipalp: *P. marginemaculatus* has a single spine between the two longest pedipalpal patellas whereas *P. viridiceps* has two small spines between the longest patellas.

ORDER SCORPIONES (Scorpions)

Typically found in the tropics and subtropics, scorpions are easily recognized by their crab-like appearance. Distinct from other arachnids by its long, segmented tail ending in a prominent, slightly curved, sharply pointed poison sting. Pedipalps are greatly enlarged and appear as a pair of large claws. Scorpions are nocturnal predators of small invertebrates and use their pincers to initially catch the prey and either crush it or inject it with venom. Although stings can be painful to humans, they are usually not serious, producing only local pain, numbness, or swelling. During the day, scorpions seek holes or shelter under rocks or tree trunks. Unlike most Arachnids, which lay eggs, female scorpions give birth to live young and carry them until they can fend for themselves. Scorpions fluoresce (glow) under ultraviolet (UV) light (a possible adaptation to avoid sunlight), which makes them easy to find at night using a UV flashlight. At least three species occur in the archipelago.

■ **FAMILY BUTHIDAE**

Small- to medium-sized species with worldwide distribution in tropical, subtropical, and temperate regions. Species found in the archipelago are small and not considered dangerous. They are frequently found under stones or bark and inside epiphytic bromeliads in many habitats.

Bahamas Bark Scorpion   *Centruroides guanensis*
Range and status: Native. Archipelago-wide; also the Caribbean.
Description: Body length 35–70 mm (1.4–2.8 in). Medium-sized, yellowish scorpion, often with a variable dusky pattern. There are dark and pale morphs, usually with two longitudinal dark bands over the abdomen separated by a thin blackish line. Claws, head, and tail are frequently spotted. It has a sharp painful sting with aftereffects lasting 1–5 hr. Two similar species are also recorded: Caribbean Scorpion (*C. testaceus*) listed for TCI; and *C. platnicki* endemic to Mayaguana (Bahamas) and Providenciales (TCI).

ORDER PSEUDOSCORPIONES (Pseudo-, False, or Book Scorpions)

Easily overlooked, very small, dark-colored arachnids (typically ≤ 5 mm [0.2 in]) with a worldwide distribution They resemble tailless scorpions, with a flat, pear-shaped, rounded abdomen and a pair of pincers as long as the body, often with a contrasting color. Found under tree bark, in leaf litter and soil, under stones, in caves, and sometimes on beaches (around the high watermark). Their small size makes field identification difficult; at least four endemic species from two families (Chernetidae and Chthoniidae) occur in the archipelago.

Family Chernetidae
Range: Northern archipelago; distribution poorly known. Currently two species described: *Antillochernes bahamensis* in Grand Bahama and *A. biminiensis* in South Bimini. Genus *Antillochernes* also recorded in Greater Antilles and south Florida.
Description: Body length c. 4 mm (0.2 in). Very small, dark-colored species. The more rigid sections of the exoskeleton (head, pincers, and central parts of the abdominal segments) are more darkly colored than the softer more flexible parts, which often gives a slightly banded appearance to the rounded abdomen. **Habits and habitat:** Typically found in caves or under stones in shaded locations.

CLASS CHILOPODA (Centipedes)

Centipedes are in the subphylum Myriapoda (which also includes the millipedes). They are fast-moving, carnivorous, venomous, nocturnal, terrestrial invertebrates with flattened, segmented bodies (each segment has a pair of legs) and long antennae. The venomous bite can be painful, and may cause a variety of symptoms, but is rarely fatal. Prey includes invertebrates, for example, insects, earthworms, spiders, slugs,

Centruroides guanensis

Antillochernes sp.

Sphendononema guildingii

and other small animals. Centipedes require a moist microhabitat because they lack the waxy cuticle of other arthropods and lose water rapidly through their epidermis; during the day they are found under rocks, logs, and leaf litter.

ORDER SCUTIGEROMORPHA (House Centipedes)

Collectively known as house centipedes, members of this group are extremely fast, agile, delicate, and short-bodied. Adults possess only fifteen pairs of very long, delicate legs, which become increasingly longer toward the end of the body. The antennae are similarly very long and whip-like.

■ FAMILY PSELLIOIDIDAE

Small family of house centipedes native to the neotropics and tropical Africa.

Sphendononema guildingii

Range and status: Native. Recorded in the northern and central archipelago (including the Biminis, Eleuthera, San Salvador), possibly archipelago-wide;

also Greater and Lesser Antilles, the Caribbean, Central and northern South America.

Description: Body length up to 35 mm (1.4 in). Medium-sized house centipede with a dark gray-black back and a median longitudinal white stripe (sometimes with a slightly greenish tinge) running the length of the body. Long delicate legs and antennae are uniformly orange-red. **Habits and habitat:** Very fast moving. Avoids direct sunlight by sheltering under rocks and tree trunks in coppice.

ORDER SCOLOPENDROMORPHA

Scolopendrons exhibit the typical centipede body form with a long dorsal-ventrally flattened segmented body with a pair of short legs associated with each of the twenty-one to twenty-three segments. The head is covered by a flat shield and bears one pair of antennae. The first pair of legs is modified to form special structures called forcipules, which are used to capture prey and inject venom. The last pair of legs point backward and are hooked at the end. This order contains some of the largest centipedes, and in addition to other invertebrates, some species feed on small vertebrates including reptiles, amphibians, and even small birds and mammals. Bites from large scolopendrons can be very painful and may result in short-term discomfort, but they are not normally life threatening. At least four species are recorded; one native and three introduced species.

■ FAMILY SCOLOPENDRIDAE

Family of large centipedes with worldwide distributon.

Scolopendra alternans

Range and status: Native. Archipelago-wide including Andros, Cat Island, Exuma, Little Inagua, New Providence, Bimini, San Salvador, Eleuthera. No data on the TCI. Also south Florida and the Caribbean.

Description: 229 mm (9 in). Large, bright orange-red species. At least three other similarly large scolopendrons occur in the archipelago.

1) *S. subspinipes* (introduced; 100–200 mm [3.9–7.9 in]) recorded throughout the northern Bahamas (and is probably archipelago-wide). The two species are difficult to distinguish but *S. subspinipes* is typically duller and has orange rather than red legs; however, the two species can only accurately be distinguished by subtle differences in the pattern of spines on top of the last pair of backward-pointing legs.

2) *S. morsitans* (introduced; 130 mm [5.1 in]) recorded from the central and southern archipelago and is generally smaller than either *S. alternans* or *S. subspinipes*; typically has yellow and black stripes on each segment and a bluish tinge to the legs.

3) The Peruvian Giant Yellow-leg or Amazonian Giant Centipede *S. gigantea* (introduced; 300 mm [12 in]) is recorded on the TCI.

Scolopendra alternans

Anadenobolus monilicorni

CLASS DIPLOPODA (Millipedes)

Millipedes are in the subphylum Myriapoda (which also includes the centipedes). They are elongated, cylindrical, and segmented with two pairs of legs per segment (centipedes have only one pair per segment), with thirty-six to 400 legs. They are typically slower moving than centipedes and unlike their carnivorous cousins, millipedes are detritivores and feed on dead plant material.

■ FAMILY RHINOCRICIDAE

Family of millipedes with a disjunct distribution, occurring throughout the Neotropics but also the South Pacific.

Yellow-banded Millipede   *Anadenobolus monilicorni*

Range and status: Probably widely distributed throughout the archipelago. Native to the Caribbean.

Description: 25–100 mm (1–4 in). Brown millipede with yellow bands and red legs.

Habits and habitat: Often found in urban and agricultural areas. If roughly handled can release a toxic irritant that will cause a dark chemical burn to the skin. Often found in leaf litter and under rocks and logs in coppice and urban habitats.

CLASS INSECTA (Insects)

Insects are arthropods with a three-part body comprising head, thorax, and abdomen. They have three pairs of jointed legs attached to the thorax, compound eyes, a pair of antennae, tough but flexible chitinous exoskeleton, and a waxy cuticle that limits water loss. They are among the most diverse groups of animals on Earth, including more than a million described species, and they represent more than half of all known animal species. The archipelago has a diverse insect fauna attributable to the proximity of the archipelago to North America and Greater Antilles. There are no endemic genera. Endemic species belong to widespread genera.

ORDER ODONATA (Dragonflies and Damselflies)

Odonata are aggressive aerial predators that capture prey on the wing. They have well-developed heads with large compound eyes; strong jaws; two pairs of long, transparent veined wings; spiny legs; and elongated abdomens. Prey includes butterflies, bees, flies, and even other smaller dragonflies. Dragonflies (suborder Anisoptera [Epiprocta]) are strong fliers with robust bodies and at rest hold their wings at right angles to the body. By contrast, damselflies (suborder Zygoptera) are usually thinner and daintier, exhibit a weaker, more fluttery flight, and when at rest most species hold their wings folded back over the abdomen. Both groups lay their eggs in or around freshwater and since the nymphs (juveniles) are primarily aquatic, adults are also typically found close to freshwater. Nymphs are aggressive predators that feed on other aquatic insects, tadpoles, and even small fish. The limited availability of surface freshwater (suitable breeding habitat) in the archipelago is associated with a more restricted fauna than on the mainland; yet, at least twenty-seven species of dragonflies and six damselflies are recorded. There are no endemic species. Many species fly year-round, but local abundance is dependent on the persistence and availability of freshwater. Restricted ranges and greater seasonal variation in detectability and breeding are expected in the drier, southern parts of the archipelago.

Dragonflies

■ **FAMILY AESHNIDAE** (Darners)

Darners are the largest dragonflies in the archipelago with long abdomens and large eyes that meet in a seam on top of the head. They appear to fly or hover continuously, but they usually hang vertically when perched. At least seven species are recorded in the archipelago.

Blue-spotted Comet Darner   *Anax concolor*
Genus: *Anax* (Green Darners)
Range and status: Archipelago-wide; also the Caribbean, Central and South America.
Description: 75–87 mm (2.9–3.5 in). Large, commonly seen dragonfly. Males have blue-green eyes, a green thorax with a few blue spots, and a brown abdomen with

Coryphaeschna viriditas, female

Anax concolor, male

large blue spots. Females are similar but have more extensive blue spotting on the thorax. **Habits and habitat:** Hunts over open ponds and clearings but also roosts in coppice habitats. **Similar species:** Green Darner (*A. junius*) (68–84 mm [2.7–3.3 in]; archipelago-wide) has a green head and thorax. Males have a thin blue abdomen marked by a central dorsal dark line, while females have a thicker yellow and brown abdomen. Both sexes have a distinctive blue and black "bulls-eye" marking on the forehead (when seen from above) just in front of their large, compound eyes, which distinguishes it from other green darners in the archipelago.

Mangrove Darner   *Coryphaeschna viriditas*

Genus: *Coryphaeschna* **(Pilot Darners)**

Range and status: Possibly archipelago-wide; also the Caribbean, Central and South America.

Description: 85–90 mm (3.2–4 in). Rare, large, green-brown dragonfly. Eyes and thorax are green, and the abdomen is green with narrow brown banding. **Habits and habitat:** Mangrove and neighboring habitats. **Similar species:** Regal Darner (*C. ingens*) (82–100 mm [3.2–3.9 in]; archipelago-wide), typical of coppice habitats, has more noticeable brown lines on the green thorax. Eye color is blue in mature females.

■ **FAMILY LIBELLULIDAE** (Skimmers)

Large group of commonly seen, brightly colored dragonflies with worldwide distribution, typically associated with standing water. At least twenty species are recorded in the archipelago.

Needham's Skimmer   *Libellula needhami*

Genus: *Libellula* **(Skimmers)**

Range and status: Northern and central archipelago, probably, southern U.S., the Caribbean, Central America.

Description: 45–58 mm (1.8–2.3 in). Distinctive, medium-sized, orange-red species with reddish-orange wing veins and a noticeable line on the leading tip of the edge of each wing (pterostigma). Mature males are typically more brightly colored than are females and young males. Face is brown in young males and females, becoming bright red in older males. Thorax is brown with pale sides. Abdomen is orange-yellow with a black mid-dorsal stripe. The front of the thorax in mature males is rusty red, and the pterostigma and abdomen are bright red. **Habits and habitat:** Frequents grassy pools. Males frequently perch on prominent vantage points in their territories. Can breed in brackish water.

Roseate Skimmer   *Orthemis ferruginea*

Genus: *Orthemis* **(Roseate or Tropical King Skimmers)**

Range and status: Archipelago-wide; also the Caribbean, southern U.S., Central and South America.

Description: 46–55 mm (1.8–2.2 in). Strong-flying, medium-sized, sexually dimorphic species. Males have two color morphs: a pinkish-blue morph and a rarer red morph; females have a brown thorax, an orange-brown abdomen, and a pale-colored stripe down the back. Wings are normally clear except for the narrow brown tips at the leading edges. **Habits and habitat:** Often flies in the midafternoon when other species are not flying. Frequents ponds and quiet water, permanent and temporary rain pools.

Band-winged Dragonlet   *Erythrodiplax umbrata*

Genus: *Erythrodiplax* **(Dragonlets)**

Range and status: Archipelago-wide; also southern U.S., the Caribbean, Central and South America.

Description: 38–47 mm (1.5–1.9 in). Commonly seen, medium-sized, sexually dimorphic species. Males are unmistakable with a dark (blue-black) thorax and abdomen and two distinctive dark bars on the fore and hindwings. Females usually lack these prominent wing bands but have dark wing tips and can occur in brown-gray (shown) or dark (male-like) forms. **Habits and habitat:** Temporary or permanent marshy ponds with emergent vegetation.

Orthemis ferruginea, male, pink morph

Orthemis ferruginea, male, red morph

Orthemis ferruginea, female or juvenile male

Libellula needhami, female or juvenile male

Erythrodiplax umbrata, male

Erythrodiplax umbrata, female

Eastern Pondhawk   *Erythemis simplicicollis*

Genus: *Erythemis* (Pondhawks)

Range and status: Archipelago-wide; also eastern North America, the Caribbean, Central America.

Description: 36–50 mm (1.4–2.0 in). Medium-sized, sexually dimorphic species: males have a green face, green and blue thorax, and a pale blue abdomen; females and juvenile males are grass green with a black-banded abdomen, with a white (pale) rectangular spot on its tip. The latter distinguishes females and juveniles from the Great Pondhawk (see below). **Habits and habitat:** Frequents many water bodies in the archipelago.

Great Pondhawk   *Erythemis vesiculosa*

Genus: *Erythemis* (Pondhawks)

Range and status: Archipelago-wide; also southern U.S., Central and South America.

Description: 56–65 mm (2.2–2.6 in). Commonly seen, large, green dragonfly with a bright green head, dull gray or brown eyes, green thorax, and black-banded green abdomen. Legs are black with greenish tips and the wings are clear. Sexes are similar in appearance, but males tend to have darker markings on the abdomen. **Habits and habitat:** Ponds, lakes, and ditches.

Vermillion Glider or Saddlebags   *Tramea abdominalis*

Genus: *Tramea* (Saddlebags)

Range and status: Archipelago-wide; also the Caribbean, Central America, southeast U.S.

Description: 41–49 mm (1.6–2.0 in). Medium-sized dragonfly with characteristic dark blotches at the base of the hindwings that give the appearance the individual is carrying saddlebags when flying. Males are red or rusty-red including eyes; females (shown) are tawny colored. Both sexes have black spots at the tip of the abdomen. **Similar species:** 1) Antillean Saddlebags (*T. insularis*) (shown; 41–48 mm [1.6–2.0 in]; archipelago-wide) has an unmarked red-brown thorax and black eyes. Face is brown in females and young males but becomes black in mature males. Top of the male's head is metallic violet. Legs are black. 2) Red-mantled Saddlebags/Glider (*T. onusta*) (41–48 mm [1.6–2.0 in]; archipelago-wide) has a large clear spot in the inner margin of each saddlebag.

Halloween Pennant   *Celithemis eponina*

Genus: *Celithemis* (Pennants)

Range and status: Archipelago-wide; also the Caribbean, eastern U.S.

Description: 30–42 mm (1.2–1.7 in). Unmistakable medium-sized species with yellowish-orange wings marked with broad, dark brown-black stripes and a red spot near the leading tip of each wing. Yellowish-green thorax with a yellowish dorsal stripe on the abdomen (which become red with age) with a black tip. Females and juveniles have yellow-green markings on the head and thorax; males

Erythemis simplicicollis, male

Erythemis simplicicollis, female or juvenile male

Erythemis vesiculosa, male

Celithemis eponina, female or juvenile male

Tramea insularis, male

Tramea abdominalis, female

have redder markings and a darker abdomen. **Habits and habitat:** The characteristic fluttering flight is similar to a butterfly. Often perched on the end of stems with the forewings held somewhat vertically and the hindwings horizontally. Frequents ponds and marshes with emergent vegetation.

Three-striped Dasher   *Micrathyria didyma*

Genus: *Micrathyria* (**Tropical Dashers**)

Range: Widely distributed through the Caribbean, southeast U.S., Central and northern South America.

Description: 32–39 mm (1.3–1.5 in). Medium-sized species with pale green thorax with three black lateral stripes and a black abdomen with pale green spots or stripes and black legs. Males (shown) have a white face, metallic blue-green eyes, and a pale blue area between the wings. **Habits and habitat:** Perches with lowered wings as shown. Flies year-round. Frequents weedy ponds and forest clearings.

Damselflies

■ **FAMILY LESTIDAE** (Spread-winged Damselflies)

Lestes spp. (subfamily Lestinae) hold their wings at approximately forty-five degrees to the body when at rest, whereas other damselflies in the archipelago hold their wings folded back over the abdomen.

Antillean Spreadwing   *Lestes spumarius*

Range and status: Archipelago-wide; also south Florida, the Caribbean.

Description: 37–44 mm (1.5–1.7 in). Dull-colored slender species. Males have brown eyes, brown thorax with pale sides, and metallic green or bronze-colored stripes between each segment along the abdomen; females have brown or blue eyes and a distinctive swollen tip to the abdomen. **Habits and habitat:** Frequents wooded areas. Often difficult to see.

■ **FAMILY COENAGRIONIDAE** (Narrow-winged or Pond Damselflies)

Large family of commonly seen damselflies with worldwide distribution, often near standing water. Delicate appearance, with narrow, stalked, clear wings and frequently patterned in black.

Citrine Forktail   *Ischnura hastata*

Genus: *Ischnura* (**Forktails**)

Range and status: Archipelago-wide, southeast U.S., Central and northern South America.

Description: 21–27 mm (0.8–1.1 in). Small, sexually dimorphic species. Males (shown) are bright yellow with green eyes (and a black cap), a green and black striped thorax, and a black-banded abdomen with a green tip. Females are duller with green eyes, brown cap, a black-gray dorsally colored thorax with white on the sides and a black-gray abdomen with dark gray banding. Immature females are orange-colored. **Habits and habitat:** Frequents ponds and densely vegetated margins along ponds.

Micrathyria didyma, male

Ischnura hastata, male

Ischnura ramburii, male

Lestes spumarius, female

Ramburs Forktail   *Ischnura ramburii*

Genus: *Ischnura* **(Forktails)**

Range and status: Archipelago-wide; also the U.S., the Caribbean, Central and South America.

Description: 27–36 mm (1.1–1.4 in). Large, brightly colored species. Males (shown) have green-black capped eyes, a green abdomen and thorax, black on the dorsal side of the abdomen with a lateral gold stripe and a blue tip. Females have two morphs: one resembles the male but differs in having a pale blue side to the thorax, while the other morph has an orange thorax and abdomen colored black dorsally and with orange on the sides. **Habits and habitat:** Most widespread species of the genus in the New World. Frequents heavily vegetated ponds and marshes. Can breed in brackish water.

Tropical Sprite   *Nehalennia minuta*

Genus: *Nehalennia* (Sprites)

Range and status: Archipelago-wide; also south Florida, the Caribbean, Central and South America.

Description: Small, slender, blue species. Males have blue eyes, a black thorax with blue on the sides, and a black abdomen with blue banding. Females are similarly colored but duller with brown eyes.

ORDER MANTODEA (Mantises)

Camouflaged predatory insects (related to termites and cockroaches; see below), with worldwide distribution in temperate and tropical habitats. Mantises sit and wait for prey (typically other insects) to come within striking distance. Their powerful jaws and two large grasping, spiked raptorial forelegs are used to capture prey. The forelegs are normally bent as if in prayer, hence their common name, Praying Mantis. They are the only insects able to turn their head without moving their body and have a high level of visual acuity. Sexual cannibalism, females eating males prior to, during, or after copulation, has been documented in many species. Unidentified species have been found in the northern and central archipelago (Abaco, Eleuthera, and Cat Island) and are probably under-recorded.

Stagmomantis **spp.**

Range and status: Uncommon in northern and central archipelago; distribution poorly known. Genus native to the Americas and the Caribbean with at least twenty-seven species recorded.

Description: 47–60 mm (2–2.5 in). Females are normally bigger than males, have thicker abdomens, and frequently have shorter wings than the males have, extending approximately three-quarters of the length of the abdomen, while those of the male extend until at least the end of the abdomen. Can exhibit considerable within-species variation in body color, e.g., green, brown or gray; nymphs (juveniles) can adjust their body color to match their environment until their final molt. **Habits and habitat:** Range of coppice, agricultural, and urban habitats. Frequently attracted to artificial lights. Photos show a green-colored female Carolina Mantis (*S. carolina*; recorded in southern Florida and possibly recorded in the northern archipelago) eating a dark-colored male and an unwinged green *Stagmomantis* sp. nymph (c. 15 mm [0.6 in]) on Eleuthera. The Grizzled or Florida Bark Mantid (*Gonatista grisea*) (Family Mantidae; not shown) is also recorded from the central archipelago.

ORDER BLATTODEA (Cockroaches and Termites)

Cockroaches are medium to large nocturnal insects with a broad flattened body, often with a shield or plate-like structure (the pronotum) covering the first segment(s) of the thorax, spiny legs, and long antennae. These insects are omnivorous scavenger insects that consume organic debris and are important in the recycling of leaf litter and feces. They have worldwide distribution but are particularly numerous in the

Stagmomantis sp., nymph

Stagmomantis carolina, female eating male

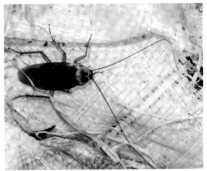

Nehalennia minuta, male

Periplaneta americana

tropics and subtropics. Most species are found in damp conditions (under stones and logs); however, several species have adapted to human environments and are considered pests because they contaminate food and transmit disease.

■ FAMILY BLATTIDAE

Family of typically large, often fast-moving, cockroaches with prominent cerci (posteriorly pointing paired sensory structures on the end the abdomen). Includes many of the commonly seen "pest" species.

American Cockroach   *Periplaneta americana*

Range and status: Introduced. Archipelago-wide; worldwide throughout the subtropics and tropics.

Description: 29–53 mm (1.1–2.1 in). Large, commonly seen, reddish-brown cockroach with a yellowish margin behind the head. **Habits and habitat:** Pest species associated with human habitation, sometimes very common. Despite its common name, it was introduced to the U.S. from Africa in the 17th century.

Eurycotis bahamensis

Range and status: Native, possibly endemic, but distribution in archipelago poorly known; currently recorded in central archipelago including Eleuthera and Andros.

Description: 25–30 mm (1–1.2 in). Distinctively colored cockroach with an orange head and thorax, with a set of three black spectacle-like markings on the pronotum and a uniformly black abdomen. **Habits and habitat:** Found in coppice under stones and logs.

■ **FAMILY BLABERIDAE** (Giant Cockroaches or Blaberids)

Medium to large, typically dull-colored (brown), slow-moving species with short legs, short antenna (often less than half of body length), and small (or absent) cerci.

Horseshoe Crab Roach   *Hemiblabera* spp.

Range and status: Limited data on distribution. At least four species are recorded in the archipelago: *H. pabulator, H. roseni, H. tenebricosa,* and *H. brunneri.*

Description: 25–58 mm (1–2.3 in). Genus of flightless cockroaches so named because of their similarity to horseshoe crabs (Family Limulidae). Both sexes have distinctively broad pronotums. **Habits and habitat:** Males court the females by hopping around and displaying their stubby wings. Little is known about this family as they tend to be shy and are found under rocks, often in coppice.

SUBORDER ISOPTERA (Termites)

Small-sized, colonial (eusocial) insects (4–15 mm [0.2–0.6 in]) with worldwide distribution. Colonies are divided into three castes: 1) reproductive individuals (at least one king and queen), 2) (sterile) workers, and 3) (sterile) soldiers, with all working collectively for the good of the colony. Colonies are typically contained in a nest made from a combination of soil, mud, chewed wood and cellulose, saliva, and feces, and it provides a climate-controlled environment that protects occupants from external environmental fluctuations. Although nests are frequently below ground, nests can also be found above ground and even high in trees. Termites are the only insects able to digest cellulose because of the microorganisms in their gut. They are important natural decomposers of plant material, but they can also be pest species and cause significant structural damage to buildings. At least twenty-seven species (representing three families and twelve genera) are recorded in the archipelago, including three endemic species.

Eastern Subterranean Termite   *Reticulitermes flavipes*

Range and status: Introduced. Northern and central archipelago. Native to eastern North America; also introduced to Europe, South America, Asia, the Caribbean.

Description: 3–5 mm (c. 0.2 in). Most common termite found in North America. A colony can include 20,000–5,000,000 individuals, the majority of which (98%) are sterile workers, the remainder are soldiers (excluding the reproducing king[s] and queen[s]). Workers are blind, wingless, soft-bodied, creamy white to grayish-white

Hemiblabera sp.

Eurycotis bahamensis

Reticulitermes flavipes, workers

Termite mound, unknown species

with a round head (shown), whereas the similar colored soldiers have a large, rectangular, yellowish-brown head with long black mandibles. An aboveground nest of an unidentified termite species is also shown. **Conservation status:** Economically one of the most important insect species in the U.S. because of the damage caused to buildings and property.

One of the largest insect orders, and it comprises the ants, bees, and wasps. Most species can sting using a modified ovipositor located at the tip of the abdomen. Many species live colonially in nests and exhibit a complex social structure, within which colony members work collaboratively.

■ **FAMILY FORMICIDAE** (Ants)

Generally small, colonial, typically red or black insects with worldwide distribution. They are easily distinguished from other insects by their elbow-shaped antennae and a slender waist (petiole) between the thorax and abdomen. Most ants are generalist predators, scavengers, and indirect herbivores. Ants can sting and bite, and some species can also spray formic acid. They control invertebrate pests, but some (introduced) ant species are also considered pests. At least ninety species are recorded in the archipelago with c. 60% of the species shared with Florida and c. 70% with Cuba. More than one-third of the species are exotics (introduced), which occur primarily in areas characterized by human disturbance. Several of the introduced species are in the genus *Solenopsis* (collectively called fire ants for the fiery pain caused by their sting and bites) and are considered threats to native insects and plant communities.

Red Imported Fire Ant   *Solenopsis invicta*

Range and status: Introduced. First recorded in The Bahamas in the early 1990s, now archipelago-wide; also introduced to Oceania, Asia, the U.S., the Caribbean. Native to South America.

Description: 2.4–6 mm (0.1–0.25 in). Small species with red-brown head and thorax, and a darker (brown-black) abdomen; workers can vary in size (see photo). Distinguishing characteristics include two triangular nodes on the petiole (waist), and ten-segmented antennae with a two-segmented club at their respective tips. **Habits and habitat:** Found in many disturbed habitats, frequently common in urban areas. Nest mounds are built of soil up to 46 cm (18 in) in diameter. Aggressive species that actively defends nests by injecting an alkaloid venom with a sting, and by biting. Diet consists of dead animals, including insects, earthworms, and vertebrates. **Conservation status:** Considered a serious pest species due to damage caused to commercial crops and property, threats posed to native plant and animal species, and health implications—stings may cause localized swelling and more severe complications.

■ **SUPERFAMILY APOIDEA** (Bees)

Insects with four wings (the hind pair typically being the smaller of the two) with a barbed stinger located in the tip of the abdomen; if used, the stinger is left in the victim and the bee dies. Bees are best known for their important role in pollination, and several species produce honey and wax. Adults feed primarily on nectar (sugar), which they access using a long proboscis, and on pollen (protein and other nutrients), which is used to feed the larvae. More than 12,000 species are described with the majority being solitary species; only c. 5% live in colonies, which typically include several fertile males (drones) and a single fertile female (queen), whereas the majority are sterile

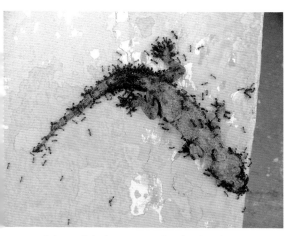

Solenopsis invicta, scavenging dead anole

Apis mellifera

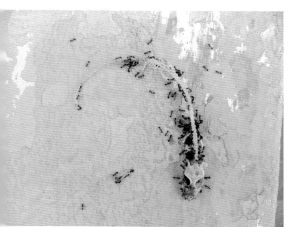

Solenopsis invicta, scavenging dead anole

workers. The largest family is the Apidae, which includes many of the commonly seen species including honey bees, carpenter bees, and bumblebees. At least six Apid species are recorded in the archipelago.

Western or European Honey Bee   *Apis mellifera*
Family: Apidae

Range and status: Introduced. Native to Europe, Asia, and Africa; introduced to North America in the 17th century and subsequently spread throughout the Americas. Recorded in north and central archipelago including Abaco, Eleuthera, New Providence, and Andros, but likely on other islands.

Description: 12–15 mm (0.5–0.6 in). Immediately recognizable red-brown species with black and orange-yellow banding on the abdomen. **Habits and habitat:** Honey bees are characterized by the production and storage of honey in wax nests, and they were one of the first domesticated insects.

Centris spp.

Family: Apidae

Range and status: At least two species recorded in the central and southern Bahamas, but distribution data are incomplete; also the Caribbean.

Description: Up to 30 mm (1.2 in). Typically large, fast-flying bees, and often active at very high ambient temperatures when many other bees are inactive. The genus contains more than one hundred species throughout the Americas. **Habits and habitat:** The females of some species collect floral oils instead of or in addition to pollen and nectar. *C. versicolor* c. 15 mm (0.6 in) recorded from the central archipelago is shown.

Carpenter Bee   *Xylocopa cubaecola*

Family: Apidae

Range and status: Southern archipelago, including Crooked Island, Mayaguana, Inagua.

Description: Up to 23 mm (0.9 in). Large, sexually dimorphic bee. Females (shown) are larger than the males and are black with a shiny abdomen; the males are yellow-brown. **Habits and habitat:** Carpenter bees are so named as they build their nests in burrows in dead wood or in wooden structures; typically solitary but often nest close to one another. Only females sting. **Similar species:** Similar in size and coloration to bumblebees (*Bombus* spp.); however, carpenter bees have shiny abdomens while bumblebee abdomens are covered with dense hair, and *Bombus* spp. are not recorded in the archipelago. At least one other *Xylocopa* spp. is recorded in the archipelago: *X. brasilianorum* (San Salvador).

■ WASPS

Wasp is the collective name of several families of winged stinging Hymenoptera characterized by a thin connection (waist) between the thorax and the abdomen; frequently considered as any member of the Hymenoptera that is not an ant or bee, but often restricted to include only yellow jackets (common wasps) and hornets. Solitary species typically do not construct nests and will instead utilize either natural cavities in the ground or in trees or will create simple egg-holding structures from mud. Colonial species build nests typically from paper pulp. Colonies typically include one or more queens; workers are sterile females produced from fertilized eggs whereas males come from unfertilized eggs and have no stingers. Adults feed on fruit and nectar, but larvae are fed insects. Wasps provide important biological control of other insects including pest species. The stinger is non-barbed and, unlike bees, can be used repeatedly.

■ FAMILY VESPIDAE (Wasps and allies)

Large family with worldwide distribution, containing both eusocial and solitary species; at least twelve Vespid species are recorded in the archipelago. Some of the most frequently encountered wasps are in the subfamily Polistinae, a group of eusocial subtropical and tropical species sometimes referred to as paper wasps. Members of

Xylocopa cubaecola, female

Centris versicolor

Polistes bahamensis

the Polistinae can be identified by a distinctive narrow waist and their characteristic profile in flight: the long legs dangle below their bodies. Individuals build a single, combed, open nest from paper pulp, often placed on buildings. All species are predatory and provide natural control of insect pests. Four species are recorded in the archipelago.

Polistes bahamensis

Range and status: Native. Throughout most of the north and central Bahamas; rare in the southern Bahamas and not recorded from the TCI. Also south Florida.

Description: 15–20 mm (0.6–0.8 in). Dull-colored red-brown species with a distinctive narrow waist with fine yellow bands on the abdomen, yellow legs, fine yellow markings on the face, and a yellow spot on the side of the thorax.

Polistes major

Range and status: Recorded from most large islands except Mayaguana and Great Inagua.

Description: 17–22 mm (0.7–0.9 in). Large, yellow *Polistes* with a yellow face, a brown dorsal patch on the yellow thorax, and a brown-striped yellow abdomen; the latter distinguishes it from other *Polistes* in the archipelago. **Habits and habitat:** Circular nests, up to 19 cm (7.5 in) in diameter, can be found in natural and urban habitats that provide shelter, e.g., under palm leaves or under the eaves of buildings. Colonies typically consist of a queen, four to nineteen sterile females, and up to four males.

Mischocyttarus cubensis

Range and status: Recorded from most islands in the north and central archipelago; rare or absent in the southern archipelago.

Description: 11–15 mm (0.4–0.6 in). Similar to *Polistes* spp. but can be distinguished by two parallel longitudinal yellow stripes on the center of the thorax and two transverse yellow stripes at the posterior end of the thorax. A noticeable yellow band occurs on the waist before the abdomen and a series of transverse yellowish stripes on the abdomen. *Polistes fuscatus* (recorded from Andros) is similar but lacks the two transverse yellow stripes at the posterior end of the thorax and yellow band on the waist.

■ **FAMILY SCOLIIDAE**

Scoliid wasps are typically black with yellow or orange markings and distinctive corrugated tips to their wings; males are thin with typically longer antennae. Adults are pollinators, but they are also parasitoids of scarab beetle larvae, with females actively seeking out the larvae, stinging them to paralyze them, and laying their eggs on the larva.

Campsomeris **spp.**

Range and status: Archipelago-wide with at least four species: *C. atrata* (San Salvador); *C. bahamensis* (southern Bahamas); *C. fulvohirta*, *C. trifasciata trifasciata*, and *C. t. nassauensis* (north and central archipelago).

Description: Medium to large, stout-bodied wasps; typically brightly marked abdomen usually with red, yellow, and white in combination with black; and long segmented antennae. **Habits and habitat:** They are parasitoids of soil-inhabiting beetle larvae.

■ **FAMILY POMPILIDAE** (Spider or Pompilid Wasps)

Family of typically dark-colored (blue or black), long-legged, solitary wasps with worldwide distribution. Their common name derives from hunting and killing spiders, often larger than themselves, as food for their larvae; adults feed on nectar.

Pepsis sp., female attacking Bahamian Tarantula (*Cyrtopholis bonhotei*)

Pepsis sp., male

Mischocyttarus cubensis

Polistes major

Campsomeris trifasciata

Pepsis spp.

Range and status: Archipelago-wide. At least three species are recorded.

Description: Up to c. 5 cm (2 in). New World genus found throughout the Americas. Commonly referred to as Tarantula Hawk Wasps, these distinctive, medium to large diurnal wasps are blue-black in color with dark rust or blue-black wings. Adults can often be observed walking rapidly on the ground while simultaneously flicking their wings. Females are larger than the males. Although both sexes feed on nectar, the females also hunt and capture tarantulas; their long legs have specialized hooked claws for grappling with their victims. The sting paralyzes but does not kill a spider, which is then transported to a burrow or nest where the female lays one egg on the spider's body. When the egg hatches the wasp larva proceeds to eat the paralyzed spider in such a way as to keep it alive as long as possible. After several weeks, the larva pupates. Following pupation, the wasp emerges to continue the life cycle. Despite their size and reputation, Hawk Wasps are not normally aggressive and rarely sting without provocation. The sting can be very painful though, especially from the larger females. Photos show an adult feeding on nectar and a female attacking a Bahamian Tarantula (*Cyrtopholis bonhotei*) (see above).

ORDER LEPIDOPTERA (Butterflies and Moths)

These distinctive and familiar insects have two pairs of large, frequently bright-colored wings covered with tiny scales. Individuals undergo a complete metamorphosis: egg to larva (caterpillar), to pupa from which a winged adult emerges. Adults feed by using their proboscis to suck nectar (but also carrion, dung, or urine), whereas caterpillars are herbivorous. Caterpillars of many butterflies are often restricted to specific plant species. Butterflies can be distinguished from moths in several ways: They have club-like antennae and moths have feathery or wiry antennae; most butterflies rest with their wings in a vertical (closed) position as opposed to outstretched as with moths; and most moths fly at night (although there are also crepuscular and day-flying species), while butterflies fly only during the day. More than seventy butterfly species and more than 500 moth species are recorded from the archipelago.

We use the following terminology to describe lepidopteran wing structure: Forewings are the front (or anterior) pair of wings; hindwings are the posterior pair of wings; upperwings refer to dorsal surface of both the fore and hindwings; and underwings refer to ventral surface of both pairs of wings.

■ **FAMILY NYMPHALIDAE (Brush-footed Butterflies)**

Large, diverse family of medium- to large-sized butterflies with worldwide distribution. Most species have a reduced pair of forelegs, and many hold their colorful wings open (horizontal) when resting. At least twenty species are recorded in the archipelago.

Subfamily Danainae (Milkweed Butterflies)

Conspicuously colored butterflies whose larvae typically feed on milkweeds (*Asclepias* spp.). These plants contain poisonous chemicals (glycosides) that are retained by both the larvae and the adults and are very distasteful to potential predators, typically birds. These species exhibit red-orange warning (aposematic) coloration to indicate their unpleasant taste. At least three species are found in the archipelago.

Danaus gilippus berenice

Danaus plexippus plexippus

Monarch   *Danaus plexippus*

Range and status: Resident (non-migratory) subspecies *D. p. megalippe* throughout the year. Probable influx of migratory subspecies *D. p. plexippus* (North American subspecies) especially in the north of the archipelago during spring and autumn migrations. Species also throughout much of the Americas.

Description: Wingspan 89–102 mm (3.5–4.0 in). Immediately recognizable species with distinctive orange-and-black patterned wings with a series of white spots along the black margins of the fore and hindwings. The body is also black with white spots. **Habits and habitat:** The North American population undergoes an annual southward late-summer into autumn migration from northern and central North America to Florida and Mexico and a corresponding northward migration in the spring. The subtle differences between the subspecies are difficult to separate in the field. Frequents many habitats including fields, meadows, urban and suburban parks, gardens, and roadsides.

Queen   *Danaus gilippus*

Range and status: Subspecies *D. g. berenice* (common to the Caribbean) archipelago-wide; species also in the Caribbean, Central America, southern U.S.

Description: Wingspan 70–88 mm (2.8–3.5 in). Similarly colored to the Monarch (*D. plexippus*) (see above) with distinctive orange-and-black patterned wings with a series of white spots along the black wing margins. The Queen is smaller and has a darker, richer shade of orange than the Monarch, lacks the black veins on its upperwings, and has white spots on its lower wings. **Similar species:** The Soldier (*D. eresimus*) (64–76 mm [2.5–3.0 in]) is probably archipelago-wide and is very similar to the Queen, but it lacks the white spots along the margin of the hindwing.

Subfamily Heliconiinae (Heliconians or Longwings)

Group of brightly colored butterflies distributed throughout the New World tropics. Typically reddish and black, with varying wing shape but forewings always elongated toward the tip. The larvae feed on passion flower vines, and both adults and larvae are distasteful to predators through their accumulation of toxic or noxious secondary compounds ingested from the host plants. Adults exhibit bright wing color patterns to signal their distastefulness. Some species roost colonially. At least four species are recorded in the archipelago.

Gulf Fritillary   *Agraulis vanillae*

Range and status: Subspecies *A. v. insularis* (common to the Caribbean) throughout the archipelago; species also in the Caribbean, Central and South America, southern U.S.

Description: Wingspan 60–95 mm (2.4–3.7 in). Commonly seen, medium-sized, brilliant red-orange butterfly with black markings and two black-encircled white dots on the leading edge of each forewing. Underwings are brown with large, elongated, iridescent silver spots. **Habits and habitat:** Northern populations are migratory, and its common name derives from migrating flights of this species sometimes seen over the Gulf of Mexico. Observed year-round in open habitats including coastal habitats, pastures, parks, gardens, and second-growth habitat. Adult food plants include sages (*Lantana* spp.), Shepherd's Needle (*Bidens pilosa*), and *Cordia* spp.; those of the caterpillar include passion flower vines (*Passiflora* spp.).

Julia   *Dryas iulia*

Range and status: Endemic subspecies *D. i. carteri* archipelago-wide; species also in Florida, Central and northern South America.

Description: Wingspan 82–92 mm (3.2–3.6 in). Distinctive bright orange butterfly with elongated forewings. Two black dots display on upper forewings with a black border on the margin of the hindwings. Males are typically brighter than the females, but with less black on forewings. **Habits and habitat:** Observed year-round. Fast flying. Observed in clearings, paths, and margins of coppice. Adult food plants include sages (*Lantana* spp.) and Shepherd's Needle (*Bidens pilosa*); those of the caterpillars include passion flower vines (*Passiflora* spp.).

Mexican Fritillary   *Euptoieta hegesia*

Range and status: Subspecies *E. h. hegesia* archipelago-wide; species also in the Caribbean, Central America, southern U.S.

Description: Wingspan 65–75 mm (2.5–3.0 in). Medium-sized, orange butterfly. Upperside of the wings is bright orange with an unmarked basal part of the hindwing (that part of the wing next to the body). A row of submarginal black spots occur on both the fore and hindwing. Underside of the wings is yellow-orange with indistinct or no submarginal black spots. **Habits and habitat:** Can be observed year-round on sunny days in open habitats. Adult food plants include sages (*Lantana* spp.), Blue Flower (*Stachytarpheta jamaicensis*), and *Turnera* spp.; those of the

Heliconius charithonia

Dryas iulia carteri

Agraulis vanillae insularis

Euptoieta hegesia hegesia

caterpillar include passion flower vines (*Passiflora* spp.), morning glories (*Ipomoea* spp.), and *Turnera* spp.

Zebra Heliconian or Zebra Longwing   *Heliconius charithonia*

Range and status: Throughout the northern and central archipelago. Subspecies *H. c. ramsdeni* on Andros and possibly New Providence; *H. c. tuckeri* on Grand Bahama and Great Abaco; species also in the Caribbean, Central and South America, southern U.S.

Description: Wingspan 40–100 mm (2.8–4.0 in). Distinctive butterfly with elongated black wings banded with narrow lemon-yellow stripes, with the lower one on the hindwing divided into dots. **Habits and habitat:** Exhibits a characteristic slow, wafting flight. Can be observed year-round across a range of coppice habitats and edge and disturbed habitats, such as fields. Adult food plants include sages (*Lantana* spp.) and Shepherd's Needle (*Bidens pilosa*); those of the caterpillar include passion flower vines (*Passiflora* spp.). Roosts in groups (up to thirty individuals) with the adults often returning to the same roost.

Subfamily Limenitidinae (Admirals and allies)

Typically, large butterflies with distinctive and bright patterns on the upperwings and cryptically colored (camouflage) underwings. The common names of certain species refer to the similarity of the upperwing markings to those on official or military uniforms (e.g., admirals), while the habit of some species interspersing their active flight with gliding is the basis of other common names (e.g., gliders).

Caribbean or Cuban Daggerwing   *Marpesia eleuchea*

Range and status: Endemic subspecies *M. e. bahamensis* throughout the north and central archipelago including New Providence, Eleuthera, Crooked Island, Andros, San Salvador; species also in the Caribbean.

Description: Wingspan 67–83 mm (2.6–3.3 in). Striking orange-brown butterfly with an elongated tip on the forewing and a long dagger-like tail on the hindwing. Three black lines run the length of both the fore and hindwings (parallel to the body); the middle line on the forewings has a distinctive point. **Habits and habitat:** Seen year-round in woodlands and clearings. Adult food plants include soldierbushes (*Tournefortia* spp.), *Cordia* spp., sages (*Lantana* spp.), and *Eupatorium* spp.

Subfamily Biblidinae (Tropical Brushfoots)

Small group of butterflies containing predominantly neotropical species.

Florida Purplewing   *Eunica tatila*

Range and status: Subspecies *E. tatila tatilista* archipelago-wide; species also in southeast U.S., the Caribbean, Central and northern South America.

Description: Wingspan 40–62 mm (1.6–2.4 in). Upperside of both wings is brown with purple iridescence at bases. Outer half of the forewing has six or seven white spots and an irregular margin (edge). Hindwing has a pale margin. Underside is gray-brown with the white spots of the forewing partially visible. **Habitat and habitats:** Coppice. Frequently sits with wings closed. Adults feed on decaying fruit and on sages (*Lantana* spp.) and *Cordia* spp.

Subfamily Nymphalinae (True Brushfoots)

Group of butterflies that sometimes includes the subfamilies Biblidinae and Limenitidinae (see above).

Cuban Crescent   *Anthanassa frisia*

Range and status: Throughout the northern and central archipelago; species also in the Caribbean, Central and South America.

Description: Wingspan 32–45 mm (1.3–1.8 in). Small, orange butterfly with black markings. Upperside of the forewing is black with orange markings with an indentation below the wing tip. The orange hindwing has a distinctive black zigzag horizontal line. **Habits and habitat:** Typically weak flight; often flies near the ground. Observed year-round in second-growth habitats, e.g., gardens, along roadsides, and in open fields. Adult food plants include sages (*Lantana* spp.); those of the caterpillar include the Shrimp Plant (*Beloperone guttata*).

Marpesia eleuchea bahamensis

Eunica tatila tatilista

Anthanassa frisia

Junonia evarete zonalis

Tropical or Caribbean Buckeye   *Junonia evarete*

Range and status: Subspecies *J. e. zonalis* archipelago-wide; species also in the Caribbean, Central and South America, southern U.S.

Description: Wingspan 50–66 mm (2–2.6 in). Brown-winged, commonly seen species with distinctive iridescent blue or lilac eyespots. Each forewing has two eyespots: a very small eyespot near the tip and a large eyespot ringed by an orange band near the lower margin, which are separated by a white lateral band. The large eyespot is banded on its outer margin by the white band. Two eyespots occur on each hindwing. **Habits and habitat:** Observed year-round. It has a fast flight and often settles on dry, open ground. Adult food plants include Blue Flower (*Stachytarpheta jamaicensis*), sages (*Lantana* spp.), and soldierbushes (*Tournefortia* spp.); those of the caterpillars also include Blue Flower (*Stachytarpheta jamaicensis*). **Similar species:** (a) Mangrove Buckeye (*J. genoveva*) (wingspan 55–63 mm [2.2–2.5 in]; archipelago-wide) typically lacks the very obvious white band between the two eyespots on the forewing and is also associated more with Black Mangroves (*Avicennia germinans*), which is the food plant of the caterpillars. (b) Common Buckeye (*J. coenia*) (wingspan 44–56 mm [1.7–2.2 in]; archipelago-wide) has the large eyespot on the forewing usually completely bordered by a creamy-white band on both sides. There is also a difference in the size of the eyespots on the hindwing: the upper (anterior) eyespot is noticeably larger than the posterior eyespot and similarly sized to the large eyespot on the forewing. *Note:* Distinguishing between these three species is sometimes difficult as some individuals exhibit characteristics of more than one species.

White Peacock   *Anartia jatrophae*

Range and status: Subspecies *A. j. guantanamo* archipelago-wide; species also in the Caribbean, Central and South America, southern U.S.

Description: Wingspan 51–70 mm (2–2.8 in). Distinctive white-gray butterfly with light brown markings and a double row of light crescents at the wing margins. Forewing has one round, black spot; the hindwing has two. Observed through-out the year: the dry season (winter) form is typically larger and lighter colored, whereas the wet season (summer form) is smaller and darker. **Habits and habitat:** Seen in many open habitats including beaches and in a range of disturbed secondary habitats (parks and fields). Adult food plants include Shepherd's Needle (*Bidens pilosa*), *Cordia* spp., sages (*Lantana* spp.), and soldierbushes (*Tournefortia* spp.); those of the caterpillar include Water Hyssop (*Bacopa monnieri*) and Ruellia (*Ruellia occidentalis*).

Painted Lady   *Vanessa cardui*

Range and status: Western Hemisphere subspecies *V. c. cardui* archipelago-wide; also the Caribbean. Species found on all continents except South America, also absent from Arctic regions.

Description: Wingspan 51–73 mm (2.1–2.9 in). Orange-brown and black butterfly with four white spots and a larger white bar in the black tips of the forewings, and black (spotted and checkered) markings on the upperwings. A row of five small black spots (sometimes with blue scales) appear on the posterior edge of the hindwing. The underwings have a black, brown, and gray pattern with four eyespots. **Habits and habitat:** Coppice and open or disturbed areas including gardens, old fields, dunes. Adult food plants include thistles (Asteraceae) and milkweeds (*Asclepia* spp.); those of the caterpillar include thistles and various legumes (Fabaceae). Painted Lady is the most widely distributed butterfly in the world and strongly migratory. North American populations originate primarily from northwest Mexico. When factors are suitable, often following heavy rains, sporadic population explosions occur in the southern populations, resulting in large-scale migrations throughout North America and the Caribbean. In the fall, northern populations may also migrate southward; individuals can overwinter in their southern breeding range.

Anartia jatrophae guantanamo

Vanessa cardui cardui

Memphis intermedia

Memphis intermedia

Subfamily Charaxinae (Leafwing Butterflies)

Leafwing butterflies are so-called because the undersides of their wings are leaf-like in color and appearance (usually with jagged edges to the hindwings). This characteristic disguises the butterfly because it resembles a dead leaf when it settles on the ground, as they keep their wings closed at rest. Adults are very robust and fast fliers.

Bahamas Leafwing   *Memphis intermedia*

Range and status: Endemic. Throughout the central and southern archipelago. Three subspecies: *M. i. intermedia* on Great Inagua Bank and the TCI; *M. i. mayaguanae* on Mayaguana Bank; *M. i. venus* on the Crooked and Acklins Bank. Species also recorded from San Salvador (subspecies not specified).

Description: Wingspan 57 mm (2.3 in). One of only three butterflies endemic to the archipelago. Upperside of the wings red-brown with darker brown at the margins. Five pale yellowish spots along the outer margin of the tip of the forewing, and the hindwing has a tail-like projection and a series of small eyespots running along its hind margin. The cryptic underside of the wing has a variety of shades and patterns of brown and gray resembling a leaf. Subspecies vary in color. **Habits and habitat:** Adults feed on Poison Wood (*Metopium toxiferum*).

■ **FAMILY LYCAENIDAE** (Hairstreaks, Blues, and Coppers)

Large family of small, blue or gray butterflies with worldwide distribution. Most species sit with their wings closed. Hairstreaks have hair-like projections on the hindwing. Many species also have a spot at the base of the tail that may function to direct predators away from the head. At least twelve species are recorded in the archipelago.

Subfamily Theclinae (Hairstreaks)

Small, fast-flying species with distinctive triangular-shaped, tailed hindwings. Their common name derives from many species having delicate hair-like lines on the underside of the wings.

Atala Hairstreak   *Eumaeus atala*

Range and status: Localized distribution throughout the archipelago; also Cuba, Cayman Islands, southeast Florida.

Description: Wingspan 38–51 mm (1.5–2.0 in). Small, unmistakable butterfly with a blue-black head, thorax, and wings and a strikingly bright red-orange abdomen. Upperside of the male wings is black with an iridescent green overlay and pale markings; the female has blue iridescence at basal areas. Underside of the wings is dull black and distinctively marked with a large red-orange spot adjoining the abdomen and three rows of irregular iridescent bluish-white spots. **Habits and habitat:** Typically seen at rest with the wings closed. Can be seen most of the year, but most commonly seen in early summer in open brushy areas and pockets of coppice in pine woodlands. Caterpillars are found only on the cycad Coontie *Zamia integrifolia* and hence geographic range limited by availability of its host plant. Threatened throughout much of its range due to habitat loss.

Martial Scrub Hairstreak   *Strymon martialis*

Range and status: Range incompletely known; also southern Florida, the Caribbean.

Description: Wingspan 28–35 mm (1.1–1.4 in). Small, blue-gray butterfly with two dark-colored tails (one long and one short) on the hindwing. Upperside of the forewing is brown-black with a violet-blue trailing edge, and the hindwing is mostly violet-blue. The more commonly seen underwings are gray-brown with a distinctive black eyespot bordered by orange near the lower tip of the wing between the origins of the two tails; the staggered postmedian black line (on both hind and forewings) is white edged. **Habits and habitat:** Typically seen February–December. Adult food plants include Bay Cedar (*Suriana maritima*), sages (*Lantana* spp.), Brazilian Pepper (*Schinus terebinthifolius*), soldierbushes (*Tournefortia* spp.), and Shepherd's Needle (*Bidens pilosa*); those of the caterpillar include Trema (*Trema micrantha*) and Bay Cedar (*Suriana maritima*).

Subfamily Polyommatinae (Blues)

Small butterflies, frequently with rounded wings and a tailless hindwing. Blue coloration is due not only to pigmentation but also to reflected light from the structure of the wing scales.

Eumaeus atala

Cyclargus thomasi bahamensis

Strymon martialis

Miami Blue or Thomas Blue *Cyclargus thomasi*

Range and status: Archipelago-wide. At least four subspecies recorded: *C. t. bahamensis* (Crooked, Acklins, Ragged, and Plana Cays); *C. t. bethunebakeri* (subspecies endemic to Florida but has been recorded in the Biminis); *C. t. clenchi* (Mayaguana, Great and Little Inagua, and the TCI); *C. t. thomasi* (Grand Bahama). Species also in the Caribbean and south Florida.

Description: Wingspan 22–30 mm (0.9–1.1 in). Small, sexually dimorphic, blue butterfly. Females are usually larger than males. Upper surface of wing is bright blue in males, while females have less blue with wide, gray wing borders and an orange-capped black spot along the outer margin of the hindwing. Underside of wing is gray in both sexes. The under hindwing has a broad, white, submarginal band and two large, dark eyespots on the outer margin. **Habits and habitat:** Scrub and pine habitats. Weak fliers. Complex life cycle involving multiple generations February–November with females laying a single, small egg. Caterpillars are frequently tended by ants.

Ceraunus Blue *Hemiargus ceraunus*
(or *H. hanno, Cyclargus ceraunus, C. hanno*)

Range and status: Archipelago-wide. Three subspecies: *H. c. antibubastus* (Grand Bahama); *H. c. ceraunus* (central archipelago); *H. c. filenus* (southern archipelago). Species also in southern U.S., the Caribbean, Central and South America.

Description: Wingspan 20–30 mm (0.8–1.2 in). Small, sexually dimorphic, blue butterfly. Upperside of the male is light blue with a darker narrow border to the wings, while that of the female is dark brown and often with blue wing bases. Underwings of both sexes are gray and have a row of dark postmedian dashes, and one or two eyespots. **Habits and habitat:** Found year-round in open and disturbed habitats including open woodland, scrub, dunes, pastures, road edges, and vacant lots. Caterpillars feed on legumes (Fabaceae), including Partridge Pea (*Cassia brachiata*), Mesquite (*Prosopis* spp.), and Rosary Pea (*Abrus precatorius*).

■ **FAMILY PIERIDAE** (White and Sulphurs)

Most species are white (Whites: Subfamily Pierinae), or yellow/orange (Sulphurs: Subfamily Coliadinae), often with black spots. Pigments that give the distinctive colors are derived from waste body products, a characteristic of the family. Both groups usually sit with their wings closed, and the upperside of the wings is rarely seen. Two species of white and at least fifteen species of sulphur butterflies are recorded in the archipelago. Some species, especially some of the small sulphurs, are difficult to distinguish in the field.

Great Southern White *Ascia monuste*

Range and status: Subspecies *A. m. evonima* throughout the archipelago; migratory subspecies *A. m. phileta* in northern islands (from southern U.S.). Species also in southern U.S., the Caribbean, Central and tropical South America.

Description: Wingspan 63–86 mm (2.5–3.4 in). Large, white butterfly. Upperside of male forewing is white with black zigzag pattern on outer margin. Female exhibits a dry and wet season form: dry season females resemble the male with heavier black zigzag pattern and a small black spot in the wing cell, whereas during the wet season the wings are darkened with black scales above and below. **Habits and habitat:** Favors coastal habitats and can be seen year-round, but most frequently from March–October. **Similar species:** At rest (wings closed) can be distinguished from the similar Florida White *Appias drusilla* (wingspan 53–770 mm [2.1–3.0 in]; subspecies *G. d. poeyi* archipelago-wide) by blue, and not just pale, clubs on the tip of the antennae.

Cloudless Sulphur *Phoebis sennae*

Range and status: Subspecies *P. s. sennae* archipelago-wide; species also in southern and eastern U.S., the Caribbean, Central and South America.

Description: Wingspan 48–65 mm (1.9–2.6 in). Commonly seen, large, yellow butterfly, so named by its apparent preference for flying on sunny days. Upperside of male wing is lemon-yellow with no markings, while that of the female is yellow or white with irregular black borders along the outer edges of both forewings and

Ascia monuste

Ascia monuste

Hemiargus ceraunus ceraunus

Kricogonia lyside

Phoebis sennae sennae

hindwings, and a black spot on each forewing. Underside of hindwing of both sexes has two pink-edged silver spots. **Habits and habitat:** Occurs year-round in disturbed and open habitats. Adult food plants include *Cordia* spp.; caterpillar feeds on *Cassia* spp., *Bougainvilla* spp., *Hibiscus* spp., sages (*Lantana* spp.), and morning glories (*Ipomoea* spp.).

Lyside Sulphur   *Kricogonia lyside*

Range and status: Archipelago-wide; also southern U.S., the Caribbean, Central and South America.

Description: Wingspan 38–60 mm (1.5–2.4 in). Small, variably patterned, sexually dimorphic sulphur, with a distinctive square-shaped tip to the forewing. Males are white above with an orange-yellow base to the forewing and sometimes a short, black bar near the tip of the hindwing parallel to the body. Females are white or yellow above with gray-black forewing tips. Underwings of both sexes vary from greenish-white to bright yellow to almost white; greener individuals have a whitish vein in the center of the hindwing and a bright yellow forewing basal patch. **Habits and habitat:** Open and scrub habitats.

Banded Yellow   *Eurema elathea*

Range and status: Subspecies *E. e. elathea* archipelago-wide. Species also in the southern U.S., the Caribbean, South America.

Description: Wingspan 32–42 mm (1.3–1.7 in). Medium-sized, sexual dimorphic sulphur. Upper forewings of the male are yellowish with a broad blackish-brown apex and a broad, straight, blackish bar along the inner margin. Hindwings of the male are white, with a broad, dark brown border. Females have white upper forewings and hindwings, both with broad, dark borders, and lack the black inner bar. **Habits and habitat:** Often in large groups in disturbed semi-open coppice edge habitats, roadsides, and agricultural land. Adult food plants include Blue Flower (*Stachytarpheta jamaicensis*), Shepherd's Needle (*Bidens pilosa*), and Pyramidflower (*Melochia pyramidata*); caterpillar feeds on legumes (Fabaceae) including *Zornia* spp. and *Stylosanthes* spp. **Similar species:** Can easily be confused with *E. daira*, but the latter has a slightly curved bar on the upperside of the forewings, rather than the straight bar of *E. elathea*.

Dina Yellow or Bush Sulphur   *Eurema dina*

Range and status: Subspecies *P. d. helios* throughout much of the archipelago; species also in southern U.S., the Caribbean, Central America.

Description: Wingspan 32–57 mm (1.3–2.2 in). Males are orange-yellow with a very narrow black border on the outer margins of the forewing, whereas females are yellow with black at the forewing tip. On the underwings of both sexes, three black spots are found on the hindwing. The wet season (summer) form is paler. **Habits and habitat:** Adults active year-round. Adult food plants include sages (*Lantana* spp.) and milkweeds (*Asclepias* spp.); caterpillar feeds on *Alvaradoa* spp.

■ FAMILY PAPILIONIDAE (Swallowtails)

Medium- to large-sized, strikingly colored butterflies, many of which have conspicuous tail-like projections on the hindwings that resemble the forked tail of a swallow. At least five species are recorded in the archipelago.

Gold Rim or Polydamas Swallowtail   *Battus polydamas*

Range and status: Subspecies *B. p. lucayus* archipelago-wide; species also in southern U.S., Central and South America.

Description: Wingspan 90–120 mm (3.5–4.7 in). Large, black butterfly with a single broad yellow band formed by distinctive spots along the hind margin of the fore and hindwings. Undersides of the forewings have the same pattern, while the hindwings have a submarginal row of red crescent-shaped markings. Unlike other

Battus polydamas lucayus

Papilio polyxenes, male

Eurema elathea elathea

Eurema dina helios

swallowtails in the archipelago, it lacks tail-like projections on the hindwing. **Habits and habitat:** Active year-round in open and disturbed areas. Adult food plants include sages (*Lantana* spp.); caterpillar feeds on pipevines (*Aristolochia* spp.).

Eastern Black Swallowtail   *Papilio polyxenes*

Range and status: Recorded from the northern Bahamas; also much of the U.S., Central and northern South America.

Description: Wingspan 80–110 mm (3.1–4.3 in). Medium-sized, predominantly black swallowtail with marked sexual dimorphism. There are two rows of yellow spots on the upper forewings following the wing margin, which continue on the hindwings. These are larger and brighter in the male with the inner spots appearing as a conspicuous, thick yellow band; the spots are duller and the inner row is almost absent in the female. On the hindwing between the two lines of yellow spots is an iridescent blue band, larger and brighter in the female. Both sexes have paired red spots with a black "bulls-eye" on each inner margin of the upper hindwings. **Habits and habitat:** Active April to October in open areas, fields, suburbs, and roadsides. Adult food plants include milkweeds (*Asclepias* spp.) and thistles (Asteraceae); caterpillar feeds on plants in the parsley family (Apiaceae).

Dusky Swallowtail   *Papilio aristodemus*

Range and status: Archipelago-wide. Three subspecies: *P. a. ponceanu*, also known as Schaus's Swallowtail (northern archipelago); *Papilio a. majasi* (central Bahamas); *P. a. bjorndalae* (recorded south of Crooked Island). Species also in the Caribbean and south Florida.

Description: Wingspan 86–130 mm (3.4–5.1 in). Dark brown swallowtail with a row of yellow submarginal spots and a broad, yellow median band. Undersides of the wings are yellow with brown markings and a broad blue and rust-colored median band. Tails are solid dark brown with a yellow border. Sexes are similar but males have yellow-tipped antennae and females are generally bigger. **Habits and habitat:** Found in dry scrub and coppice habitats; difficult to approach. **Conservation status:** Schaus's Swallowtail found in southern Florida and the northern Bahamas is listed as critically endangered.

Bahamian Swallowtail   *Papilio andraemon*

Range and status: Archipelago-wide (except on Turks bank); also Cuba, Jamaica, Cayman Islands.

Description: Wingspan 96–102 mm (3.8–4.0 in). Similar to Dusky Swallowtail (see above). Distinguished by the yellow bar at the tip of the forewing and the yellow-filled long tails of the hindwing. **Habits and habitat:** Typically seen from April–October. Found in scrub and coppice. Caterpillars feed on plants in citrus family (Rutaceae), for example, *Ruta* spp., and *Zanthoxylum* spp.

■ **FAMILY HESPERIDAE** (Skippers)

Typically, small, stout-bodied, short-winged butterflies with the tip of the antennae forming a hook-like projection. Their common name derives from their darting, skipping flight. At least sixteen species are recorded in the archipelago.

Subfamily Pyrginae (Spread-wing Skippers)

Large subfamily of butterflies with worldwide distribution. Wings are held wide open when basking in the sun but held closed at rest. Males can be territorial and utilize prominent perches within their territories. Some species also feed on fluids from dung, carrion, and rotting fruit.

Mangrove Skipper   *Phocides pigmalion*

Range and status: Archipelago-wide. Two subspecies: *P. p. batabano* (Andros); *P. p. batabanoides* (most other islands in the archipelago). Species also in the Caribbean, Central and South America, south Florida.

Description: Wingspan 48–70 mm (1.9–2.8 in). Distinctive blue-black butterfly with iridescent blue scaling on the upperwings, and blue lines and dashes on the abdomen. Hindwing tapers into a small, stubby tail and has a submarginal row of faint, light blue spots. Underside of the forewing is a dull brownish black, while that of the hindwing is black with a blue sheen and several faint pale blue bands. **Habits and habitat:** Adults can be found year-round especially in coastal habitats

Papilio aristodemus ponceanu, male

Phocides pigmalion batabanoides

Papilio andraemon

Epargyreus zestos zestos

including mangroves. Adult food plants include *Bougainvillea* spp., sages (*Lantana* spp.), Shepherd's Needle (*Bidens pilosa*), Redgal (*Morinda royoc*), and Bushy Lippia (*Lippia alba*); caterpillar food plants include the Red Mangrove (*Rhizophora mangle*). Subspecies *P. p. batabanoides* is shown.

Zestos Skipper   *Epargyreus zestos*

Range and status: Archipelago-wide. Two subspecies: *E. z. zestos* (northern archipelago); *E. z. inaguarum* (Crooked Island, Great Inagua, the TCI). Species also in the Caribbean, Central and northern South America.

Description: Wingspan 45–60 mm (1.8–2.4 in). Small, red-brown butterfly. Upperside is reddish-brown with a transparent gold stripe across the middle of the forewing that can also be seen when the wings are closed. Underside of hindwing is reddish-brown and usually has no gold mark. **Habits and habitat:** Adults are associated with coastal scrub. Adult food plants include sages (*Lantana* spp.) and *Bougainvillea* spp.; caterpillar feeds on plants in pea and bean family (Fabaceae).

Hammock Skipper   *Polygonus leo*

Range and status: Subspecies *P. l. histrio* throughout most of the archipelago; species also in Caribbean, Central and South America, southern U.S.

Description: Wingspan 43–59 mm (1.7–2.3 in). Commonly seen, small, brown butterfly similar to the long-tailed skipper (*U. proteus*; see below) but lacks the long tail. Upperside of the fore and hindwings is black-brown with three square, white markings on the forewing. Underside of the hindwing is brown with a blue sheen and a black spot near the base. Hindwing has a noticeable lobe. **Habits and habitat:** Adults frequently rest upside down on the underside of leaves. Typically active on sunny days, but fly mostly in the shade. Often seen in openings in coppice. Adult food plants include Blue Flower (*Stachytarpheta jamaicensis*), sages (*Lantana* spp.), Shepherd's Needle (*Bidens pilosa*), and *Cordia* spp.; caterpillar feeds on Jamaican Dogwood (*Piscidia piscipula*) and other plants in the pea and bean family (Fabaceae).

Long-tailed Skipper   *Urbanus proteus*

Range and status: West Indian subspecies *U. p. domingo* throughout the archipelago, except the Turks Islands. Species also in the Caribbean, southern North America, and South America.

Description: Wingspan 40–60 mm (1.6–2.4 in). Widely distributed, small, dark butterfly with a characteristic, very long, tail. Upperside of the fore and hindwings is dark (blackish) brown with a V-shape of pale spots on the forewing and a similar-colored band along the trailing edge of the fore and hindwings. Body and wing bases are iridescent blue-green. **Habits and habitat:** Observed year-round in open habitats, such as brushy fields, road edges, coppice, gardens, and other secondary habitats. Adult food plants include *Bougainvillea* spp., sages (*Lantana* spp.), Blue Flower (*Stachytarpheta jamaicensis*), *Cordia* spp., soldierbushes (*Tournefortia* spp.), and Shepherd's Needle (*Bidens pilosa*); caterpillars feed on plants in pea and bean family (Fabaceae).

Florida Duskywing   *Ephyriades brunnea*

Range and status: Subspecies *E. b. brunnea* throughout the archipelago; species also in south Florida, Greater and Lesser Antilles.

Description: Wingspan 42–48 mm (1.7–1.9 in). Distinctive fast-flying, dark-colored skipper. Sexually dimorphic. Upperside of males is black with a silky sheen and a circle of white transparent spots near the apex of the forewing. Females are less black and more lilac with dark bands and larger spots on the forewings. **Habits and habitat:** Typically associated with pine–sabal palm scrub and coastal habitats. Adults feed on a variety of shrubs including Shepherd's Needle (*Bidens pilosa*), *Croton* spp., and sages (*Lantana* spp.); caterpillar feeds on Key Byrsonima (*Byrsonima lucida*) and Barbados Cherry (*Malpighia glabra*).

Urbanus proteus domingo

Ephyriades brunnea brunnea

Polygonus leo histrio

Cymaenes tripunctus tripunctus

Subfamily Hesperiinae (Grass Skippers)

Typically small, orange and/or brown species with pointed forewings. Individuals can hold or position their wings in a triangle-like shape, with the forewings held upright and the hindwings folded flat, which may facilitate warming while basking in the sun.

Three-spotted Skipper   *Cymaenes tripunctus*

Range and status: Subspecies *C. t. tripunctus* throughout most of the archipelago; species also in the southern U.S., the Caribbean, South America.

Description: Wingspan 29–35 mm (1.1–1.4 in). Small, dark-colored skipper. Wings are dull dark brown. Upperside of forewing has three tiny transparent white spots on the leading edge near the tip and two to three spots in the middle of the wing. Underside of the hindwing is yellow-brown with faint pale spots near the center. **Habits and habitat:** Typically found in grassy habitats. Adult food plants include Blue Flower (*Stachytarpheta jamaicensis*), Tropical Whiteweed (*Ageratum conyzoides*), Coral Vine (*Antigonon leptopus*), and Jack in the Bush (*Chromolaena odorata*); caterpillar feeds on Guinea grass (*Panicum maximum*) and Sugarcane (*Saccharum officinarum*).

Fiery Skipper   *Hylephila phyleus*

Range and status: West Indian subspecies *H. p. phyleus* throughout most of the archipelago; species also in North America, the Caribbean, Central and South America.

Description: Wingspan 32–38 mm (1.3–1.5 in). Small, sexually dimorphic skipper with short antennae. Males have a golden-orange upperside bordered by a black outer margin, while females are brown-black with less orange. Underwings of both sexes are pale brown with paler checks, but male has scattered black spots. **Habits and habitat:** Found in open habitats including fields, lawns, gardens, roadsides, and second-growth scrub. Adult food plants include Shepherd's Needle (*Bidens pilosa*), Blue Flower (*Stachytarpheta jamaicensis*), and sages (*Lantana* spp.); caterpillars feed on grasses (Poaceae).

■ **SUPERFAMILY NOCTUOIDEA** (Noctuid or Owlet Moths)

Noctuids are a large lepidopteran family of typically dull-colored species, often with lines or spots on their wings. Most species are nocturnal or crepuscular and are often attracted to artificial light. At rest, adults often hold their wings in a roof-like form above the body. Caterpillars frequently feed on toxic plants. This superfamily comprises the majority of moth species in the archipelago, with more than 190 species recorded.

■ **FAMILY EREBIDAE** (Erebid Moths)

Very large, diverse family of moths with worldwide distribution containing some of the largest macromoth species (wingspan up to 127 mm [5 in]).

Subfamily Erebinae (Erebine Moths)

Typically medum- to large-sized moths with worldwide distribution, but particularly diverse in the tropics; also includes several pest species. Certain species have cryptically colored (camouflage) forewings, which conceal brightly colored hindwings that are suddenly revealed when the moth is disturbed, possibly distracting would-be predators. Like other Noctuoids, erebine moths can detect the calls of echolocating bats or other approaching predators using hearing organs (tympana), which are among the most sensitive in the superfamily.

Black Witch or Erebus Moth   *Ascalapha odorata*

Range and status: Archipelago-wide; also southern U.S., the Caribbean, Central and northern South America.

Description: Wingspan 90–170 mm (3.5–6.7 in). Large, commonly seen, dark-colored nocturnal moth, but seen during the day when flushed from roosting sites. The large wings are dark brown. Both the fore and hindwings are crossed by series of alternating light and dark undulating lines and bands, and there is often an iridescent blue cast over the wings. The diagnostic mark is a small green or orange spot on each upper forewing shaped like a number nine or a comma. The larger females

Ascalapha odorata, male

Ascalapha odorata, female

Hylephila phyleus phyleus, male

Melipotis famelica

have pinkish-white bands across the middle of both upperwings, which males lack. **Habits and habitat:** Roost in dense shade during the day, e.g., in thickets and under the eaves of houses. In The Bahamas, locally known as Money Moths or Moneybats: The legend is that if they land on you, you will come into money. Caterpillars feed on woody legumes (Fabaceae), e.g., *Acacia* spp., *Cassia* spp., and Mesquite (*Prosopis* spp.).

Melipotis famelica

Range and status: Probably archipelago-wide; also southern U.S., the Caribbean, Central America.

Description: Wingspan 42 mm (1.7 in). Small, dark-colored, gray-brown moth. Upperside of forewing is dark with a lighter or pinkish wide band across the center and along the hind margin with a large spot between. Upperside of hindwings dark with a white slash radiating from the body and a thinly defined white margin. Undersides of wings similar but lighter with less defined patterns. **Habits and habitat:** Caterpillars known to feed on False Tamarind (*Leucaena latisiliqua*).

Subfamily Arctiinae (Tiger and Lichen Moths)

Large and diverse family of brightly colored, often day-flying moths. Caterpillars are hairy and are often referred to as wooly bears or worms. The most distinctive feature of the family is a specialized tymbal organ on the hind section of the thorax used to produce ultrasonic sound that is used in mating and predator defense. Caterpillars of many species acquire distasteful or poisonous chemicals from their host plants and retain the chemicals in the adult or moth form. Moths advertise these defenses with bright "warning" (aposematic) coloration, unusual postures, odors, or, in adults, ultrasonic vibrations. Some mimic other moths that are poisonous or wasps that sting.

Faithful Beauty or Uncle Sam Moth   *Composia fidelissima*

Range and status: Archipelago-wide; also the Caribbean, southeast U.S.

Description: Wingspan 48–64 mm (1.9–2.5 in). Unmistakable, brightly colored, day-flying moth with blue-back forewings and dark blue hindwings and abdomen, frequently with an iridescent sheen. Leading edge of the forewing next to the head is red with a series of three paired white spots. There are also white spots along the outer edge of both the fore and hindwings. Head is almost black with a series of white spots arranged in three transverse lines, the body is iridescent blue, and legs are black with white rings. Coloration between populations can be surprisingly variable. **Habits and habitat:** Shrub and coppice habitats year-round. Caterpillars feed on Bay Bean (*Canavalia rosea*).

Spotted Oleander Caterpillar Moth   *Empyreuma affinis*

Range and status: Archipelago-wide; also southern U.S., the Caribbean.

Description: Wingspan 43–48 mm (1.7–1.9 in). Strikingly colored, day-flying moth with a dark brown body with metallic blue highlights and small white dots on the thorax and abdomen. Forewings are light chocolate brown or red with a border fringe of deeper brown. Area between the costal margin and subcostal veins on the forewing is carmine red. Hindwings are entirely carmine red with a deep brown border fringe. **Habits and habitat:** The solitary caterpillars are orange with silver spots and long reddish-brown hairs, and they feed on Oleander (*Nerium oleander*), a highly toxic ornamental shrub. Caterpillars are immune to the plant's toxin, which they store in their bodies, and their bright (aposematic) coloration serves as a warning to potential predators.

Polka Dot Wasp or Oleander Moth   *Syntomeida epilais*

Range and status: Archipelago-wide; also the Caribbean, Central and northern South America, southeast U.S.

Description: Wingspan 45–51 mm (1.8–2.0 in). Unmistakable day-flying, wasp-like, metallic blue moth with distinctive white polka dots on the wings, white spotting on the thorax and abdomen, and a red tip on the abdomen. **Habits and habitat:** The gregarious caterpillars are orange, with dark tubercles (bumps) bearing long, black, non-stinging hairs. They feed on Oleander (*Nerium oleander*), a highly toxic ornamental shrub. Caterpillars are immune to the plant's toxin, which they store in their bodies, and their bright (aposematic) coloration serves as a warning to potential predators.

Utetheisa ornatrix

Empyreuma affinis

Empyreuma affinis, caterpillar

Composia fidelissima

Syntomeida epilais

Ornate Moth or Bella Moth   *Utetheisa ornatrix*

Range and status: Archipelago-wide; also southern U.S., the Caribbean, Central and northern South America.

Description: Wingspan 33–46 mm (1.3–1.8 in). Small, distinctively pale-colored moth with light pink, light orange, or yellow forewings with white bands, each containing a row of black dots. The extent of patterning can be variable. Hind-wings are bright pink with an irregular marginal black band. **Habits and habitat:** Commonly seen diurnal species. Caterpillars feed on a variety of plants including rattlepods (*Crotalaria* spp.), which contain poisonous alkaloid compounds that are also retained by adults to deter predation.

■ **SUPERFAMILY BOMBYCOIDEA** (Silk, Emperor, and Sphinx Moths)

Medium- to large-sized species with worldwide distribution. Contains some of the more striking moths found in the archipelago.

■ **FAMILY SATURNIIDAE** (Wild Silk-worm Moths)

The Saturniids include many of the largest species of moth. Adults are characterized by large size, heavy bodies covered in hair-like scales, lobed wings, reduced mouth-parts, and small heads. Often brightly colored. Males can generally be distinguished by their large, broad antennae.

Io Moth   *Automeris io*

Range and status: Resident subspecies *A. i. lilith* recorded in the north of the archipelago; species also in the Caribbean, Central America, eastern North America.

Description: Wingspan 50–87 mm (2.0–3.5 in). Medium-sized, colorful moth with a large black to bluish eyespot, with white in the center, on the upperside of each hindwing. On the mainland U.S., adult males are mostly yellow, but males of the resident subspecies have orange-brown forewings (over a yellow ground color), whereas females (shown) have brown forewings. **Habits and habitat:** Adults are associated with coppice and feed on more than one hundred plant species. Green caterpillars are covered in urticating or "stinging" hairs, and contact with them can result in tingling, rashes, and nausea.

■ **FAMILY SPHINGIDAE** (Hawk and Sphinx Moths)

Medium- to large-sized (28–175 mm [1.1–6.9 in]) robust, fast-flying moths capable of flying long distances. Distinctive narrow wings and streamlined abdomen are adaptations for rapid flight, which can exceed 20 kmh (12 mph). Their ability to travel long distances enhances their role in long-distance flower pollination. Includes nocturnal, crepuscular, and daytime species. Some species feed like hummingbirds using a long proboscis, which is rolled up when not in use. Caterpillars have harmless hooks or horns at their posterior end, and they are sometimes called hornworms. At least sixty species have been recorded in the archipelago.

Grote's Sphinx   *Cautethia grotei*

Range and status: Resident subspecies *C. g. grotei* archipelago-wide; species also in the Caribbean, eastern U.S.

Description: Wingspan 28–40 mm (1.1–1.6 in). Medium-sized species. Upperside of the forewing is pale silvery gray or brown with dark (brown or black) markings, and the upperside of the hindwing is deep yellow-orange with a wide black or brown border. **Habits and habitat:** Adults fly at dusk. Larvae feed on plants of the coffee family (Rubiaceae) including Snow Berry (*Chiococca alba*) and Black Torch (*Erithalis fruticosa*).

Hemaris thysbe

Cautethia grotei grotei

Automeris io lilith, female

Hummingbird Clearwing   *Hemaris thysbe*

Range and status: Northern archipelago; also most of North America.

Description: Wingspan 40–55 mm (1.6–2.2 in). Variably colored, day-flying migratory species. Thorax is typically olive to golden-olive on upper side and white-yellow below. Abdomen is dark-colored (burgundy or black) with light olive to dark golden patches on the upper side. Wings are transparent with visible veins and a reddish-brown border. Light-colored legs and the lack of striping on underside of the thorax distinguish it from other species in the genus. **Habits and habitat:** Found in open and second-growth habitats, gardens, and settlements; adults feeds on variety of flowers including sages (*Lantana* spp.). Moths in this genus are known as clearwing moths (because of colored veins outlining clear wing patches) or hummingbird moths because they hover in front of flowers using their long proboscis (19–21 mm [0.7–0.8 in]) to collect nectar from flowers in a manner similar to hummingbirds.

Half-blind Sphinx or Coffee Sphinx   *Perigonia lusca*

Range and status: Resident subspecies *P. l. lusca* throughout the archipelago; species also in the Caribbean, Central and South America, south Florida.

Description: Wingspan 55–65 mm (2.2–2.6 in). Medium-sized, gray-brown species with distinctive continuous dark median band running over the upperwing and the body, and a black band along the tip of the wings margined in gray. Upper hindwing typically has a yellow-orange band along the leading edge and a partial yellow margin or spot along the trailing edge. **Habits and habitat:** Coppice and edge/urban habitats. Caterpillar feeds on plants in the coffee family (Rubiaceae).

Tetrio Sphinx Moth   *Pseudosphinx tetrio*

Range and status: Archipelago-wide; also the Caribbean, South and Central America.

Description: Wingspan 127–140 mm (5–5.5 in). Large, brown moth with white and gray markings on the forewings and darker hindwings. Body is striped with narrow gray-white bands separated by wider black bands. The distinctive large caterpillar (c. 150 mm [5.9 in]) is conspicuously colored black with narrow yellow bands and a red-orange head, with an orange bump with a black horn roughly 20 mm (0.8 in) long at the posterior end. Legs are orange with black spots. **Habits and habitat:** Adult food plants include Madagascar Periwinkle (*Catharanthus roseus*) and Pequi (*Caryocar brasiliense*); caterpillars feed on Frangipanis (*Plumeria* spp.) and Golden Trumpet (*Allamanda cathartica*). The caterpillars' consumption of toxic plants makes them distasteful to most predators. In addition, they are coated in barbed urticating hairs (a skin irritant), and they can bite.

Tersa Sphinx Moth   *Xylophanes tersa*

Range and status: Resident subspecies *X. t. tersa* throughout the archipelago; species also in the Caribbean, Central and South America, southeast U.S.

Description: Wingspan 60–80 mm (2.4–3.2 in). Large, distinctive, pale-brown species. Upperside of the forewing is pale brown with lavender-gray at the base with dark brown lengthwise lines throughout and a darker central line, making each forewing look like a leaf. Upperside of the hindwings is dark brown with a band of whitish, wedge-shaped marks. **Habits and habitat:** Active around sunset. Often attracted to lights.

ORDER DIPTERA (Flies and Mosquitoes)

Collectively known as flies, this large order comprises c. 120,000 described species with worldwide distribution. They are characterized by large compound eyes, a large moveable head, and sucking mouthparts, which have been modified in some species to pierce skin to suck blood. Diptera are dull or metallic colored and possess only one pair of wings; the second pair of wings has been modified to form a pair of club-like balancing sensory organs called "halteres." The characteristic buzzing of flies is caused by the high frequency of wing beats, c. 300 beats per second. Larvae lack legs and wings and can be terrestrial or aquatic. More than 250 species are recorded in the archipelago; some are important disease vectors.

Pseudosphinx tetrio,
caterpillar

Pseudosphinx tetrio

Perigonia lusca

Xylophanes tersa

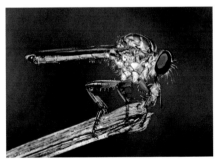

Efferia sp.

Family Asilidae (Robber Flies)

Range and status: Archipelago-wide; at least eight species are recorded.

Description: Medium- to large-sized, 30–50 mm (1.2–2 in), predatory flies. They have stout, spiny legs, a dense moustache of stiff bristles on the face, three simple eyes (ocelli) in a depression between their two large compound eyes, and frequently long tapered abdomens. **Habits and habitat:** They feed on a wide range of invertebrates including beetles (Coleopterans), butterflies (Lepidopterans), grasshoppers (Orthopterans), ants and wasps (Hymenopterans), and even spiders (Arachnids). Prey are stabbed with the short, stout proboscis and injected with saliva and toxins, with the subsequent digested liquid contents being sucked out. *Efferia* spp., a Puerto Rican species from one of the same genera found in the archipelago, is shown.

Family Bombyliidae (Bee Flies)

Range and status: Archipelago-wide; at least thirteen species recorded.

Description: Medium- to large-sized, stout-bodied, hairy, and often bright or distinctively colored species, so named because of their similarity to bees. Wings often have dark markings and are frequently held outstretched at rest. Individuals fly similarly to hover flies (see below). **Habits and habitat:** Adults generally feed on nectar and pollen, but larval stages are predators or parasites of eggs and ground-living larvae of other insect species (e.g., Coleoptera, Hymenoptera, Lepidoptera, and Diptera). Two species are shown: *Ligyra cerberus* (c. 15 mm [0.6 in]; archipelago-wide) and *Chrysanthrax* sp. (c. 15 mm [0.6 in]; north and central archipelago, possibly archipelago-wide).

Family Ceratopogonidae (No-see-ums, Midges, and Sandflies)

Range and status: Archipelago-wide; at least twelve species recorded.

Description: 1–4 mm (< 0.1–0.2 in). Small, slender, two-winged flies with long legs and a long proboscis. They are typically most active at dawn and dusk (crepuscular). Often seen in large swarms near aquatic or semi-aquatic habitats; the larvae require moist conditions and can utilize freshwater, saline water (salt-marsh and mangrove swamps), muddy or sandy substrates, rotting fruit, and water-holding plants, e.g., bromeliads. **Habits and habitat:** Males and females feed on nectar, but the females require blood for their eggs to mature. Females have cutting "teeth" on elongated mandibles on their proboscis to facilitate blood-sucking, and bites on human skin can cause intensely itchy, red welts that can persist for more than a week. Smaller species are tiny enough to pass through typical insect screening (c. 4900/in^2) and a finer mesh size of 10,000/in^2 is required to block all no-see-ums. As with mosquitos (see below), some species are important vectors for both human and animal diseases, and they cause significant discomfort for residents and tourists. Because of their potential economic importance, they are well studied in the archipelago. *Culicoides furens* is shown.

Family Culicidae (Mosquitoes)

Range and status: Archipelago-wide; at least thirty-four species recorded.

Description: Small flies, typically most active at dawn and dusk (crepuscular), are often seen in large swarms near freshwater. Most species are associated with damp and wet conditions; the eggs, larvae, and pupae are all aquatic and the adult emerges from the mature pupa as it floats on the water. **Habits and habitat:** Although many species feed on nectar, the females of some species feed on blood by piercing the skin with their specialized needle-like proboscis and injecting an anticoagulant to facilitate blood flow. They are key vectors for human, domestic, and wild animal diseases. Important vector species include *Aedes aegypti*, responsible for transmitting diseases including dengue fever, yellow fever, and Zika

Culicoides furens

Ligyra cerberus

Chrysanthrax sp.

Aedes aegypti

Lucilia sp.

fever. *Anopheles* spp. may transmit malaria, a disease not endemic to the archipelago, through visitors from Hispaniola where it remains endemic. The Yellow Fever Mosquito, *Aedes aegypti* (4–7 mm [< 0.02 in]), is shown. Introduced to the Americas from Africa, it is now pantropical and archipelago-wide. Distinguished from other mosquitos by white markings on legs and a lyre-shaped marking on the upper surface of its thorax.

Family Calliphoridae (Blowflies, Blue-Green Bottles, Carrion Flies)

Range and status: Throughout most of the archipelago; at least five species are recorded.

Description: Typically shiny flies, often with a metallic blue, green, or black thorax and abdomen. **Habits and habitat:** Larvae of most species are scavengers on carrion and dung; adults can detect dead animal matter from up to 1.6 km (1 mi) away. Associated with unsanitary conditions and are potential vectors of many diseases in humans and livestock. However, larvae are used for cleaning non-healing wounds, and the sequence and timing of larval development is important in forensic medicine. *Lucilia* sp. is shown. It is a North American species from one of the same genera found in the archipelago.

Family Sarcophagidae (Flesh Flies)

Range and status: Archipelago-wide; at least thirty-four species recorded.

Description: Medium- to large-sized flies, similar to Blowflies (Family Calliphoridae; see above) but lack their metallic coloring. Instead, individuals have gray bodies with black and gray longitudinal stripes on the thorax, red eyes, and a bristled, red-tipped abdomen. **Habits and habitat:** They differ from most flies in that they are ovoviviparous and lay hatched or hatching maggots instead of eggs on carrion, dung, decaying material, or open wounds of mammals, hence their common name. *Sacrophaga* sp. is shown (15–20 mm [0.6–0.8 in]; recorded at least in the north and central archipelago; probably archipelago-wide).

Family Syrphidae (Hover and Flower Flies)

Range and status: Archipelago-wide; at least twelve species are recorded.

Description: Thickset, typically brightly colored, flies with spots, stripes, and bands of yellow or brown covering their bodies often mimicking bees or wasps (Batesian mimicry). **Habits and habitat:** They hover in front of or on flowers from which they consume nectar and pollen; many species are important flower pollinators. Many larvae are insectivores and prey on aphids and other plant-sucking insects and therefore are potentially important in biological control. *Copestylum eugenia* (c. 15 mm [0.6 in]; recorded at least in the northern archipelago) mimics the female carpenter bee (*Xylocopa cubaecola*) (see above).

Family Tabanidae (Horse and Deer Flies)

Range and status: Archipelago-wide; at least eight species recorded.

Description: Medium- to large-sized diurnal species with prominent compound eyes, typically found in direct sunlight. Individuals are brightly colored, and the head and thorax are covered in short hairs. Membranous forewings can be clear, uniformly shaded gray or brown, and/or patterned. **Habits and habitat:** Both sexes feed on nectar and pollen. Females bite animals to obtain blood to facilitate egg production; their mouthparts are formed into a stout stabbing organ with two pairs of sharp cutting blades and a sponge-like structure to lap up the blood that flows from the wound. Larvae are predatory and require (semi) aquatic habitats. Five of the recorded species are of genus *Tabanus*; a North American species of this genus is shown.

Family Tipulidae (Crane Flies)

Range and status: Throughout most of the archipelago; at least eight species recorded.

Description: Medium- to large-sized, long-legged, delicate flies, with a slender brown-gray body and stilt-like legs, commonly referred to as Daddy-long-legs. **Habits**

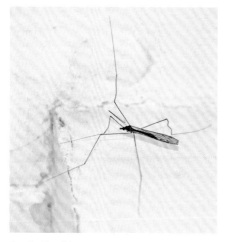

Family Tipulidae

Tabanus sp.

Sacrophaga sp.

Copestylum eugenia

and habitat: They resemble giant mosquitoes but do not bite humans or animals. Adults are short-lived and do not feed, but the larvae, termed "leatherjackets," can be agricultural pests, because they feed on roots of commercially valuable plant species, such as fruits and vegetables.

ORDER COLEOPTERA (Beetles)

More beetle species occur than any other insect order, and they comprise c. 40% of all described insect species. They are distinguished from other insects by hard forewings (elytra). Their hard exoskeleton is made up of numerous plates called "sclerites," separated by thin sutures, which provides armored defenses while maintaining flexibility. Beetles are widespread and found in almost all types of habitats. The beetle fauna of the archipelago is well described; approximately 1000 species from seventy-four families are known.

■ **FAMILY SCARABAEIDAE** (Scarab Beetles)

A family of beetles with more than 27,000 species worldwide. Large, stout-bodied, typically nocturnal beetles sometimes with bright metallic colors; distinctive, clubbed antennae made up of small plates that can be compressed into a ball or fanned out like leaves; and front legs that often exhibit modifications for digging. Males frequently have distinctive ornamentation on the head, such as highly developed horns. Includes some of the heaviest (100 g [3.5oz]) and largest beetles (up to 160 mm [6.5 in]). The characteristic large, pale-colored, C-shaped larvae live underground or under debris. Many species are scavengers of dung, carrion, or decaying plant material, while some diurnal species feed on flowers (pollen, nectar, or their petals) and are referred to as flower chafers or scarab beetles. At least thirty-nine species recorded in the archipelago, including two endemics.

Euphoria sepulcralis
Subfamily: Cetoniinae (Fruit and Flower Chafers)
Range: North and central archipelago, range incompletely known; also southeast U.S.
Description: 10–14 mm (0.4–0.6 in). Small, dark-colored (dark brown or black), diurnal beetle, with a metallic bronze or green sheen, noticeable small indentions, and tiny white spots along the otherwise smooth and shiny elytra. **Habits and habitat:** Often seen in flowers or foliage of flowering plants.

Strategus spp.
Subfamily: Dynastinae (Rhinoceros Beetles)
Genus: *Strategus* (Ox Beetles)
Range: Archipelago-wide; at least five species recorded.
Description: 35–50 mm (1.4–2.0 in). Large-sized, nocturnal, typically dark-colored species, including some of the largest beetles in the archipelago. Common name refers to the characteristic horn(s) borne only by the males, which are used in fighting during breeding and for digging. *Strategus tapa* (male; species recorded in the central archipelago); and *S. anachoreta* (female; recorded in the northern archipelago).

■ **FAMILY TENEBRIONIDAE** (Darkling Beetles)

Large family of beetles with c. 20,000 species worldwide (10–80 mm [0.4–3.2 in]), frequently with black or metallic-colored elytra, often with longitudinal lines or markings, and distinctive eleven-segmented antennae. Diet includes both fresh and decaying vegetation. At least sixty-eight species are recorded in the archipelago.

Blapstinus spp.
Range and status: Archipelago-wide; at least nine species recorded.
Description: 5–10 mm (0.2–0.4 in). Typically small-sized, shiny, metallic dull-colored genus. A North American species of this genus is shown.

Strategus tapa, male

Euphoria sepulcralis

Strategus anachoreta, female

Blapstinus sp.

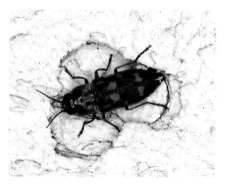

Melanophila notata

■ **FAMILY BUPRESTIDAE** (Jewel or Metallic Wood-boring Beetles)

Large family of beetles with c. 15,000 species with worldwide distribution (3–80 mm [0.2–3.2 in]), whose common name refers to their frequently glossy iridescent colors. Often bullet-shaped and seen on prominent locations, such as trunks, logs, and flowers in direct sunlight. Brightly colored specimens are prized by collectors. At least thirty species recorded in the archipelago.

Melanophila notata

Range and status: Northern archipelago, possibly archipelago-wide; also the Caribbean, Central America, North America.

Descripton: 15–20 mm (0.6–0.8 in). Medium-sized, black beetle with a coarse body texture and three or four yellow-orange spots on the elytra (wing case). *Melanophila* spp. are commonly known as fire beetles.

■ **FAMILY CERAMBYCIDAE** (Long-horn Beetles)

Large family of frequently large-sized beetles with c. 25,000 species worldwide, with extremely long antennae, often as long as or longer than the body. Both adults and larvae feed on dead wood. Although the majority are dull or cryptically colored, some species are mimics of hymenopterans (ants, bees, and wasps). At least sixty-three species are recorded in the archipelago.

Hardwood Stump Borer   *Stenodontes chevrolati*

Range and status: Throughout northern and central archipelago, but range poorly known; also south Florida and the Caribbean.

Description: 42–60 mm (1.7–2.4 in). Large-sized, sexually dimorphic, black beetle. Males have massively enlarged mouthparts. **Habits and habitat:** Range of coppice habitats.

Mesquite Borer   *Placosternus difficilis*

Range and status: Archipelago distribution poorly known; also southern U.S., the Caribbean, Central America.

Description: 19 mm (0.75 in). Medium-sized wood-boring, brown-black species with distinctive gold-yellow markings resembling a wasp (Batesian mimicry). **Habits and habitat:** Adults feed on nectar and pollen; larvae feed on wood.

Lagocheirus araneiformis

Range and status: Archipelago distribution poorly known; also the Caribbean, Central and northern South America.

Description: 20–28 mm (0.8–1.1 in). Cryptically colored, gray-brown species with very long antennae. **Habits and habitat:** Often attracted to artificial lights. A pest of Cassava (*Manihot esculenta*) and Sugarcane (*Saccharum* spp.).

Plinthocoelium **spp.**

Range and status: Archipelago distribution poorly known; also southern U.S., the Caribbean, Central and South America.

Description: 20–25 mm (0.8–1.0 in). Dark-colored (black) head and body with orange-red upper legs and dark-colored (black) lower legs.

■ **FAMILY CARABIDAE** (Ground Beetles)

Large family of shiny black or metallic-colored, typically predatory beetles with c. 40,000 species worldwide. They have ridged wing covers and paired glands in the lower back of the abdomen that produce noxious and caustic secretions used to deter predators. Large-sized species are unable to fly. Nocturnal species are typically black, but diurnal species can be brighter colored. Frequently found under the bark of trees, under logs, or among rocks. More than ninety species recorded in the archipelago.

Stenodontes chevrolati, female

Stenodontes chevrolati, male

Plinthocoelium sp.

Lagocheirus araneiformis

Placosternus difficilis

Harpalus pensylvanicus

Pennsylvania Dingy Ground Beetle   *Harpalus pensylvanicus*

Range: Northern archipelago, but range poorly known; common through much of North America.

Description: c. 14 mm (0.6 in). Small-sized, dark-colored species; head, thorax, and abdomen brown-black with reddish legs. **Habits and habitat:** Unusual for the Carabid family: This species is omnivorous.

Diurnal Marginated Tiger Beetle *Cicindela (Ellipsoptera) marginata*

Range: Northern and southern Bahamas; range poorly known.

Description: 11–14 mm (0.3–0.6 in). The diurnal Tiger beetles are characterized by large bulging eyes; metallic sometimes bright coloration; long, slender legs; and large, curved mandibles. **Habits and habitat:** Typical of beaches and sandy areas. Active predators; very fast running, feeding on other invertebrates. Controlling for body size, they are one of the fastest animals on the planet. To avoid obstacles while running, they hold their antennae rigidly and directly in front of the body and mechanically sense their environment.

■ **FAMILY MELOIDAE** (Blister Beetles)

Small family of typically diurnal beetles named for their defensive secretion of a chemical agent (cantharidin) that causes blistering of the skin. Many are conspicuously colored, which serves as a warning (aposematic) coloration to predators. Larvae are insectivorous and/or parasites of other insects whereas adults sometimes feed on flowers and leaves of plants. At least two species recorded from north and central archipelago, but distribution and number of species incompletely known.

Nemognatha punctulata

Range and status: North and central archipelago; also Caribbean, southeast U.S.

Description: c. 13 mm (0.5 in). Conspicuously colored beetle with an orange-red head, thorax, and legs and orange and/or black elytra. **Habits and habitat:** Often on flowers.

■ **FAMILY ELATERIDAE** (Click Beetles)

Small family of beetles with c. 9300 species worldwide, comprising small- to medium-sized, dull to metallic-colored, nocturnal species. Named for the unusual click mechanism they possess, it can bounce the beetle into the air to avoid predation and to right itself when on its back. At least thirty-six species are recorded in the archipelago.

Chalcolepidius spp.

Range and status: Range poorly known but at least north and central archipelago.

Description: 16–45 mm (0.6–1.8 in). Medium- to large-sized, robust genera with an arched metallic-colored body, frequently with lateral stripes on the pronotum (a shield or plate-like structure covering the front of the thorax) and/or elytra (wing cover). **Habits and habitat:** Little is known about adult behavior. Observed in a variety of habitats. Larvae are predatory, feeding on wood-boring larvae and termites. *Chalcolepidius silbermanni* is shown.

Cicindela (Ellipsoptera) marginata

Photinus sp.

Nemognatha punctulata

Chalcolepidius silbermanni

■ **FAMILY LAMPYRIDAE** (Fireflies)

Small family of typically brown nocturnal beetles comprising c. 2000 species, found in temperate and tropical climates often close to aquatic habitats, such as marshes or wet, wooded areas. Nocturnal species can bioluminesce; they possess a chemical (luciferin) in the segments at the end of the abdomen that emit light when oxygen reacts with the chemical. Light is emitted primarily to attract mates but also in defense and to deter predators. Both sexes can emit light; males often emit light on the wing whereas females remain stationary on vegetation when displaying. Diurnal species do not usually produce light. They are also known as lightning bugs or lamplighters. At least three species recorded in the north and central archipelago.

Photinus **spp.**

Range and status: Distribution incompletely known. At least one species, *P. manni*, recorded in the central Bahamas (Andros and Eleuthera).

Description: c. 10 mm (0.4 in). Small, typically dull-colored terrestrial genera; often displays on or near the ground. Light display frequently begins around dusk and

223

involves an intense period of c. 30 minutes, which decreases in intensity thereafter. Body contains chemicals to deter predation. Females of a co-existing firefly genus *Photuris* (recorded on Andros) are known to prey on *Photinus* spp. males by mimicking the light response of *Photinus* spp. females. A North American *Photinus* sp. is shown.

ORDER HEMIPTERA (Bugs)

Distinguished from other insect groups by their specialized piercing mouthparts, typically used to penetrate plant tissues and suck out the sap. This group is sometimes split into two suborders: Heteroptera (includes stink bugs, assassin bugs, and water bugs) and the Homoptera (includes leafhoppers, cicadas, and aphids). The difference between the two suborders lies in the structure of the wings. In homopterans, all four wings are entirely membranous, for example, cicadas, whereas the base of the forewings of heteropterans (referred to as true bugs) are hardened (leathery). In addition to feeding on the sap of plants, some heteropterans also feed on other animals, primarily insects. *Note:* After hatching from an egg, juveniles develop by molting through a series of instar (nymph) forms before reaching maturity. The appearance of these instar forms can be markedly different from that of adults. More than 170 species are recorded in the archipelago.

Family Belostomatidae (Giant Water Bugs)

Range and status: Archipelago-wide. At least two species are recorded: *Lethocerus americanus* and *L. medius*.

Description: 52–60 mm (2–2.4 in). Large-sized, predatory species (the biggest heteropterans) with a large, elongate, leathery brown body. Front legs are twice as thick as other legs, usually held in front of head, and are used for grasping prey while the thinner middle and hind legs point toward rear and are used for swimming. **Habits and habitat:** Cannot breathe under water; breathes through a snorkel-like tube at the end of the abdomen. They ambush their prey, which includes large invertebrates and small vertebrates, such as fish and frogs. Prey is held by the powerful forelegs and injected with venom that also aids digestion. Although mostly encountered in or near freshwater, they are strong fliers and are often attracted to bright lights at night. Their bite is considered one of the most painful insect bites (they are commonly referred to as toebiters), but there are no long-term medical consequences.

Family Coreidae (Squash or Leaf-footed Bugs)

Range and status: Archipelago-wide; at least nine species are recorded.

Description: Herbivorous insects, whose common names reflect that several species are pests of squash (gourd) plants, such as pumpkins, cucumbers, and watermelons, and also refer to the leaf-like expansions some species have on their hind legs. Variable body shape, typically oval but can be more elongate and slender. They have four-segmented antennae, forewings that are numerously veined and basally leathery with membranous tips (hemielytra), a stink gland at the end of the thorax, and frequently an enlarged hind tibia. *Chondrocera laticornis* is

Lethocerus sp.

Chondrocera laticornis

Jadera haematoloma

Loxa viridis

Euthyrhynchus floridanus

shown (16–18 mm [0.6–0.7 in]; range poorly known but recorded from northern archipelago).

Family Rhopalidae (Scentless Plant Bugs)

Range and status: Archipelago-wide.

Description: Similar to Coreidae (see above) but typically smaller in size and have a reduced scent gland opening and numerous veins on the hemielytra. Typically feeds on fruit and seeds, with some species considered pests. The distinctively colored Red-shouldered or Soap Berry Bug (*Jadera haematoloma*) is shown (10–14 mm [0.4–0.6 in]; probably archipelago-wide; also throughout the U.S., the Caribbean, Central and northern South America).

Family Pentatomidae (Stink Bugs)

Range and status: Archipelago-wide; at least eight species are recorded.

Description: c. 13 mm (0.5 in). Typically, small green or brown species characterized by five-segmented antennae and a well-developed, frequently triangular scutellum (the hardened extension of the thorax over the abdomen). **Habits and habitat:** Glands in the thorax between the first and second pair of legs produce a foul-smelling liquid (cyanide with a rancid almond scent) used to deter potential predators, hence their common name. This family includes many commercially important herbivorous pest species, but also includes some predatory species. Two species shown: *Loxa viridis*, a herbivorous species (12–17 mm [0.5–0.7 in]; probably archipelago-wide) and Florida predatory stink bug (*Euthyrhynchus floridanus*) (12–17 mm [0.5–0.7 in]; north and central archipelago, possibly archipelago-wide).

Family Scutelleridae (Jewel or Shield-backed Bugs)

Range and status: Archipelago-wide; at least five species are recorded.

Description: Commonly referred to as jewel bugs or metallic shield bugs because of their often brilliant coloration and the enlargement of the last section of their thorax into a continuous shield over the abdomen and wings. They can easily be distinguished from stink bugs (Pentatomidae; see above) because the shield completely covers the abdomen and the wings. They are closely related to stink bugs and may also produce an offensive odor when disturbed. *Sphyrocoris obliquus* is shown (12–17 mm [0.5–0.7 in]; probably archipelago-wide; also the southern U.S.).

Family Reduviidae (Ambush and Assassin Bugs)

Range and status: Probably archipelago-wide, but range incompletely known.

Description: Predatory insects with long legs and an elongated head, distinct narrowed neck, and a prominent, segmented tube (rostrum) for feeding. Most species are dark colored with hues of brown, black, red, or orange. **Habits and habitat:** Tip of the rostrum fits into a groove on the underside of the thorax, where it can be rasped against ridges to produce sound. This ability is often used to deter predators. If the threat continues, the rostrum can be used to deliver a painful bite. Predatory species use the rostrum to inject lethal saliva that liquefies the insides of the prey, which are then sucked out. A few species suck blood and transmit disease, e.g., Chagus disease. The Milkweed Assassin Bug *Zelus longipes* is shown (16–18 mm [0.6–0.7 in]; probably archipelago-wide; also southern U.S., Central and northern South America).

Family Cicadidea (Cicadas)

Range and status: At least two species recorded in the archipelago.

Description: Cicadas live in temperate to tropical climates and are recognizable by their large size and unique sounds. Adults have large prominent eyes set wide apart on the sides of the wide, blunt head; short antennae; and distinctive transparent membranous forewings. Male cicadas have specialized paired membranous drums on the sides of the abdomen called "tymbals," which produce clicking sounds as the membrane is tightened and relaxed. Each species produces a unique "song." **Habits and habitat:** The complex life cycle involves the female laying her eggs in the bark of a twig, and after hatching, the nymphs drop to the ground where they burrow underground to depths of 0.3–2.5 m (1–8.5 ft) and remain there for most of their lives (typically two to five years) feeding on sap from roots. During this time, they go through additional nymph stages, and when they finally re-emerge metamorphose into an adult. Duration of the respective stages of the life cycle is species-specific. *Diceroprocta bonhotei* (29–35 mm [1.1–1.4 in]), possibly endemic and recorded in the northern and central archipelago and typical of coppice and disturbed (including urban) habitats is shown as an adult and a nymph. A smaller species, *Ollanta caicosensis* (18–22 mm [0.7–0.9 in]), is found in the southern archipelago.

Sphyrocoris obliquus

Zelus longipes

Diceroprocta bonhotei, nymph

Diceroprocta bonhotei, adult

ORDER ORTHOPTERA (Grasshoppers, Locusts, and Crickets)

Herbivorous insects with a cylindrical body, large compound eyes, powerful mouth-parts (mandibles), elongated well-developed hind legs, and excellent jumping abilities. Males of many species mechanically produce distinctive chirping noises by either rubbing their wings against each other or against their legs to attract females. Most species have two sets of wings. Forewings are stiff and leathery and not suitable for flight, and the hindwings are membranous. Males have a single unpaired plate at the end of the abdomen; larger females have two pairs of valves (ovipositors) at the end of the abdomen that are used to dig sand during egg laying. At least twenty-three species are recorded in the archipelago.

Family Acrididae (Short-horned Grasshoppers)

Range and status: Archipelago-wide; at least ten species are recorded.

Description: Medium- to large-sized species with short, stout antennae (normally half the length of the body), well-developed wings, a specialized drumhead-like structure (tympani) on the side of the first abdominal segment, and large well-developed hind legs. Their calls are generated by rubbing their legs against their wings. The commonly seen American Bird Grasshopper (*Schistocerca americana*) is shown (39–55 mm [1.5–2.2 in]; archipelago-wide; also southeast U.S., Central America). Adults are medium- to large-sized, yellow-brown, with fully developed pale wings with large brown spots, while the smaller juveniles are green, yellow, or red, usually with a lateral pattern of black markings. **Habits and habitat:** Typical of open areas, including dunes, scrub habitats, and agricultural areas. Frequently flies when disturbed.

Family Tettigoniidae (Katydid or Bush-crickets)

Range and status: Archipelago-wide; at least seven species are recorded.

Description: Distinguished from other grasshoppers by their long, thin antennae, which frequently exceed the length of their body. Most species are camouflaged with their body shape and color matching host plant species. **Habits and habitat:** Nocturnal. Often attracted to artificial lights at night. Calls are generated by rubbing their front wings together. Two species shown: a) Broad-winged or Greater Angle-wing Katydid (*Microcentrum rhombifolium*) (52–63 mm [2.1–2.5 in]; northern archipelago, also U.S.) and b) Giant Katydid (*Stilpnochlora couloniana*) (c. 65 mm [2.6 in]; northern and central archipelago; also Florida, the Caribbean).

Family Gryllidae (Crickets)

Range and status: Archipelago-wide; at least twenty-five species are recorded.

Description: Crickets are frequently confused with grasshoppers with long antennae but can be distinguished by their long tail-like projections at the end of the abdomen. **Habits and habitat:** Nocturnal, hiding under cover during the day. Their loud, persistent chirping is made by rubbing their wings together. Cuban Ground Cricket (*Neonemobius cubensis*) is shown (20–25 mm [0.8–1.0 in]; probably archipelago-wide).

ORDER PHASMATODEA (Stick and Leaf Insects)

Phasmids are herbivorous insects found predominantly in the tropics and subtropics, and they resemble sticks or leaves, hence their common names. Primarily nocturnal, remaining inactive on vegetation during the day when their camouflage makes them difficult to see. Females are usually larger than males. At least three genera from three different families are recorded.

Malacomorpha androsensis

Family: Pseudophasmatidae

Range and status: Possibly endemic. North and central archipelago, perhaps on other islands.

Neonemobius cubensis

Microcentrum rhombifolium

Malacomorpha androsensis, female

Stilpnochlora coulomana

Schistocerca americana

Description: Male: 24–27 mm (0.9–1.1 in); female: 37–49 mm (1.5–1.9 in). Thickset, flightless species with a cylindrical granulose body and a dark mottled coloration (light brown to a dark reddish-brown). Females are longer and thicker than the males (female shown). **Habits and habitat:** Coppice. Can spray a toxic chemical from glands on the end of the thorax that can cause an intense burning irritation of the eyes and mouth of potential predators.

Mayer's Walkingstick   *Haplopus mayeri*

Family: Phasmatidae

Range and status: Northern and central archipelago; also Florida, the Caribbean.

Description: Male: 88–118 mm (3.5–4.6 in); female: 119–153 mm (4.7–6.0 in). Large-sized, less widely recorded species. Females are larger than the males and are mottled gray, while the males (shown) have a brown head and thorax, striped green and brown abdomen, and green legs. Both sexes have a pair of unequal, black-tipped, spinose horns on the head and smaller spines along the length of the thorax. **Habits and habitat:** Coppice.

Clonistria **spp.**

Family: Diapheromeridae

Range and status: Northern and central archipelago; poorly known in the archipelago. Genus is widespread through the Caribbean and South America.

Description: Medium-sized, wingless genus characterized by a smooth, elongate head and body. Sexual dimorphism is pronounced. The smaller male is shown. **Habits and habitat:** Coppice.

ORDER NEUROPTERA (Ant Lions)

The term "ant lion" applies to the larval form, sometimes referred to as "doodlebug" in North America because of the distinctive spiraling trails it leaves in the sand while looking for a location to build its pit-like trap.

Family Myrmeleontidae

Range and status: Archipelago-wide; at least five species recorded.

Description: Adults resemble a damselfly (see above) with two pairs of long, narrow, multi-veined wings, but differ by having long, clubbed antennae. *Vella fallax* (48–68 mm [2–2.5 in], archipelago-wide) is shown. Larvae, which are not normally seen, have an enormous pair of sickle-like jaws. **Habits and habitat:** After searching for suitable sandy habitat, often leaving distinctive trails in the substrate, the larva excavates a funnel-like pit and buries itself at the bottom with the large jaws just below or sometimes above the surface. Any small insects that venture over the edge of the pit fall to the bottom and are eaten. If the prey attempts to scramble up the walls of the pit, the larva flicks loose sand, which undermines the sides of the pit, causing them to collapse and brings the prey down with them to the awaiting predaceous larva. The pit-like traps are commonly seen in sandy soils.

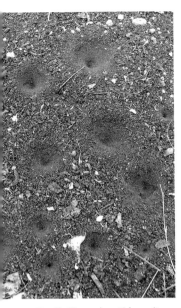

Vella fallax, adult

Family Myrmeleontidae, larva

Family Myrmeleontidae, pits

Haplopus mayeri, male

Clonistria sp.

Fish

Extant fish are divided into three groups: 1) jawless fish (superclass Agnatha) including lampreys and hagfish; 2) cartilaginous fish (superclass Chondrichthyes) including sharks and rays; and 3) bony fish (superclass Osteichthyes). Bony fish are further split into lobe-finned fish (Class Sarcopterygii), a small group of fish with soft, fleshy, paired fins, such as lung fish whose ancestors gave rise to the first land animals; and ray-finned fish (Class Actinopterygii) whose fins are webs of skin supported by bone and comprise 99% of the c. 30,000 extant species of fish. Given the terrestrial focus of this guide, we have only highlighted several of the endemic freshwater species and several commercially or culturally important marine species (all are in class Actinopterygii).

The archipelago has limited freshwater, and the freshwater fish fauna includes approximately 36 species; many of these can tolerate brackish water; 7 are introduced and at least 7 are endemic. By contrast, the archipelago has an extensive coral reef system and a large marine fish fauna; more than 600 marine species are recorded (see suggested guides in the reference section).

Freshwater Fish

FAMILY CYPRINODONTIDAE (Pupfish)

Small-sized fish with a flat head, an extendable mouth, generally large scales, and the absence of a well-developed lateral line (the visible line along the side of a fish containing sense organs to detect movement and pressure). The common name derives from the males' puppy-like aggressive behavior of encircling each other to mark out territories during the mating season. Many species are ovoviviparous (females retain their eggs internally and give birth to live young).

■ *CYPRINODON* SPP.

Almost fifty species are described from freshwater habitats to hypersaline lagoons throughout Central America, the Caribbean, southern U.S., and northern South America. Generally small in size, they feed on algae and organic detritus. Most species do not co-occur. At least four species are recorded in the archipelago, including three endemic species.

Durophage Pupfish   *Cyprinodon brontotheroides*

Range and status: Endemic. San Salvador.

Description: Length up to 36 mm (1.4 in). Small, sexually dimorphic species. Females (and immature males) are silvery-tan with pale gray vertical banding or a complete absence of markings. Adult males possess metallic-blue speckling on the top-front region of the body and usually a black margin on the tail, and breeding males sometimes are orange below. Females and juveniles have black and white eye-like spots on the dorsal fin. Distinguished from other Cyprinodon species in the archipelago by a protruding nasal region encasing its upper jaw. **Habits and habitat:** Specialist predator of small, hard-shelled invertebrates; its powerful jaw is modified to crush shells. Restricted to a few hypersaline lakes on San Salvador.

Scale-eating Pupfish   *Cyprinodon desquamator*

Range and status: Endemic. San Salvador.

Description: Length up to 30.7 mm (1.2 in). Small, sexually dimorphic species. Females and immature males are silvery-tan with irregular dark, wide bars along the length of the body. The body of adult males is almost completely black, which distinguishes them from other pupfish. In addition, both sexes lack the black terminal margin to the tail but do have black on the median fins. Both sexes have a greatly enlarged lower jaw (and a bulldog-like face) and a more elongated body than other pupfish in the archipelago. **Habits and habitat:** Scale-eating specialist that bites scales from co-existing fish species. Restricted to six hypersaline lakes on San Salvador. Unusual for pupfish, both *C. brontotheroides* and *C. desquamator* co-exist with one another, and also with the Sheepshead Pupfish *C. variegatus* (30–75 mm [1.2–3.0 in]; native to the eastern coasts of North and Central America and the Caribbean). The fourth species of Bahamian pupfish, *C. laciniatus* (52–60 mm [2–2.5 in]) is recorded only from New Providence.

FAMILY POECILIIDAE (Guppies, Swordtails, and allies)

Family of small-sized, subtropical and tropical, typically freshwater (and brackish) fish species, many of which are widely used in the pet trade. Originally distributed in the tropical Americas (southern U.S. and northern South America) and Africa. Deliberate introductions for mosquito control and release of unwanted pets, however, has resulted in Poeciliids being found in almost all tropical and subtropical areas of the world. African species typically lay eggs whereas species in the Americas give birth to live young (viviparous).

■ GENUS *GAMBUSIA*

More than forty species are described throughout the eastern and southern U.S., the Caribbean, and Central and northern South America. These fish are small (40–200 mm [1.6–7.9 in]) with an elongate body, flattened head, short snout, and spineless fins. The anal fin (unpaired fin on the lower posterior part of the body between the anus and tail) of males is modified into a thin, elongate intromittent organ (an external organ specialized to deliver sperm during copulation). Adult males are usually smaller

Cyprinodon brontotheroides, male

Cyprinodon brontotheroides, female

Cyprinodon desquamator, female

Cyprinodon desquamator, male

Gambusia sp., female

Gambusia hubbsi, two males

and more brightly colored than the females. Females give birth to live young and have a characteristic distended abdomen when pregnant. They feed on small invertebrates (including insects) and aquatic vegetation. Some species have been widely introduced outside their native range to control aquatic larvae of insect vectors of disease (hence their common name of mosquito fish), but they can sometimes become a threat to native aquatic species. *Gambusia* are very popular in the pet trade.

Bahamian Mosquito Fish   *Gambusia* **spp.**

Range and status: Endemic. Three species recorded in the northern and central archipelago: *G. hubbsi* (Great Bahama Bank including Andros and New Providence); *G. manni* (Great Bahama Bank, including Eleuthera and Long Island, but not known to occur with *G. hubbsi*); and *Gambusia* sp. (unnamed species) (restricted to Little Bahama Bank).

Description: Length 20–32 mm (0.8–1.3 in). Small, delicate, silvery, elongate fish with a variable red-green wash and a dark scale outlining, which gives the body a cross-hatched appearance. The small, variably colored (reddish-orange to pale yellow) dorsal fin, which can be extended during displays, is situated posteriorly. Small, dark spots occur on the tail, dorsal fin, and sporadically along the lateral line. Males are more conspicuously colored and distinctively marked than females, with brighter coloration on the dorsal fin, more obvious spotting on the fins and tail, and a more pronounced dark mark under the eye. **Habits and habitat:** Occupies many aquatic habitats including tidal creeks, blue holes, and inland freshwater marshes.

Marine Fish

FAMILY MEGALOPIDAE (Tarpon)

Family comprises two species of large predatory fish, both in genus *Megalops*, one in the Indo-Pacific Ocean and the other in the Atlantic Ocean.

Atlantic Tarpon   *Megalops atlanticus*

Distribution: Widely distributed in salt, brackish, and freshwater habitats along the western Atlantic coast, Gulf of Mexico, and throughout the Caribbean; typically restricted to waters between 22–28 °C (72–82 °F).

Description: Length 1.2–2.4 m (4–8 ft); weight 27–127 kg (60–280 lb). Large, scaled, silvery fish with bluish or greenish back, large eyes, and broad mouth with a prominent lower jaw. The swim bladder is used not only to control buoyancy but to "breathe"—can tolerate water with low oxygen content by coming to the surface and gulping in air. It is an obligate "air breather" and if unable to access the water surface, individuals can die. Juveniles are typically found in freshwater habitats and adults more frequently found in marine habitats. **Conservation:** Highly prized game fish (typically catch and release), not only because of their large size but also because of the fight they put up when caught. An essential part of the economically important sand flat fishery in the archipelago, which also includes Western Atlantic Bonefish (*Albula vulpes*, see below), Snook (*Centropomus undecimalis*), and Cobia (*Rachycentron canadum*).

FAMILY ALBULIDAE (Bonefish)

Small family of pantropical, medium- to large-sized, slender, silver fish. The mouth is located under the down-turned snout, which is an adaptation for feeding on organisms found on the sea floor (i.e., benthic species) including crabs, shrimp, worms, and other invertebrates. Can often be seen in shallow water with just their tail (caudal fin) protruding out of the water as they pick food from the bottom substrate. Typical of marine habitats but can also be found in brackish water. They can breathe air through a modified swim bladder.

Megalops atlanticus

Albula vulpes

Western Atlantic Bonefish   *Albula vulpes*

Range and status: Archipelago-wide. Pantropical. Common in south Florida, Bermuda.

Description: Length 770–900 mm (31–35 in); weight 5.9–8.6 kg (13–19 lb). Large, silver fish with a blue-green back and slight shading on the sides giving the impression of fine lateral lines running from the gills to the tail. Bases of the pectoral fins (behind the gills) are sometimes yellow. **Habits and habitat:** Predominately a coastal inshore species, commonly found in intertidal sand flats with abundant beds of seagrass and in mangrove areas. Found in shallow water during the rising tide, retreating into deeper water during a falling tide. Bottom-feeding species that consumes worms, fry, crustaceans, and mollusks. **Conservation:** Bonefishing (typically catch and release) is a popular sport and is an essential part of the economically important sand flats fishery in the archipelago, which also includes Tarpon (*Megalops atlanticus*, see above), Snook (*Centropomus undecimalis*), and Cobia (*Rachycentron canadum*).

FAMILY ISTIOPHORIDAE (Sailfish, Marlin, and allies)

Istiophorids are in the Order Istiophoriforme (Billfish). Billfish are large-sized, fast-swimming, predatory, open ocean species, and so named because of their prominent spear-like snout or bill (rostra), which is used to slash at, stun, and/or occasionally spear their prey. Diet includes fish, cephalopods (e.g., squid), and crustaceans. The order and family contain many popular gamefish.

Blue Marlin   *Makaira nigricans*

Distribution: Subtropical and tropical Atlantic, Pacific, and Indian Oceans; migratory, but typically restricted to waters 22–31 °C (72–88 °F).

Description: Length 3.4–4.3 m (11–14 ft); weight 91–821 kg (200–1810 lb). Females are larger than males. Large-sized, open water species with a long elongated body, a spear-like snout (c. 20% of body length), and a long, rigid dorsal fin that can be extended forward to form a crest. Very fast swimmers, reaching speeds of about 80 kmh (50 mph) in the open ocean. Like all billfish, can change color but is typically blue-black along the back with a silvery white underside, with rows (c. 15) of pale, vertical, cobalt-colored stripes along each side of the body. Fins are blue-black or brown-black. **Conservation:** Considered one of the premier game fish. Commercial long-line fishing considered a threat to this species and it is listed as Vulnerable. It is the national fish of The Bahamas.

FAMILY SERRANIDAE (Sea Bass and Grouper)

Large family (400+ species) with worldwide distribution in subtropical and tropical reefs and coastal habitats. Typically, brightly colored, thickset, with relatively large (toothed) mouths, often with a pair of noticeable, canine-like teeth projecting from the lower jaw. Typically ambush predators feeding on fish and crustaceans (e.g., crabs and shrimp). Many of the larger species are commercially important. The family also includes some of the largest bony fish species, such as Giant Grouper (*Epinephelus lanceolatus*), which can reach 2.7 m (8.9 ft) in length and weigh 400 kg (880 lb). Many species begin their lives as females and change to males later in life.

Nassau Grouper   *Epinephelus striatus*

Range and status: Archipelago-wide; also Florida, the Caribbean.

Description: Length 300–1200 mm (1–4 ft); weight 4.5–9 kg (10–19.8 lb), occasionally up to 25 kg (55 lb). Medium- to large-sized, shallow water, coral reef species, but up to depths of 60 m (197 ft). Buff-tawny colored, with five dark brown vertical bars along the body and fins. Distinct dark markings on the face and a characteristic fork-shaped mark on top of the head continues as a dark line running from the tip of the snout through each eye. Dark markings are punctuated by a series of irregular pale spots and blotches (over both the head and body). **Habits and habitat:** Adults feed mainly on fish, crustaceans, and mollusks. Juveniles feed mostly on crustaceans and are found in shallow coastal waters, often close to seagrass beds. It is the most important grouper species for the commercial fishery in the Caribbean and is threatened by overfishing. Usually solitary, but huge schools

Makaira nigricans

Epinephelus striatus

involving thousands of fish form during mass spawning events in December and January, around the time of the full moon, and always in the same locations thus making them vulnerable to overfishing. It is a slow breeder, requiring five years to reach reproductive maturity. Because its populations have been depleted, fishing restrictions or harvest bans occur throughout its range.

Amphibians

These vertebrates do not generate their own body heat, and hence are dependent on the environment to maintain their body temperature (ectothermic). Most species undergo metamorphosis from gill-bearing aquatic larvae to terrestrial adults that use lungs to breathe air. The three modern orders of amphibians (Class Amphibia) are 1) frogs and toads (Order Anura), 2) salamanders and newts (Order Caudata), and 3) caecilians (limbless amphibians that resemble snakes; Order Gymnophiona), comprising approximately 6500 species.

Amphibian skin is thin, moist, and permeable to water in contrast to the impermeable scaly skin of reptiles. Because of this thin, moist skin, amphibians must remain moist to prevent dehydration and to facilitate respiration; therefore, most amphibians require freshwater to live and breed. Saltwater is not easily tolerated by most amphibians because of their water-permeable skin and eggs. Water loss and breakdown of critical functions occur when they are in saline environments. This water loss is a result of the process of osmosis in which fluids pass through a permeable membrane from low saline (internal) to high saline (external) environments. Therefore, physiologically, the ocean is a desert for most amphibians, with water loss being a constant problem, and saltwater therefore poses a significant barrier to amphibian dispersal and colonization. Amphibian distribution in the archipelago has also been restricted by the absence of historic land bridges (with North America or the Greater Antilles), historic fluctuating sea levels (and the corresponding effects on land area), and the scarcity of permanent surface freshwater. Only eight species occur, and all are frogs or toads (Order Anura). Of these, one is endemic, two are native (all exhibit adaptations to limited access to freshwater), and the other five species were introduced in the 20th century and have a limited distribution in the wetter northern islands. **Important:** Avoid contact with the Cuban Tree Frog and the Cane Toad; both secrete toxic mucous if handled, which can cause a range of medical problems.

The amphibians described are listed in the following order: Eleutherodactylids (Eleutherodactylidae); tree frogs (Hylidae); narrow-mouthed frogs (Microhylidae); true frogs (Ranidae); and true toads (Bufonidae).

Family Eleutherodactylidae (Eleutherodactylids)

Family of frogs native to the tropical Americas and the Caribbean. Two species (genus *Eleutherodactylus*) are recorded in the archipelago. This genus consists of small, dainty, cryptically colored frogs and are also referred to as "rain frogs" or "tink frogs" because of their sharp, high-pitched, insect-like calls. They typically exhibit direct development: Eggs hatch directly into small frogs, completely bypassing the water-dependent tadpole stage of a typical amphibian, which is an adaptation to habitats with limited permanent freshwater. One species is endemic to The Bahamas, while the other also occurs outside the archipelago.

Cuban Flathead Frog or Greenhouse Frog   *Eleutherodactylus planirostris*

Range and status: Throughout the Little and Great Bahama Banks. Introduced to the TCI including Providenciales, Grand Turk, North and Middle Caicos. Also Cuba, The Cayman Islands, and introduced to Guam, Jamaica, southeast U.S.

Description: 13–32 mm (0.5–1.3 in). Very small, nocturnal, terrestrial, reddish-brown frog, with golden-flecked brown eyes. Occurs in two morphs: one with a longitudinal (head to tail) striped pattern along the back, the other with a mottled pattern with irregular light or dark markings. Underside is normally pale. Feet are distinctive having three unwebbed toes with adhesive discs on their tips. **Habits and habitat:** Most conspicuous at night and/or during wet weather, rarely seen during the day. Found in natural or urban habitats; frequently associated with damp conditions including coppice, cave entrances, banana or sinkholes, under rocks, and in gardens. **Vocalizations:** One to six bird-like chirps.

Bahamas Flathead Frog   *Eleutherodactylus rogersi*

Range and status: Endemic. Restricted to several of the Exumas (including Darby Island and Bell Island), Long Island, and San Salvador.

Description: 13–32 mm (0.5–1.3 in). Appearance and behavior is very similar to *E. planirostris* (see above). Best identified by distribution.

Family Hylidae (Tree Frogs)

Many species show adaptations for an arboreal lifestyle, including forward-facing eyes that provide binocular vision and adhesive pads on the fingers and toes. In the non-arboreal (terrestrial or semi-aquatic) species, these features may be greatly reduced or absent. Two species are found in the archipelago: one native and one introduced.

Cuban Tree Frog   *Osteopilus septentrionalis*

Range and status: Throughout the Little and Great Bahamas Banks and most islands in the southern Bahamas. Introduced to the TCI; recorded on Providenciales, Grand Turk, North and Middle Caicos. Also Cuba, the Lesser Antilles, and introduced to southeast U.S.

Osteopilus septentrionalis at night

Osteopilus septentrionalis during the day

Eleutherodactylus planirostris

Eleutherodactylus rogersi

Description: 76–140 mm (3–5.5 in). Large-sized, nocturnal, arboreal species with variably colored rough skin ranging from olive-brown to bronze to gray-white. Color varies with temperature and environment. **Habits and habitat:** An excellent climber, normally sleeping above ground during the day. A voracious predator that feeds on invertebrates, other frogs, small reptiles, small crustaceans, and even nestling birds. Skin on the head is fused to the skull, resulting in fewer blood vessels to reduce water loss. Prolific breeder, producing up to 4000 eggs during the breeding season in batches laid in freshwater. Eggs can hatch in less than thirty hours, and the tadpoles can fully develop in a month. Found in damp, shady habitats in coppice, wetlands, agricultural, and urban areas. A highly invasive species and because of its broad diet is considered a threat to native species where it is introduced. **Vocalizations:** Repetitive, variably pitched, and slightly rasping; has been compared to a squeaking door or a "snoring rasp." **Warning:** Secretes a toxic mucus from their skin when handled, which can cause an allergic reaction, asthma, or an intense burning sensation should it come in contact with the eyes and nose.

Squirrel Tree Frog   *Hyla squirella*

Range and status: Introduced. Grand Bahama, San Salvador. Native to southeast U.S.

Description: 38 mm (1.5 in). Small-sized, nocturnal, shiny-skinned tree frog; typically green but can also be varying shades of yellow or brown, sometimes with white or brown blotching; and with a distinctive blunt nose. **Habits and habitat:** Found in wetlands and coppice. **Vocalizations:** Two distinct calls: a nasal trill produced during breeding and a rapid chatter in vegetation during humid or wet weather.

Family Microhylidae (Narrow-mouthed Frogs)

Widely distributed, diverse family of primarily small-sized frogs with typically stout hind legs, short snouts, and spherical teardrop bodies. They can be arboreal or terrestrial; some live close to water. The two main shapes for the microhylids are wide bodies and narrow mouths or the more normal frog proportions. Species with narrow mouths generally eat termites and ants, and the others have diets typical of most frogs. One species recorded in the archipelago.

Eastern Narrowmouth Toad   *Gastrophryne carolinensis*

Range and status: Introduced. Grand Bahama, Abaco, New Providence. Native to the southeast U.S.

Description: 22–35 mm (0.9–1.4 in). Small-sized, nocturnal, variably colored species; brown, red-black, and gray-green with a distinctive narrow head (with a fold of skin that runs immediately behind their small eyes). They have short legs, unwebbed feet, and lack eardrums (tympana). Males have a dark throat. **Habits and habitat:** Found in wetlands and coppice. **Vocalizations:** A long, nasal, high-pitched *waaaaaaaaaaaaah*. Has been compared to the *baaaing* of sheep.

Family Ranidae (True Frogs)

True frogs are widely distributed and are found on all continents except Antarctica. They are characterized by prominent eyes; strong, long, frequently webbed hind feet that are adapted for leaping and swimming; smooth or slimy skin; and usually lay eggs in clusters. Three species are recorded in the archipelago, all are introduced.

Rana grylio

Gastrophryne carolinensis

Hyla squirella

Pig Frog   *Rana grylio*

Range and status: Introduced. Great Bahamas Bank including New Providence, Andros. Native to the southeast U.S.

Description: 82–139 mm (3.3–5.5 in). Medium-sized, nocturnal, gray-green frog with dark (brown or black) blotching, a distinctive granular skin, a pointed nose, and visible eardrums (tympana). Hind feet are webbed. **Habits and habitat:** Primarily aquatic, it is found in and around freshwater wetlands. Varied diet including crustaceans, insects, fish, and even other frogs. **Vocalizations:** The common name of the species derives from its pig-like grunt.

Green Frog   *Rana clamitans*

Range and status: Introduced. Great Bahamas Bank, including Abaco, Andros, New Providence. Native to the northeastern U.S.

Description: 50–100 mm (2–4 in). Diurnal, green or brown frog with a yellow throat and two noticeable longitudinal ridges of skin that run the length of the back (from behind each eye). Hind feet are partially webbed. Males have larger eardrums (tympana) than females. **Habits and habitat:** Primarily aquatic, found in and around freshwater wetlands. **Vocalizations:** Distinctive call sounds like a plucked banjo string, usually given as a single note, but sometimes repeated.

Southern Leopard Frog   *Lithobates sphenocephalus*

Range and status: Introduced. Grand Bahama. Native to southeast U.S.

Description: 90 mm (3.5 in). Medium-sized, nocturnal green or light brown frog, with dark (brown or black) blotching, a pointed nose, and long legs. Two noticeable longitudinal ridges of skin run the length of the back (from behind each eye). An excellent jumper with unwebbed hind feet. **Habits and habitat:** Found in freshwater wetlands. **Vocalizations:** Guttural trill.

Family Bufonidae (True Toads)

Toads are native to every continent except Australia and Antarctica. More terrestrial than most frogs, and they may be found in drier environments than frogs. They have stubby bodies with short hind legs (for walking instead of hopping), warty dry skin, and a pair of parotoid (or poison) glands behind the eyes. These glands contain an alkaloid poison (bufotoxin) that can be very toxic to some species and is usually excreted when the toad is stressed. In contrast to frogs, toads lay eggs in long chains. One introduced species is recorded in the archipelago.

Cane Toad   *Rhinella marina*

Range and status: Introduced. First recorded in 2013 on New Providence. Native to Central and South America.

Description: 100–150 mm (3.9–5.9 in). Very large terrestrial toad, up to 380 mm (15 in) and 2.7 kg (5.8 lb). Warty skin can be a variety of colors including gray, yellowish, red-brown, or olive-brown, with varying patterns; juveniles lack the warty skin of adults and have smooth, dark skin. Distinct ridges above the eyes, which run down the snout. Toes have webbing at their base, while the fingers are web-free. **Habits and habitat:** Capable of inflating their lungs, puffing up, and lifting their body off the ground to appear larger than they actually are. Has a diverse diet; captures surprisingly large prey, including small rodents, reptiles, other amphibians, and birds. Unlike many other amphibians, they also eat plant material and carrion. Because of its voracious appetite, the toad was introduced throughout the Pacific and the Caribbean to control pests of sugarcane (hence its common name). Found in open grassland, woodland, as well as in urban areas

Rana clamitans

Lithobates sphenocephalus

Rhinella marina

including gardens. Requires freshwater to breed. Individuals can breed when six months old and produce batches of 8000–25,000 eggs in long distinctive strings up to 20 m (66 ft) long. Considered a serious threat due to predation on native species, competition for food and/or breeding sites, and poisoning of toad predators. **Warning:** Secretes highly toxic mucous and should not be handled.

Reptiles

Three (of the four) extant orders of reptiles have been recorded in the archipelago: Crocodilia (crocodiles and related species), the Squamata (lizards and snakes), and the Testudines (turtles and tortoises). Most species lay eggs, and all have scaly, watertight skins (made of keratin, the structural protein of hair and nails), which minimize water loss. They are commonly referred to as cold-blooded (ectothermic) as their internal temperature varies with ambient temperature and as a result their movements become sluggish in cold conditions.

The current reptile fauna of the archipelago, which is primarily West Indian in origin, lacks the historical "giants" of the recent past (crocodiles and giant tortoises), but the largest native predators in the archipelago are still reptiles: a species of snake (boa, 2.3 m [7.5 ft]) and a lizard (iguana, 1.5 m [5 ft]). More than sixty species are currently recorded from the archipelago including (at the time of publication) thirty-one endemic species.

We use the following terminology: carapace (shell) length (CL), total length (TL; tip of nose or snout to the tip of the tail), and snout-vent length (SVL; the vent is the posterior opening for the intestinal, reproductive, and urinary tracts), which generally excludes the tail in lizards, geckos, and skinks and the tip of the tail in snakes. The reptiles are listed in the following order: terrapins and turtles, lizards, and snakes.

Order Testudinata (Tortoises and Turtles)

Testudines are the oldest living reptiles and are unique among vertebrates as the pelvic (hip) and pectoral (chest) girdles are contained within a shell made of ribs and fused with bone. The upper shell is termed the "carapace" and the lower shell, which encases the belly, is termed the "plastron." The outer layer of the shell is covered by large modified scales called "scutes," which overlap the seams between the shell bones and add strength to the shell. This group lacks teeth, and many can withdraw their head and limbs, partially or fully, into their shells. Clutches of eggs are buried in sand. After hatching, the young fend for themselves. Three species of freshwater turtles (terrapins or sliders) breed in The Bahamas and one species occurs in the TCI. Five of the seven extant species of marine turtle are found in the archipelago, and three of these breed regularly in the archipelago. There are no tortoises on the islands currently.

FAMILY EMYDIDAE (Cooters, Sliders, and allies)

Emydids are a diverse family consisting primarily of aquatic or semi-aquatic species, with a typically domed and oblong carapace, and are widely distributed throughout the Americas, Europe, and Africa. The most common turtle species sold in the pet trade are in this family.

Cat Island Terrapin or Jamaican Slider   *Trachemys terrapen*

Range and status: Native, possibly endemic. Throughout the Great Bahama Bank including Eleuthera, Exuma, Andros, New Providence, Cat Island.

Description: CL: 200–270 mm (7.9–11 in). Oval-shaped, slightly domed carapace varies from dark or grayish-brown to yellowish-olive and is wider and its edge more serrated toward the rear of the animal. No markings on the cream or yellow-colored plastron. Head and body are usually gray to olive; no distinctive markings on the skin but for the occasional exception of faded pale-colored lines on the chin, neck, and forelimbs and/or a whitish moustache under the nostrils. Distinguished from other *Trachemys* species in the archipelago by the combination of its blunt and rounded snout, a distinct terminal notch in the upper jaw, the wrinkled appearance of the scutes, and the absence of any pattern on the carapace. **Habits and habitat:** Primarily feeds on plants but also small fish, frogs, aquatic invertebrates, and carrion. Found in most fresh to brackish wetlands throughout its range. Breeding season for the Jamaican population is February–September, and clutch size varies from three to eight eggs with three to four clutches per year. Bahamian populations, however, may have a more restricted breeding season because of the limited availability of freshwater. **Conservation status:** An estimated 60% of the Bahamian population occurs on Cat Island, hence its common name. Currently unclear whether this species is endemic to The Bahamas or is conspecific with the Jamaican terrapin (which is listed as Vulnerable). Limited suitable habitat (fresh or brackish water) suggests the Bahamian population is small.

Great Inagua Terrapin or Central Antillean Slider   *Trachemys stejnegeri*

Range and status: Native. Endemic subspecies *T. s. malonei* on Great Inagua. Introduced to Caicos Bank and sporadically recorded on Pine Cay. Two other subspecies occur, one on Puerto Rico and the other on Hispaniola.

Description: CL: 240 mm (9.4 in). Oval, slightly domed carapace is widest around its midpoint, and its edge is serrated toward the rear of the animal. In adults, the carapace is dark-colored varying through gray, brown, and olive, to black. Yellow streaks may occur on juveniles. Plastron is typically yellow. Head is short, and the snout blunt or pointed, while the upper jaw is medially notched. Head is gray to olive with cream to yellow stripes and its supratemporal stripe (behind and above the eye) is reddish-brown. Neck, limbs, and tail are gray to olive with cream to yellow stripes. **Habits and habitat:** May spend considerable time out of water. **Conservation status:** Small estimated population of about 2000 is largely restricted to the west side of Great Inagua. As with other terrapins, introduced mammalian predators threaten this turtle. Habitat loss has negatively affected native and endemic *Trachemys* populations in The Bahamas, including hurricane-

Trachemys terrapen

Trachemys scripta elegans

Trachemys stejnegeri malonei

induced saltwater contamination of the freshwater ponds. Continued importation of introduced Red-eared Slider (*T. scripta elegans*, see below) for the pet trade and their subsequent release is considered a significant genetic threat to native *Trachemys* sp. Composition of the *Trachemys* populations on New Providence and Exuma is unclear but may include hybrids with introduced species.

Red-eared Turtle or Slider   *Trachemys scripta*

Range and status: Introduced. New Providence, Grand Bahama, and a few records from Providenciales. Native to southern U.S. and northern Mexico.

Description: CL: 200–330 mm (8–13 in). Individuals are typically brightly colored, and the oval carapace, neck, and limbs are striped in shades of green and yellow. A characteristic red mark around the ears from which it derives its common name distinguishes it from other sliders found in the archipelago. Older individuals sometimes have a melanistic (dark) coloration of a dark grayish-olive green, with nondescript shell and body markings, thus potentially leading to confusion with native *Trachemys* species where they co-exist. **Habits and habitat:** Breeding usually occurs from March–July. Introduced through pet trade. (Of three subspecies,

T. s. elegans is the archetypal pet turtle with millions of turtles cultivated each year.) Other introduced species of terrapin recorded in the archipelago are the Eastern box turtle (*Terrapene carolina*) and the Hispaniolan slider (*Trachemys decorata*).

SUPERFAMILY CHELONIOIDEA (Marine Turtles)

Representatives from both families of marine turtle are found in the archipelago: leatherbacks (Family Dermochelyidae) and sea turtles (Family Cheloniidae). All are large, long-lived, have a streamlined shell and large flippers, and a low reproductive rate, only reaching sexual maturity at 10–25 years old. Breeding season for marine turtles in the archipelago is April–September. Mating occurs offshore, and females come ashore on sandy beaches, typically at night, and bury their eggs in pits dug above the high-water mark; each species produces a unique set of tracks in the sand. After fifty to seventy days, the young emerge and return to the sea. Most species are declining in numbers because of the loss of nesting beaches, pollution, overharvesting (adults, juveniles, and eggs), and unintentional death as commercial fishing bycatch. Since 2009, sea turtles, their eggs, and nests have been fully protected in The Bahamas. Since 2014, export of turtle products and the harvesting of nesting females and eggs are prohibited in the TCI, although green and hawksbill turtles can be harvested at sea (other turtle species are fully protected), but individuals must be of a minimum size and landed alive to check biometrics.

In addition to the species described here, the Kemp's or Atlantic Ridley turtle (*Lepidochelys kempii*; CL: 580–660 mm [23–26 in], weight 32–49 kg [70–108 lb]) is occasionally encountered in the archipelago but is not known to nest. This marine turtle is the rarest and smallest, with a heart-shaped, dark green-gray carapace, typically wider than it is long, and a white or yellowish plastron. Listed as Critically Endangered.

Green Turtle   *Chelonia mydas*

Range and status: Pantropical and subtropical distribution. Scattered breeding records of the Atlantic subspecies *C. m. mydas* throughout the archipelago. Foraging adults and juveniles recorded throughout the archipelago.

Description: CL: 830–1140 mm (33–45 in), weight: 110–190 kg (240–420 lb). Oval carapace is usually a patterned olive-brown; skin is dark (brown but also gray to black), although head scales may be edged with yellow. Plastron is pale (white or yellow). Distinguished from other sea turtles by having only a single pair of prefrontal scales (scales in front of its eyes) instead of two. Common name derives from the green fat usually found beneath the carapace. **Habits and habitat:** Typically herbivorous, feeding primarily on sea grass and algae. **Conservation status:** Endangered.

Hawksbill Turtle   *Eretmochelys imbricata*

Range and status: Pantropical distribution. Breeding of the Atlantic subspecies *E. i. imbricata* recorded in the central and southern islands. Foraging adults and juveniles found archipelago-wide, especially in central and southern islands.

Chelonia mydas mydas

Eretmochelys imbricata imbricata

Description: CL: 710–890 mm (28–35 in), weight: 46–70 kg (101–154 lb). Easily distinguished from other species by its sharp, pronounced, curving beak; distinctively colored carapace, with black and brown mottling on an amber background; and a characteristic elliptical carapace with a serrated saw-like edge. Front flippers have two visible claws whereas other species have only one. **Habits and habitat:** Occurs year-round; most commonly found in shallow lagoons and coral reefs. Feeds primarily on sea sponges, sea anemones, and jellyfish, including the dangerous Portuguese Man O' War (*Physalia physalis*). Because of ingestion of "toxic" prey, hawksbill meat can also be toxic if eaten. **Conservation status:** Historically, hawksbill shells were the primary source of turtle shell used for decorative purposes. The worldwide population is estimated to have declined more than 80% in the past hundred years. Listed as Critically Endangered.

Loggerhead Turtle   *Caretta caretta*

Range and status: Worldwide. Scattered breeding records of Atlantic subspecies *C. c. caretta* throughout the archipelago. Foraging adults and juveniles occasionally encountered.

Description: CL: 800–1100 mm (31–43 in), weight: 70–170 kg (155–375 lb). Largest of hard-shelled turtles with characteristic heavy head and jaws. Head and heart-shaped carapace ranges from a yellow-orange to a reddish-brown, while the plastron is typically pale yellow-brown. **Habits and habitat:** Typically carnivorous; feeds on a variety of bottom-dwelling invertebrates including mollusks (cephalopods, gastropods, bivalves), crustaceans (isopods, brachiopods, decapods), sponges, corals, worms, sea anemones, echinoderms (sea urchins, starfish, sand dollars), barnacles, fish, and even hatchling turtles. **Conservation status:** Vulnerable.

Leatherback Turtle   *Dermochelys coriacea*

Range and status: Worldwide. Rarely encountered, with sporadic occasional breeding records in the northern Bahamas (Abaco).

Description: CL: 1000–2000 mm (39–78 in), weight: 250–700 kg (550–1500 lb). Largest of all living sea turtles, differentiated from other species by its lack of a bony (hard) carapace. Instead, its large, pear-shaped carapace is covered by dark leathery skin and oily flesh with white spots and has seven distinct ridges running along its length. Front flippers can grow up to 2.7 m (9 ft) long. **Habits and habitat:** Feeds primarily on jellyfish, tunicates, squid, and octopus. This species is one of the deepest diving marine animals, having been recorded up to 1280 m (4199 ft) deep. Primarily an open ocean (pelagic) species. The most significant regional nesting sites are in Trinidad and Tobago and east coast of Florida. **Conservation status:** Vulnerable.

Order Squamata (Lizards and Snakes)

Squamates are the most diverse of all the reptile groups with more than 10,000 species of snakes and lizards. They differ from other reptiles in that they shed their skin periodically, either in one piece or in patches; other reptiles either shed scales one at a time, for example, crocodiles, or they add new layers from beneath, such as tortoises and turtles. Squamates also have a unique joint between the skull and the jaw, which enables them to open their mouths very wide to consume large prey and to produce a powerful bite, especially noticeable in snakes. They are highly visual, often with well-developed color vision. Visual cues are important for locating prey and communication. Visual displays involving specific postures, sometimes in conjunction with coloration, are used in territorial and/or mating displays by some Squamata.

Caretta caretta caretta

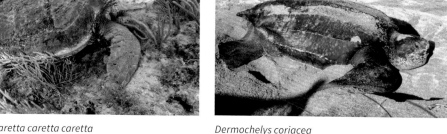

Dermochelys coriacea

LIZARDS

More than 6000 species are recorded worldwide and are classified into four general suborders: Iguania (iguanas, chameleons, agamids, and New World lizards); Gekkota (geckos); Amphisbaenia (worm lizards); and Autarchoglossa (skinks and whiptails). Most are four-legged (quadrapedal) and diurnal, but geckos are nocturnal. They do not normally swim. Most species lay eggs, and the majority are carnivorous (often adopting a "sit and wait" strategy). Males are frequently territorial, exhibiting visual displays to attract females and to intimidate rival males. In addition to visual displays, some species use pheromones (chemical signals) to communicate, and many species can detach their tails to escape from predators ("autotomy"). They differ from snakes by having external ears and eye lids. At least thirty-six species are recorded in the archipelago; twenty-two endemic, eight native, and six introduced species. *Note:* Color is not always a reliable means of species identification as there can be considerable differences in coloration between individuals of the same species, especially between different subspecies and sexes. Even within the same (sub)species, individuals of some groups (e.g., Anolis lizards and Sphaerodactylus geckos) can significantly alter their coloration in response to temperature, social interactions, the dominant colors in their environment, and other factors. Consequently, the distribution and presence of co-existing species on the respective limestone banks and island groups are given in species accounts to facilitate identification.

FAMILY TEIIDAE (Whiptail Lizards)

Teiids lizards are in the suborder Autarchoglossa (Skinks and Whiptails) and are found throughout the Americas and the Caribbean. They are collectively called "whiptails" (because of their long tails) and/or race/jungle-runners (because of their ability to move at speed). These fast-moving, medium-sized, ground-dwelling, diurnal lizards differ from other lizards in the archipelago by their size; long, straight tails; and characteristic quick jerky movements. The head and forked tongue continually probe the substrate. Many species dig burrows and are restricted to areas with sandy soils. They feed primarily on arthropods but also small lizards. Two species occur in the archipelago, both in The Bahamas.

Blue-tailed or Cuban Whiptail   *Pholidoscelis auberi*

Range and status: Native. Widely distributed on the Great Bahama Bank. Twelve subspecies recorded: *P. a. behringensis* (Andros); *P. a. bilateralis* (Ragged Island); *P. a. felis* (Cat Island); *P. a. focalis* (northern Exuma Cays); *P. a. kingi* (Andros [Gibson Cay, Big Wood Cay]); *P. a multilineata* (Berry Islands); *P. a. obsoleta* (Long Island, Exuma Cays); *P. a. parvinsulae* (Green Cay); *P. a. richmondi* (Bimini Island); *P. a. sideroxylon* (Andros); *P. a. thoracica* (New Providence, Rose Island, Eleuthera [and surrounding cays], Little San Salvador); and *P. a. vulturnus* (Andros). Species also in Cuba.

Description: SVL: 115–136 mm (4.5–5.4 in). Comparatively large, slim, terrestrial lizard, with a long, blue-gray tail. Color and patterning on the back and sides is variable between the twelve recorded subspecies but can be divided into two general groups: 1) subspecies with three longitudinal stripes along their back comprising a medial (central) stripe and two dorsoventral stripes separated by light through to dark brown coloring; and 2) subspecies in which the dorsal stripes are absent or faded. Most subspecies have dark sides, usually black or brown, while the throat color ranges from cream or yellow to black. **Habits and habitat:** Adults and juveniles (miniature versions of the adults) are unmistakable because of their distinctive, quick, jerky movements. Frequently seen in rocky or sandy areas or among leaf litter. Photo of *P. a. thoracica* characteristic of New Providence, Eleuthera, and Little San Salvador is shown.

Inagua Whiptail   *Pholidoscelis maynardii*

Range and status: Endemic. Great and Little Inagua. Three subspecies: *P. m. maynardii* (northwest Great Inagua); *P. m. uniformis* (east, south, and west Great Inagua); and *P. m. parvinaguae* (Little Inagua).

Description: SVL: 70–72 mm (2.8 in). Medium-sized, slim, terrestrial lizard. Coloration varies with subspecies. *P. m. maynardii* is distinctively marked with three longitudinal dark (black) stripes and two tan or yellow stripes along the back, and a fawn or grayish-brown tail. *P. m. uniformis* lacks this distinctive dorsal striping and is uniformly light grayish-brown (sometimes with an indistinct brown-black lateral line), a pale bluish-gray or white underside, and a grayish-brown tail. Intermediate forms between these two subspecies are found along the boundaries of their ranges. *P. m. parvinaguae* (Little Inagua; not shown) is uniform pale brown, or gray- or reddish-brown.

Pholidoscelis maynardii maynardii

Pholidoscelis auberi thoracica, male

Pholidoscelis maynardii uniformis

Spondylurus caicosae

FAMILY SCINCIDAE (Skinks)

Skinks are diurnal lizards with indistinct necks, short legs, and long tails. Most species are terrestrial or burrowing and about 45% of species are viviparous (give birth to live young). Two species of the genus *Spondylurus*, collectively called Four-lined Skinks and so-named because of typically four (occasionally three to six) dark dorsolateral stripes, are found in the archipelago, both in TCI.

Caicos Island Skink   *Spondylurus caicosae*

Range and status: Endemic. Widely distributed throughout the Caicos Islands.

Description: SVL: 73 mm (2.9 in). Medium-sized, brownish-gray lizard with small, triangular-shaped dark brown spots distributed in thin lines on the back and tail. Narrow, dark brown dorsolateral stripes extend from the top of head along the first third of body, under which runs a broad, dark brown, lateral stripe with whitish spots (extending from the snout to the first third of body). A thin, pale, lateral stripe also extends from tip of snout to first third of body. **Habits and habitat:** Unlike many skinks, this species is a good climber and often found near stone walls. **Conservation status:** Considered at risk due to the loss of habitat and threats posed by alien predators.

Turks Island Skink   *Spondylurus turksae*

Range and status: Endemic. Turks Bank; rare on Grand Turk Island, Cotton Cay, Gibbs Cay.

Description: SVL: 79 mm (3.1 in). Medium-sized lizard colored similarly to the Caicos Island skink (*S. caicosae*; see above) with a greenish-brown dorsal color, but it lacks the small, dark brown spots on the back and tail. A pale gray dorsolateral stripe extends from behind the eyes along the first third of the body, and a dark brown lateral stripe runs from the snout to the hind limbs, usually becoming less distinct around the mid-body. A pale greenish-brown stripe extends from top of head to first third of body. **Habits and habitat:** Typically found in scrub habitats in leaf litter or under stones. **Conservation status:** Critically Endangered due to a low population density, very restricted range, and threats posed by alien predators.

FAMILY LEIOCEPHALIDAE (Curly-tailed Lizards)

Curly-tailed or lion lizards are in the suborder Iguania (New World Lizards and allies). They are medium-sized, stocky, diurnal lizards (TL: c. 18 cm [7 in]) distributed widely throughout the Caribbean especially in dry coastal and rocky areas but also shrub and pinelands and even populated areas. The characteristic that separates them from other lizards in the archipelago is their distinctive tail that normally, but not always, curls above and over their back and gives the family its common name. These alert, fast lizards are typically terrestrial but can be good climbers. Males are often found basking in sunlight on the top of rocks, logs, or other vantage points. Coloration and patterning along the body is variable both within and between species, but they are often brown or gray with darker patterning along the back and sides. Males are often larger, brighter, and more distinctively colored than the females and can be aggressive, especially during the breeding season (February–October), when they exhibit a variety of territorial displays, such as head bobbing, tail curling, strutting, and inflation of the sides of the neck, which can escalate to full-blown fights with conspecifics. Arthropods (including roaches, mosquitoes, ants, spiders, and small crustacea) make up a significant part of their diet, which can be supplemented with flowers, seeds, small fruits, and even *Anolis* lizards. Five species are recorded in The Bahamas and one in the TCI. The species do not overlap in distribution; location is the primary factor in their identification.

Cuban or Northern Curly-tailed Lizard   *Leiocephalus carinatus*

Range and status: Native. Widely distributed across the Great and Little Bahama Bank. Four subspecies: *L. c. coryi* (western Great Bahama Bank including the Bimini Islands, Andros, and the Berry Islands); *L. c. hodsdoni* (eastern and southern Great Bahama Bank including Little San Salvador, Cat Island, Long Island, Guana Cay, Pinders Cay, Cay Verde, and the Ragged Islands); *L. c. virescens* (northern and eastern Great Bahama Bank including Green Cay, Eleuthera, and the Exuma Cays); *L. c. armouri* (Grand Bahama and Little and Great Abaco). Introduced on New Providence. Species also in Cuba, Cayman Islands, and introduced to Florida.

Spondylurus turksae

Leiocephalus greenwayi, male

Leiocephalus carinatus virescens, male

Description: SVL: 103–130 mm (4.1–5.1 in). Large-sized, dull-colored, brownish-gray with dark and pale markings giving the impression of brown-black and white banding on the face, back, sides, and tail. Subspecies vary in color and the intensity of their markings. Males are typically larger than females and more prominently marked, especially around the head. **Habits and habitat:** Tail often curled over their back and/or the lizards perch on prominent positions. *L. c. virescens* is shown.

East Plana Curly-tailed Lizard   *Leiocephalus greenwayi*

Range and status: Endemic. East Plana Cay.

Description: SVL: 71–75 mm (2.8–3.0 in). Medium-sized, dark to light brown, possibly green lizard with darker brown median spots and light flecks, and a distinct black patch above each shoulder and a smaller one above each hind leg. Brown tail has a series of rings along its length.

Inagua Curly-tailed Lizard   *Leiocephalus inaguae*

Range and status: Endemic. Great Inagua.

Description: SVL: 74–90 mm (2.9–3.5 in). Medium-sized, gray-brown, sexually dichromatic species. Males have a tan back with approximately ten transverse, typically rectangular, indistinct markings separated by pale stripes along its length. Along the sides are bands corresponding to the bands on the dorsum. Throat-streaks are black, and the transverse bars on the abdomen are tinted pink. A pink wash is also found on the undersurface of the tail and thighs. Three conspicuous black spots appear on the sides of the body in the shoulder region, which continue along the sides of the body as a series of smaller dark spots. The paler gray-brown female is less distinctly marked with a series of twelve or more transverse, frequently faded bars of dark brown down the back and a dark, lateral stripe along the side. These bars are continued on the sides of the abdomen as a series of narrow transverse stripes broadly flecked with green or white. Underparts are white except for these transverse bars on each side of the abdomen and a series of irregular streaks of dark brown on the throat.

San Salvador or Rum Cay Curly-tailed Lizard   *Leiocephalus loxogrammus*

Range and status: Endemic. San Salvador, Conception Island, Rum Cay. Two subspecies: *L. l. loxogrammus* (Rum Cay); *L. l. parnelli* (San Salvador).

Description: SVL: 70–90 mm (2.8–3.5 in). Medium-sized, brown, sexually dichromatic species. Males have light or dark brown upper parts marked with black diagonal lines, which are sometimes flecked with yellow, while the lateral fold along the side of the body is creamy to pale brown. Three or four black neck bars and a series of dark markings run from the eye onto the neck. Throat and underside are bright yellow-orange with an orange-red wash on the tail. Females are less distinctly marked and duller than males, with light brown upper parts with two broad yellow lines and a series of dark cross bands or chevrons along the back. Uppersides of the body are dark brown with black streaks, while the lateral line is pale brown. In contrast to males, females have a cream-colored throat. *L. l. parnelli* is shown.

Spotted or Crooked-Acklins Curly-tailed Lizard   *Leiocephalus punctatus*

Range and status: Endemic. Crooked-Acklins Bank, Samana Cay.

Description: SVL: 61–80 mm (2.4–3.1 in). Dull-colored species with limited dichromatism. Head, body, and flanks range from brown to charcoal with lighter cream to off-white and dirty gray spots. Throat is gray with irregular indistinct dark markings, while the tail has a series of dark and pale rings.

Turks and Caicos Curly-tailed Lizard   *Leiocephalus psammodromus*

Range and status: Endemic. Throughout the TCI. Six subspecies: *L. p. psammodromus* (Turks Island); *L. p. aphretor* (Long Cay); *L. p. apocrinus* (Big Ambergris and Little Ambergris Cays [Caicos Bank]); *L. p. cacodoxus* (Fort George Cay, The Providenciales, Sugar Loaf Island [Caicos Bank]); *L. p. hyphantus* (Pine Cay, Water Cay, Stubb Cay [Caicos Bank]); *L. p. mounax*

Leiocephalus inaguae, male

Leiocephalus inaguae, female

Leiocephalus loxogrammus parnelli, male

Leiocephalus loxogrammus parnelli, female

Leiocephalus psammodromus, male

Leiocephalus psammodromus, female

Leiocephalus punctatus, male

Leiocephalus punctatus, male

(Long Cay off Cockburn Harbour, South Caicos [Caicos Bank]). In addition, individuals not attributed to a specific subspecies are recorded on the Caicos cays.

Description: SVL: 83–105 mm (3.3–4.1 in). Medium-sized, yellow-brown, sexually dichromatic species, with dark barring crossing the body (ten to fifteen transverse stripes), although this patterning is often obscured by black mottling or stippling. Gray-blue belly and tail. Males are larger, brighter, and more distinctively marked than females. Locally known as bugwally.

FAMILY POLYCHROTIDAE (Anolis lizards)

Anolis lizards are in the suborder Iguania (New World Lizards and allies). They are small, primarily arboreal, diurnal, insectivorous lizards found throughout the Americas and the Caribbean. They are one of the most species-rich and diverse genera of terrestrial vertebrates with more than 400 recorded species (c. 130 in the Caribbean). They have the ability to rapidly change color, often to match the background colors of their environment. Coloration can be very variable within a species, which makes identification difficult. Anoles climb vertical surfaces, such as walls, fence posts, trees, and leaves by means of the special "adhesive" pads on each toe. Males and females have a throat fan or dewlap, made of erectile cartilage that extends from the neck and throat area. Dewlaps are used mostly by males in a variety of visual displays, including encounters with territorial rivals, potential mates, and predators. Dewlap displays are often accompanied by head bobbing and push-up displays, and the erection of dorsal crests on the head, neck, and body. The color and pattern of the dewlap, as well as the specific pattern of head movements, is species-specific.

Caution should be taken while handling *Anolis* lizards as the tail can break off when they are escaping predators or during fights. The tail continues to wriggle strongly for some minutes after detaching, presumably to distract the predator as the tailless anole escapes.

Eight species of anoles and a suite of subspecies are found across the archipelago, including three species endemic to The Bahamas and one species endemic to the archipelago as a whole. They are the most widespread and frequently seen reptiles in the archipelago. Several species have overlapping distributions; co-existing species are given in the species accounts.

Cuban Twig Anole *Anolis angusticeps*

Range and status: Native. Endemic subspecies *A. a. oligaspis* widely distributed on The Great Bahama Bank; species also in Cuba.

Description: SVL: 47–53 mm (1.9–2.1 in). Elongated snout, flattened head and body, and short limbs. Dorsal coloration is grayish to yellowish-brown, but occasionally dark brown with irregular small markings that are dark brown and black. A series of alternate dark and light brown bands is present on the tail; the dark bands are approximately twice as broad as the light ones. Chin, throat, belly, and undersurface of the limbs are cream-colored with faint, irregular dark mottling laterally. Underside of the tail is irregularly dark and light brown. Dewlap is pale pink or orange-yellow. **Habits and habitat:** Arboreal, matching their coloration

Anolis brunneus

Anolis brunneus

Anolis brunneus

Anolis angusticeps oligaspis, dark coloration

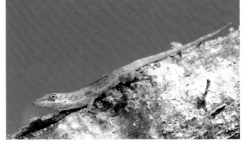

Anolis angusticeps oligaspis, light coloration

to mimic that of the bark of the host tree, which makes them difficult to see. Co-exists with *A. sagrei*, *A. distichus*, and *A. smaragdinus* throughout its range. Can be distinguished by a (dorsal-ventral) flattened head and body, and shorter limbs.

Crooked Island Anole   *Anolis brunneus*

Range and status: Endemic. Crooked and Acklins Bank, Plana Cays.

Description: SVL: 70–76 mm (2.8–3.0 in). Distinctive elongated snout and dorsal crest (not always erected). Variably colored ranging from dark brown to various shades of gray to gray-green, often with a barred patterning on the flanks and a dark line through the eye running onto the neck. Dewlap is pink, magenta, or ruby-red.
Habits and habitat: Arboreal; seen on trunks or branches. Elongated snout and arboreal habitat distinguishes it from *A. sagrei* on Crooked-Acklins Bank and *A. scriptus* on the Plana Cays.

Bark or Hispaniolan Gracile Anole   *Anolis distichus*

Range and status: Native. Great Bahama Bank. Five subspecies: *A. d. distichus* (Great Bahama Bank, except Bimini Islands, Andros, Berry Islands, Eleuthera); *A. d. biminiensis* (Bimini Islands); *A. d. dapsilis* (Eleuthera and surrounding cays, Little San Salvador, and introduced on Grand Bahama); *A. d. distichoides* (Andros, Berry Islands); *A. d. ocior* (Rum Cay, San Salvador, surrounding cays). Introduced on Grand Bahama; species also in Cuba, Hispaniola.

Description: SVL: 50 mm (2.0 in). Variably colored, including gray, a range of browns, greens, and yellows, and frequently with marbled coloration. Dewlap color is variable but is typically pale including white, yellow-orange, or red. **Habits and habitat:** Species perches relatively close to the ground and is typically seen on tree trunks but also on ground or on rocks. Co-exists with *A. sagrei* (Great Bahama, Little Bahama, Rum Cay, San Salvador Banks, respectively); *A. angusticeps* (Great Bahama Bank); and *A. smaragdinus* (Great and Little Bahama Banks). Similar to the Brown Anole (*A. sagrei*) but lacks chevrons or blotches on its back, and the scales on the belly are smooth, whereas those of *A. sagrei* have a noticeable crease in the center, visible under magnification. More terrestrial and has a shorter, more rounded snout than either *A. angusticeps* or *A. smaragdinus*.

Cay Sal Anole   *Anolis fairchildi*

Range and status: Endemic. Cay Sal Bank.

Description: SVL: 76 mm (3.0 in). Elongated snout frequently green, spotted with white (one spot per scale) with dark green lateral streaking. Tail has a series of fine, white, transverse lines and the dewlap is pink. Distinguished from co-occurring *A. sagrei* by its more elongated snout and more arboreal habits.

Cuban Knight Anole   *Anolis equestris*

Range and status: Introduced. Small populations on Abaco, New Providence, Grand Bahama, Providenciales (TCI). Native to Cuba.

Description: SVL: 90–190 mm (3.5–7.5 in). Largest anole with a characteristic triangular head with distinctive bony ridges. Typically, green with yellow (or red-green, dark green, or brown) markings around the head and shoulders. Dewlap pink. **Habits and habitat:** Arboreal, aggressive species with a distinctive threat display: individuals stand erect on all fours, gape menacingly, turn green, and may bob their head. Male will extend its dewlap, and both sexes can "puff themselves up" with air. Commonly sold in the pet trade, which accounts for most introductions. Combination of large size and aggressive behavior potentially poses a predatory threat to a range of native species, including other anoles, geckos, and even adult birds. Because of its size, sometimes mistaken for an immature Green iguana (see below). Co-exists with *A. angusticeps, A. distichus,* and *A. sagrei* (New Providence); *A. carolinensis/A. smaragdinus,** A. distichus,* and *A. sagrei* (Grand Bahama); and *A. scriptus* (Providenciales). Adults distinguished from other species by their large size. *The introduced *A. carolinensis* (native to southeast U.S.) is possibly recorded from Grand Bahama (Little Bahama Bank) and is very similar in appearance to *A. smaragdinus*.

Anolis distichus dapsilis

Anolis distichus dapsilis

Anolis equestris

Anolis fairchildi

Anolis sagrei ordinatus, male

Anolis sagrei ordinatus, male, dewlap extended

Anolis sagrei ordinatus, female

Cuban Brown Anole   *Anolis sagrei*

Range and status: Native. Endemic subspecies *A. s. ordinatus* recorded throughout the Little and Great Bahama Banks, Crooked-Acklins Bank, Conception Bank, Rum Cay, San Salvador, and Cay Sal Bank. Introduced Providenciales. Native to Cuba. Introduced elsewhere in the Caribbean, southeast U.S., Asia.

Description: SVL: 50–65 mm (2.0–2.5 in). Most commonly seen and widely distributed anole in central and northern Bahamas. Dark to light brown, but can also exhibit paler colors, with lighter-colored transverse markings and/or chevrons on the back. Males have dark brown diamond shapes along the backbone and frequently a pattern of beige-yellow dots, whereas females have a light yellow dorsal stripe and two brown stripes on the sides. Males have a large reddish-brown dewlap with a yellow margin and a non-serrated crest on the neck and back (only occasionally erected). **Habits and habitat:** Frequently observed low on tree trunks and on or near the ground. Co-existing species are *A. distichus* (Great Bahama, Little Bahama, Conception Banks); *A. angusticeps* (Great Bahama Bank); *A. smaragdinus* (Great Bahama, Little Bahama, Rum Cay, San Salvador Banks); and *A. brunneus* (Crooked-Acklins Bank). Shorter, more rounded snout and more terrestrial than *A. fairchildi, A. angusticeps, A. smaragdinus,* and *A. brunneus.* Scales on belly are different from *A. distichus* (see above).

Silver Key Anole   *Anolis scriptus*

Range and status: Endemic. Four subspecies found in the southern archipelago. *A. s. sularum* (Samana Cay, including Booby Cay, West Plana Cay); *A. s. leucophaeus* (Great Inagua, including Sheep Cay, Little Inagua); *A. s. mariguanae* (Mayaguana Island, Booby Cay); *A. s. scriptus* (TCI).

Description: SVL: 64–76 mm (2.5–2.9 in). Most common anole species in the southern archipelago. Color varies from brown to pale gray. In bright light, adults are usually yellowish or grayish-white, finely speckled with dark gray, whereas in the shade individuals are frequently dark brown. Throat is often yellow, and there may be a yellowish tinge on the head and occasionally on the back. Dewlap is orange-yellow with lines of pale spots. **Habits and habitat:** Usually found low on tree trunks and on or near the ground. Possibly co-exists with *A. equestris* on Providenciales (Caicos Bank).

Bahamian Green Anole   *Anolis smaragdinus*

Range and status: Endemic. Two subspecies recorded in the archipelago. *A. s. smaragdinus* (Great Bahama Bank [except islands with other subspecies], including Little San Salvador, Cat Island, Long Island); *A. s. lerneri* (Bimini Islands, Berry Islands, Andros; also Conception Bank [subspecies not described]).

Description: SVL: 51–64 mm (2.0–2.5 in). Green (sometimes brown) anole with a distinctive elongated snout and a rose-pink dewlap. **Habits and habitat:** Difficult to find; mostly found high in trees. Co-exists with *A. distichus, A. sagrei, A. angusticeps* (Great Bahama Bank), and *A. sagrei* (Conception Bank). *Note:* The introduced *A. carolinensis* (native to southeast U.S.) is possibly recorded from Grand Bahama (Little Bahama Bank) and is very similar in appearance to *A. smaragdinus.*

Anolis smaragdinus lerneri   *Anolis scriptus scriptus*

FAMILY IGUANIDAE (Rock Iguanas)

West Indian rock iguanas (*Cyclura* spp.) are in the suborder Iguania (New World Liz-ards and allies). They are large, diurnal, typically ground-dwelling lizards and are one of the most threatened groups of reptiles in the world because of loss of habitat and the negative impact of nonnative species. Most species survive in remnant populations that are restricted to islands and cays that lack nonnative predators (dogs and cats). Three endemic species are scattered throughout the archipelago. Adults are easily dis-tinguished from other lizards in the region by their large size 70–150 cm (2.2–4.9 ft) and a prominent dorsal crest that runs down the back. Body colors are highly variable, between and within species, depending on age, sex, and environmental conditions. Males are typically larger, more colorful, and have more prominent dorsal crests than the females; juveniles, by contrast, are usually solid brown or gray with faint darker stripes. Well-developed dewlaps help regulate their body temperature and are used in courtship and territorial displays. Iguanas are primarily herbivorous, feeding on leaves, flowers, and fruits, supplemented with insects, mollusks, crustaceans, arach-nids, lizards, and carrion. Individuals reach maturity at six to nine years of age and live at least twenty-five years. Mating occurs in May and June, with clutches of three to ten eggs usually laid in June or July in nests excavated in pockets of soil exposed to the sun. Subtle differences occur between subspecies in their coloration and scale morphology but ranges do not overlap so misidentifying a species is highly unlikely when their island of origin is known. All endemic rock iguanas are protected, and it is illegal to harvest, possess, purchase, or sell iguanas, their parts, and eggs or to interfere with their nests.

Northern Rock Iguana   *Cyclura cychlura*

Range and status: Endemic. Andros, The Exumas. Three subspecies: Andros Rock Iguana (*C. c. cychlura*), Andros; Allen Cays Rock Iguana (*C. c. inornata*), on several cays in the northern Exumas; and Exuma Rock Iguana (*C. c. figginsi*), on seven small disjunct cays throughout the central and southern Exumas.

Description: SVL: 465–475 mm (18.3–18.7 in). Largest iguana in the archipelago, reaching a total length of c. 1.5 m (c. 5 ft).

1) *C. c. cychlura* (shown). Dark gray to black, with yellowish-green or orange-red tinged scales on the head, dorsal crest, and legs. **Habits and habitat:** Unique among iguanas, females lay their eggs within excavated termite mounds and guard them until hatching occurs. **Conservation status:** Only iguana in The Bahamas that is not currently confined to small predator-free cays; has an estimated population of 2500–5000. Listed as Endangered.

2) *C. c. inornata.* Dark gray-black with tan, pink, or orange-colored mottling; lacks the horn-like frontal or prefrontal scales. **Conservation status:** Because of its restricted range and very small population of 400–500, it is listed as Endangered.

3) *C. c. figginsi.* Coloration is highly variable between populations. Adults from Bitter Guana and Gaulin Cays are dull gray-black with diffuse pale gray spots with either a white or a light red crest and the head scales are black-tinged with orange on the snout and lower jaw. Adults from Guana Cay are dull black with diffuse, pale white ventral and throat coloration with light blue patterning of the head and neck, and gray (with a red tinge) or intense scarlet dorsal crest scales. **Conservation status:** Despite an estimated population of 1000–1200, this subspecies is considered the most threatened of the Northern Rock Iguanas due to a decreasing and highly disjunct (fragmented) population. Listed as Critically Endangered.

Turks and Caicos Rock Iguana   *Cyclura carinata*

Range and status: Endemic. Two subspecies. Bartsch's iguana or Booby Cay iguana (*C. c. bartschi*), restricted to Booby Cay (near Mayaguana Island) in southern Bahamas; Turks and Caicos rock iguana (*C. c. carinata*), widely distributed throughout the TCI.

Description: SVL: 285–335 mm (11.2–13.2 in). Smallest iguana in the archipelago. Typically gray, brown, or dull green and often patterned by darker markings on the sides. Coloration varies by island. This species lacks the large scales on the upper surface of the head. Males have larger dorsal spines than other iguana species, whereas females lack dorsal spines.

1) *C. c. bartschi.* Greenish to brownish-gray, with a yellow dorsal crest, a faint yellow-brown net-like pattern on the body, a golden iris, and nine to ten dark vertical stripes on the sides of the body wall that fade with age. **Conservation status:** Species is particularly prone to predation by introduced predators, e.g., dogs and cats, because of its small size. This, combined with a very restricted range, and an estimated population of only 200–300, has resulted in the subspecies being listed as Critically Endangered.

Cyclura cychlura cychlura, female

Cyclura cychlura cychlura, male

Cyclura carinata carinata, female

Cyclura carinata carinata, male

2) *C. c. carinata* (shown). **Conservation status:** Despite a relatively large population (c. 30,000) scattered on 50–60 cays in the TCI, this subspecies is considered Critically Endangered since the majority of the population is located on only three large, predator-free islands: Big Ambergris, Little Ambergis, and East Bay cays and is vulnerable to habitat loss and the introduction of alien predators.

San Salvador Rock Iguana or Central Bahamian Rock Iguana   *Cyclura rileyi*

Range and status: Endemic. Three subspecies: San Salvador Rock Iguana (*C. r. rileyi*), San Salvador; Acklins Rock Iguana (*C. r. nuchalis*), restricted to two natural populations in the Acklins Bight and an introduced population on a cay in the northern Exumas; White Cay iguana (*C. r. cristata*) occurs on several cays in the southern Exumas.

Description: SVL: 320–395 mm (12.6–15.6 in). Most colorful iguana in the archipelago and one of the most threatened West Indian iguanas.

1) *C. r. rileyi* (shown). Largest and most colorful of the subspecies. Adult back color is highly variable and can range from red, orange, or yellow to green, brown, or gray, usually patterned by darker markings, with males having the brightest colors. Dorsal crest scales on the neck are always higher than on the body. **Conservation status:** Estimated population of 500–1000. Critically Endangered.

2) *C. r. nuchalis.* Strikingly colored with orange-yellow highlights on a darker gray to brown background. Can be distinguished from the San Salvador and White Cay iguanas by a combination of several scale features. **Conservation status:** Estimated population of c. 13,000. Endangered.

3) *C. r. cristata.* Smallest and least brightly colored of the subspecies. Dorsum of adults is usually gray with brown to orange-brown wavy lines, and the dorsal crest scales, forelimbs, and portions of the head and face are typically highlighted in orange. **Conservation status:** Estimated population of c. 2000. Critically Endangered.

Green or Common Iguana   *Iguana iguana*

Range and status: Introduced. Sporadic reports throughout the archipelago, including Stirrup Cays, Cat Cay, Bimini (The Bahamas); and Grand Turk, Providenciales (TCI). Native to Central and South America, the Caribbean.

Description: SVL: 300–420 mm (11.8–16.5 in). Large, frequently arboreal species. Despite their common name, the body color ranges from blue, black, green, red, and even pink depending on the subspecies, typically with few markings on body, and a banded tail. **Conservation status:** Commonly sold in the pet trade. Escapees and released pets are the primary source of introductions. Highly invasive and a potential competitor with native endangered iguana species. Increasing in some areas.

Iguana iguana

Cyclura rileyi rileyi, male

Cyclura rileyi rileyi, female

SUBORDER GEKKOTA (Geckos)

Geckos are small- to average-sized, soft-skinned, primarily nocturnal lizards found in warm climates worldwide. They are unique among lizards in their social vocalizations. Unlike other lizards, most geckos appear to have no eyelids (they are fused) and instead have a transparent membrane, which they lick to clean. The pupils of the eye are vertical. Many species are well-known for their specialized toe pads that enable them to climb smooth and vertical surfaces and their ability to change their colors to match their environment. They feed on insects, spiders, and other small invertebrates.

■ *HEMIDACTYLUS* SPP.

Hemidactylus geckos are in the family Gekkonidae and are commonly referred to as House Geckos because of their ability to adapt to and co-exist with humans. They can be very vocal, emitting squeaking or barking noises during territorial disputes or to deter predators. Two species are found in the archipelago, both introduced.

Indo-Pacific Gecko   *Hemidactylus garnotii*

Range and status: Introduced to The Bahamas in 1990s. Recorded on Abaco and New Providence. Native to southeast Asia.

Description: SVL: 59 mm (2.3 in). Small, nocturnal gecko with a long, narrow snout and a distinct fold of skin along the flanks and another bordering the posterior edge of the hind limbs. During the day, typically seen as dark gray or brown with light markings, whereas at night it is a pale, translucent color. Belly is orange or yellow. **Habits and habitat:** Associated with human habitation and at night are seldom found far from buildings with outdoor lights. By day, individuals hide in cracks, crevices, and under tree bark. In addition to sexual reproduction, viable young (clones of the mother) can be produced by the development of unfertilized eggs (termed parthenogenesis). Its adaptability and ability to undergo a unisexual mode of reproduction has made it a very successful invader and colonizer.

Tropical House Gecko   *Hemidactylus mabouia*

Range and status: Introduced. Archipelago-wide; also the Americas, the Caribbean. Native to Africa.

Description: SVL: 68 mm (2.7 in). Pale, nocturnal gecko with a pink-gray, gray-brown, pale white, or tan body, usually with dark chevrons on back and tail. **Habits and habitat:** Associated with human habitation and at night are seldom found far from buildings with outdoor lights. By day, individuals hide in cracks, crevices, and under tree bark typically exhibiting cryptic darker body coloration and often matching their body color to that of the background. The eyes, which are black at night, often appear metallic (bronze) colored during the day. Also known as Common House Gecko.

■ *TARENTOLA* SPP.

Geckos in this genus are in the family Phyllodactylidae and are commonly referred to as wall geckos. They are thickset, nocturnal lizards with noticeably heavier scaling than other geckos. Many species have rows of bumps, spine-like projections, or skin flaps on the body to increase the camouflage. All species have wide toes and are good climbers; climbing pads are present on both the pads of the feet and on the toes. Males are usually larger than females. One species is recorded in the archipelago.

American Wall or Cave Gecko   *Tarentola americana*

Range and status: Native. Endemic subspecies *T. a. warreni* recorded throughout The Great Bahama Bank including Andros, North Eleuthera, Great Exuma, Long Island; species also in Cuba.

Description: SVL: 111–120 mm (4.4–4.7 in). Large, thickset, dull-colored gecko (gray to brown) with a series of darker broad transverse bands running the length of body and tail. **Habits and habitat:** Occurs on cliffs, in caves, around Coconut Palms, and on buildings.

Hemidactylus mabouia at night

Tarentola americana warreni

Hemidactylus mabouia during day

Aristelliger barbouri

Hemidactylus garnotii

■ **ARISTELLIGER SPP.**

Geckos in this genus are in the family Sphaerodactylidae and are commonly referred to as croaking geckos. They are restricted to the Caribbean; two species are found in the archipelago.

Inagua or Striped-Caribbean Gecko *Aristelliger barbouri*

Range and status: Endemic. Great Inagua.

Description: SVL: 46–50 mm (1.8–2.0 in). Small, dull-colored, gray-brown gecko with a broad, dark brown streak on either side of the head, merging with a pair of large brown blotches on either side of the neck and a series of six irregular transverse streaks on the body. The anterior half of each streak is dark brown while the posterior half is paler. Tail is encircled above with approximately nine broad bands of yellow or dark brown. Underside of the head and body are white. **Habits and habitat:** Nocturnal, secretive, and difficult to see.

Caicos Croaking Gecko   *Aristelliger hechti*

Range and status: Endemic. TCI.

Description: SVL: 75–90 mm (2.9–3.5 in). Large gecko with a short, rounded snout, typically dull gray-brown and frequently mottled, sometimes with a prominent irregularly edged dorsal band along the back. A pair of dark eyespots, sometimes with a pale central spot may adorn the shoulder region. A dark line runs from the eye to the back of the head. Tail has approximately seven dark bands alternating with narrow paler bands that broaden nearer the tip, which is often black.

■ *SPHAERODACTYLUS* SPP.

Sphaerodactylids (Family Sphaerodactylidae) can be distinguished from other geckos in the archipelago by their small size 24–41 mm (0.9–1.6 in) in length, round rather than vertical eye pupils, and the presence of a single adhesive pad at the end of each digit from which they derive their common name: Sphaero = round, dactylus = finger (*Gr*). The genus comprises approximately ninety species, the majority of which are endemic to the Caribbean. The soft skin has a velvet-like texture. Many species exhibit sexual dichromaticism (sexes are colored differently), while species that frequent dry areas appear to be duller in color than species in wetter habitats. They are found in leaf litter and under stones, dead wood and plant material. Females usually lay a small clutch size, frequently one disproportionately large egg. A combination of their small size and their secretive and nocturnal behavior means that they frequently go undetected; their variable coloration also makes species difficult to identify. Nine species are recorded in the archipelago: seven in The Bahamas (including two introduced species) and three in the TCI (including one species found also in The Bahamas). Several species have overlapping distributions. Co-existing species are noted in the species accounts.

Ocellated Gecko or Jamaican Stippled Sphaero   *Sphaerodactylus argus*

Range and status: Introduced. *S. a. argus* found on North Bimini and New Providence; species also in the Caribbean, Florida, Central America.

Description: SVL: 24–27 mm (0.9–1.1 in). Very small, sexually monochromatic species. Both sexes are light brown with a pointed snout and yellow-white dots or flecking, which can give the impression of lines, particularly on the head, neck, and to a lesser degree on the back. Tail is orange or red and also flecked. **Habits and habitat:** Found in a variety of natural habitats and human habitations including leaf litter, coastal vegetation, and rubbish piles. Co-exists with *S. notatus* (North Bimini).

Caicos Least Gecko   *Sphaerodactylus caicosensis*

Range and status: Endemic. Throughout the Caicos islands including the Providenciales and Caicos cays.

Description: SVL: 32 mm (1.3 in). Small, sexually dichromatic species. Both sexes have a pointed snout, but males have a gray body and tail with dark flecking and an unmarked dull yellow head; females have a gray or tan body with seven or eight brown transverse stripes, an orange tint, longitudinal dark stripes on the head that

Aristelliger hechti

Sphaerodactylus argus

Sphaerodactylus caicosensis, female

Sphaerodactylus caicosensis, male

Sphaerodactylus copei cataplexis, male

Sphaerodactylus copei cataplexis, female
(lacking tail)

run through the eyes and merge into the body pattern, and frequently an orange
tinge to the tail.

Cope's Least Gecko or Haitian Big-scale Sphaero   *Sphaerodactylus copei*

Range and status: Introduced. *S. c. cataplexis* on Andros, Eleuthera, New
    Providence. Native to Hispaniola.

Description: SVL: 40–41 mm (1.6 in). Relatively large, sexually dichromatic species
    with more conspicuous scales than other Sphaeros in the archipelago. Males are
    variably colored ranging from light to dark gray, brown or green, with orange-red
    patterning and variable head coloring ranging from gray, yellow to orange, with
    dark flecking or spots. Females are gray-brown with one to four transverse some-

times indistinct body bands, a black collar with four pale spots, and a head pattern with a series of green, brown, or orange lines on a gray or blue-gray background, often with a distinctive yellow spot on top of the head. Co-exists with *S. nigropunctatus* and *S. notatus* throughout its range.

Rum Cay Least Gecko or Central Bahamas Sphaero   *Sphaerodactylus corticola*

Range and status: Endemic. Central Bahamas. Four subspecies: *S. c. aporrox* (East Plana Cay); *S. c. campter* (Crooked-Acklins Bank); *S. c. corticola* (Rum Cay); *S. c. soter* (San Salvador).

Description: SVL: 37–39 mm (1.5 in). Small, terrestrial species with limited sexual dichromatism. Both sexes have a dark, typically medium-brown or purple-brown body with variable amounts of dark and/or white flecking, the latter sometimes having a salt-and-pepper appearance. Spotting along the body can fuse to form longitudinal markings. Head pattern is also variable and can be absent or also composed of dark spotting. Co-exists with *S. notatus, S. copei,* and *S. nigropunctatus* (only on the Rum Cay Bank).

Inagua Least Gecko or Inagua Sphaero   *Sphaerodactylus inaguae*

Range and status: Endemic. Inagua.

Description: SVL: 26.5 mm (1.0 in). Small, sexually dichromatic species. Males are pale brown with a series of large, dark brown spots on the head and the body interspersed with a few smaller and less distinct spots. Females have three elongated spots or longitudinal stripes of dark brown on the head and a large dark spot on the shoulder surrounded by a narrow dark line to form a rectangular "target" pattern. Remainder of the body has a few small, dark blotches or indistinct cross-bands, while the tail is more heavily spotted. **Habits and habitat:** Frequently found in a range of natural and urban coastal habitats. Co-exists with *S. notatus* (Great Inagua Bank). Local name Bubatani or Snail Lizard.

Mayaguana Least Gecko or Southern Bahama Sphaero   *Sphaerodactylus mariguanae*

Range and status: Endemic to the southern islands in the archipelago: Mayaguana Island and Booby Cay (The Bahamas) and introduced to Grand Turk (TCI).

Description: SVL: 41 mm (1.6 in). Relatively large species with limited sexual dichromatism exhibiting variable coloration and pattern varying from light to dark brown, and can be unspotted, lightly to heavily spotted, or alternatively exhibit a mottled pale and dark pattern. Head pattern is also variable including a dark stripe through the eye. Co-exists with *S. underwoodi* (Turks Bank).

Black-spotted Least Gecko or Three-banded Sphaero   *Sphaerodactylus nigropunctatus*

Range and status: Native. Five subspecies recorded throughout the northern and central Bahamas. *S. n. nigropunctatus* (eastern Great Bahama Bank, including New Providence, Eleuthera, Cat Island); *S. n. decoratus* (Rum Cay);

Sphaerodactylus corticola

Sphaerodactylus inaguae, male

Sphaerodactylus mariguanae

Sphaerodactylus nigropunctatus, female

Sphaerodactylus nigropunctatus gibbus, male

S. n. flavicauda (western Great Bahama Bank, including Andros and The Berry Islands, and Cay Sal Bank); *S. n. gibbus* (Exuma Cays, Long Island, Little San Salvador, Green Cay, intergrades with *S. n. nigropunctatus* on northern most Exuma Cay); *S. n. porrasi* (Ragged Island); not recorded in the TCI. Species also in Cuba.

Description: SVL: 40 mm (1.6 in). Relatively large, sexually dichromatic species. Males are beige to gray, typically with dark spots, although depending on the subspecies the spots may extend from head to the tail, occur only on the back, or be missing altogether. Head and tail are typically yellow-orange. Females are beige-colored with a black collar and approximately five black transverse body bands. Typically, three black dorsal bands occur between the front and the hind legs,

277

with a similarly black-banded tail. Juveniles show the same coloration as females. Co-exists with *S. notatus* (throughout the Great Bahama and Rum Cay Banks), *S. copei* (Andros, New Providence, Eleuthera), *S. corticola* (Rum Cay), and *S. argus* (Bimini Islands).

Reef Gecko or Brown-spotted Sphaero   *Sphaerodactylus notatus*

Range and status: Native. Three subspecies. *S. n. amaurus* (Great Bahama Bank); *S. n. atactus* (Cuban subspecies introduced on Great Inagua); *S. n. peltastes* (Little Bahama Bank); species also in Cuba, Hispaniola, Florida, Honduras.

Description: SVL: 33–34 mm (1.3 in). Widely distributed terrestrial species with limited sexual dichromatism. Both sexes are brown with extensive darker brown mottling and a yellow-brown tail. A series of dark brown-black stripes frequently run longitudinally from the head along the first third of the body, although head patterning is absent for some individuals. Male head color is light brown and is sometimes patterned with darker scales; females often have three dark lines on a light brown head. Juveniles are similar to adults but can have a white tip to their tail. Co-exists with the sometimes similarly colored but slightly larger *S. corticola*, but only on Rum Cay (at the boundary of their respective distributions), and *S. nigropunctatus* (on the Great Bahama Bank).

Underwood's Dwarf Gecko   *Sphaerodactylus underwoodi*

Range and status: Endemic. Grand Turk and surrounding cays.

Description: 30–32 mm (1.2 in). Medium-sized, sexually dichromatic species. Males are dark brown, sometimes with a salt-and-pepper pattern, or yellow-orange with brown makings and a yellow-orange head. Females are dark brown with gray stippling with a pair of stripes extending from behind the eyes onto the neck and the top of the back. Co-exists with *S. mariguanae* (Turks Bank).

SNAKES

This widely distributed and diverse predatory group comprises more than twenty families and more than 3000 species. Snakes differ from lizards by the lack of eyelids and external ears, and most species shed their skin in one piece. The majority are non-venomous and either swallow prey alive or kill by constriction. Venomous species use their poisonous bite primarily to catch and kill their prey, rather than for self-defense. Snake vision is less acute than that of lizards and at best they can detect and follow movements. Although some species detect the body heat of prey by special sensors at the front of the head, most snakes use smell to track their prey. The characteristic forked tongue continually moves in and out of the mouth, collecting chemicals from the air, ground, and water, which are then passed to the vomeronasal organ (or Jacobson's organ) in the mouth. This organ provides the snake with the ability to assess prey or predators; differences in the chemical composition between the left and right fork of the tongue provides a directional sense of smell. In addition, snakes use vibrations, detected through the air and on the ground, to sense their environment.

At least fifteen species of native snakes occur in the archipelago, representing five

Sphaerodactylus underwoodi, female   *Sphaerodactylus underwoodi*, male

Sphaerodactylus notatus

families: boas (Boidae), racers (Colubridae), wood snakes (Tropidophiidae), worm snakes (Typhlopidae), and blindsnakes (Leptotyphlopidae). In addition, several North American species have been introduced to the islands during the 20th century.

Snakes are feared throughout much of the archipelago and frequently are intentionally killed on sight. Only one species, however, the Brown Racer (see below), is potentially mildly venomous, but it is shy, non-aggressive, and poses no threat to humans unless handled. The bite has no long-term medical consequences. Snakes are very important in controlling native and non-native pest species.

FAMILY COLUBRIDAE

Large and diverse group comprising more than 60% of all snake species worldwide. One endemic and one established introduced species are recorded in The Bahamas. In addition to the species described below, at least six other Colubrid snakes are sporadically recorded in the archipelago: the Cuban Racer (*Cubophis cantherigerus*, native; recorded only once on Cay Sal Banks) and at least five introduced species including the Common Garter Snake (*Thamnophis sirtalis*), Ribbon Snake (*T. sauritus*), Northern Brown Snake (*Storeria dekayi*), Eastern Rat Snake (*Pantherophis alleghaniensis*), and Rough Green Snake (*Opheodrys aestivus*).

Brown Racer   *Cubophis vudii*

Range and status: Endemic. Five subspecies. *C. v. vudii* (Great Bahama Bank, except Bimini Islands, Cat Island, Little San Salvador); *C. v. aterrimus* (Little Bahama Bank); *C. v. picticeps* (Bimini Islands); *C. v. raineyi* (Crooked-Acklins Bank); *C. v. utowanae* (Great Inagua). Not recorded on TCI.

Description: SVL: 765 mm (30.2 in). Commonly seen, medium-sized, shy, non-aggressive, diurnal slender snake recorded throughout The Bahamas except in San Salvador and Mayaguana. Coloration and patterning can be highly variable both between and within subspecies and can range from red-brown to darker brown or even black and may be pattern-less or have faint lateral stripes, irregular banding or even a combination of flecking and banding. **Habits and habitat:** Found in many natural habitats and human settlements. Frequently observed basking in the sun. If threatened, individuals give a defensive display of extending their throats like a hood (shown), similar to a cobra. Fast-moving and actively pursue prey on the ground, in trees, and even in water, but they also ambush prey. Prey includes lizards, frogs, and other snakes. **Important:** Potentially slightly venomous, but since individuals are small, rear-fanged/toothed, and possess only a small amount of mild toxin, unless handled they pose no significant threat and will avoid human contact whenever possible. The bite has no long-term medical consequences.

Corn Snake   *Pantherophis guttatus*

Range and status: Introduced. Grand Bahama, New Providence, Abaco, Grand Turk. Native to southeast and central U.S.

Description: SVL: 1200–1800 mm (47–71 in). Distinctively colored constrictor species with black-margined, reddish blotches on a background of gray, yellow, orange, and/or tan. Belly is boldly checkered in black on white. Juveniles are similar to adults but with duller colors. **Habits and habitat:** Frequently seen on the ground but also an excellent climber. Usually moves slowly. Usually freezes when threatened. Not venomous and typically docile.

FAMILY BOIDAE (Boas)

Boas (boa constrictors) are heavy-bodied, typically arboreal snakes that ambush their prey and kill by constriction. Originally it was assumed that after the prey had been grasped and restrained, sufficient pressure was applied to prevent it from inhaling after each exhale and then prey eventually succumbed due to asphyxiation. Recent research now indicates that after the prey has been captured, the applied pressure disrupts the circulation system and results in the cessation of blood flow and the supply of oxygen to the vital organs.

The archipelago has five species of West Indian boas (WIBs), which are frequently referred to locally as fowl snakes (caused by observations of them preying upon domestic chickens). These distinctive non-venomous snakes are typically mottled gray-brown with striking regular patterns in dark gray or black along their body. They are the largest snakes found in the archipelago (up to 2.5 m [c. 8 ft]); females are normally larger than males. WIBs are very adaptable and occur in natural and

Cubophis vudii vudii

Cubophis vudii vudii

Cubophis vudii vudii, hood showing

Pantherophis guttatus

urban habitats. Prey is dependent on the size of the snake: young feed mainly on small insects and lizards, whereas larger snakes consume bigger prey including rodents (rats and mice), frogs, large lizards (even young iguanas), small birds, and even other snakes. Mating takes place in early spring, and live young are born in early fall. There is no parental care. WIBs are needlessly feared throughout the archipelago and are frequently unnecessarily killed.

Northern Bahamas or Abaco Island Boa   *Chilabothrus exsul*

Range and status: Endemic. Little Bahamas Bank; Grand Bahama; Great and Little Abaco.

Description: SVL: 810 mm (32 in). Long, slender, gray snake with an unpatterned head and a series of dark markings along the body and tail, frequently with an iridescent reddish sheen. **Habits and habitat:** Found in coppice, pine, and disturbed habitats.

Southern Bahamas or Turk's Island Boa   *Chilabothrus chrysogaster*

Range and status: Endemic. Two subspecies: *C. c. relicquus* (Great Inagua, Sheep Cay); *C. c. chrysogaster* (TCI, including Turks Island, Grand Turk).

Description: SVL: 800–1780 mm (32–70 in). Three color morphs of this large, nocturnal, gray snake: spotted, striped, and non-patterned (not shown). The spotted morph has a series of irregular dark markings on a light gray background and is the most commonly seen morph. The striped morph has two dark dorsolateral lines running from the neck to the tip of the tail, while the very uncommon pattern-less snakes are mostly solid gray with a few indistinct dark markings. All morphs have a gray head and a dark line from behind the eye and along the cheek. Juveniles are orange-red, long, and thin. **Habits and habitat:** Found in a range of coppice with many large, flat rocks.

Bahamian Striped Boa   *Chilabothrus strigilatus*

Range and status: Endemic. North and central archipelago. Five subspecies: *C. s. ailurus* (Alligator Cay, Cat Island); *C. s. fosteri* (Bimini Islands); *C. s. fowleri* (Andros and Berry Islands); *C. s. mccraniei* (Ragged Island); *C. s. strigilatus* (New Providence, Eleuthera, Long Island, the Exuma Cays, Andros).

Description: SVL: 2330 mm (91.7 in). Large, typically gray-brown snake with distinctive dark-colored barring and mottling but with considerable variation in the extent and intensity of the patterning both within and between subspecies. For example, *C. s. ailurus* on Cat Island (similar to *C. s. strigilatus* on Eleuthera) has somber tones of chestnut and dusky brown while the Bimini subspecies, *C. s. fosteri*, has a ground color of shimmering black, decorated by a series of irregular gray blotches along the length of the back and a contrasting creamy white belly. **Habits and habitat:** Found in many natural habitats including coppice and mangroves as well as in human settlements, where it may enter buildings. Typically nocturnal and frequently arboreal, both hunting and sleeping or basking in trees. Diet includes other reptiles, birds, and small introduced mammals.

Chilabothrus chrysogaster chrysogaster, striped morph

Chilabothrus chrysogaster chrysogaster, spotted morph

Chilabothrus exsul

Chilabothrus strigilatus strigilatus

Chilabothrus strigilatus strigilatus with White-crowned Pigeon (*Patagioenas leucocephala*)

Conception Bank Silver Boa   *Chilabothrus argentum*

Range and status: Endemic. Conception Island Bank.

Description: SVL: 927–1029 mm (36.0–40.5 in). Medium-sized, pale-colored arboreal species exhibiting significantly reduced coloration as compared with other boas in the archipelago. Dorsal ground color is silver gray to a very light tan, occasionally with a very faint gray dorsal stripe extending the length of the spine with jagged edges and occasional interruption. A few scales are darker gray to brown, scattered across the dorsum either singly or in clusters of three to eight. Underside is pure cream-white with no markings or other coloration. Discovered in 2016 and little more is known.

Crooked-Acklins Boa   *Chilabothrus schwartzi*

Range and status: Endemic. Crooked and Acklins Bank. Rare.

Description: SVL: c. 785 mm (c. 31 in). Juveniles (< 150 mm [6 in]) have an orangish-red dorsal coloration with darker blotches across the body (shown), whereas adults have less distinct or completely lack the saddle-like markings and appear to be primarily gray dorsally, with a rufous wash on the head, and a cream-colored venter. **Habits and habitat:** Juveniles are frequently arboreal and feed on Anoles. Adult behavior is poorly known. Elevated to species status in 2017 (formerly considered a subspecies of *C. chrysogaster* [the Southern Bahamian Boa]; see above) but little is known.

FAMILY TROPIDOPHIIDAE (Wood Snakes)

Small, nocturnal, terrestrial (ground-dwelling) snakes (300–408 mm [11.8–23.6 in]). They are significantly smaller and daintier than the large, more robust Bahamian boas (see above); many individuals can easily fit in the palm of your hand. Most can change color (darker when inactive and paler at night). The tip of the tail is yellow-orange in juveniles and darker-colored (often black) in adults; at first glance, juveniles resemble a small rattlesnake. Diet is mostly sleeping lizards but also frogs and small birds. They remain underground or under vegetation during the day and are rarely seen, only emerging at night or when it rains. When threatened, they coil up into a tight ball. The local name on Andros Island (The Bahamas) is "shame snake" and may refer to this behavior. A more extreme defensive behavior is the ability to voluntarily bleed from the eyes, mouth, and nostrils (autohemorrhage). All give birth to live young. Three species are recognized. Color and size varies among subspecies.

Northern Bahamas Trope or Pygmy Boa   *Tropidophis curtus*

Range and status: Endemic. Great Bahama, Cay Sal Banks. Three subspecies. *T. c. androsi* (Andros); *T. c. barbouri* (Eleuthera, Cat Island, Exumas, Long Island, Ragged Island); *T. c. curtus* (Bimini Islands, Cay Sal Bank, New Providence).

Description: TL: 408 mm (15 in). Small, dull-colored snake with a dorsal color consisting of various shades of brown with darker brown blotches and stripes. Ventral side is spotted. Locally known as Thunder Snake or Shame Snake.

Chilabothrus argentum

Chilabothrus schwartzi, juvenile

Tropidophis curtus barbouri

Tropidophis canus

Inaguan Trope   *Tropidophis canus*

Range and status: Endemic. Great Inagua.

Description: TL: 408 mm (15 in). Similar to Pygmy Boa (*T. curtus*; see above) with a pale grayish, yellow-brown background color with dark-colored spots and stripes along its back, and speckled underside.

Caicos Pygmy Boa   *Tropidophis greenwayi*

Range and status: Endemic. Caicos Islands. Two subspecies. *T. g. lanthanus* (South Caicos, Middleton Cay, Long Cay, North Caicos, Middle Caicos, Providenciales); *T. g. greenwayi* (Big Ambergris Cay). Not recorded on the Turks islands.

Description: SVL: 301–313 mm (11.9–12.3 in). Terrestrial, dull-colored snake with a variable color and pattern but usually with a series of dark brown blotches on a light brown background and a uniform dark brown on the back of the head, becoming slightly paler on the neck and body. **Habits and habitat:** Found mainly under limestone rocks in areas of rocky coppice or dense scrub.

SUPERFAMILY TYPHLOPOIDAE (Blind or Thread Snakes)

Thread snakes are nocturnal, burrowing, non-venomous snakes with narrow, long bodies (hence their common name). They have distinctive glossy or shiny scales, a rounded indistinct head with vestigial (near sightless) eyes, and resemble shiny, unsegmented earthworms. A specialized (rostral) scale at the very tip of the snout on the upper jaw overhangs the mouth to form a shovel-like burrowing structure, and the tail ends with a horn-like scale, which gives it a pointed appearance. Rarely seen, spending most of their time underground, they usually emerge only when it rains or when the water table rises. They feed on ants and termites and are often found near their colonies. Most species are oviparous. Six species are recorded in the archipelago.

In addition to the blindsnakes described below, there is the endemic Inaguan Blindsnake (*Typhlops paradoxus*, TL: 363 mm [14.3 in]). This is a poorly known, long, slender species with a brown dorsum and yellow-white venter and was formerly considered a subspecies of Bahamian Slender Blindsnake (*T. biminiensis*).

San Salvador Blindsnake   *Epictia columbi*
Family: Leptotyphlopidae (Slender Blindsnakes or Thread Snakes)
Range and status: Endemic. San Salvador.

Description: SVL: 125–175 mm (5–7 in). Small, very thin, brown snake with a blunt head and shiny scales, similar in appearance to an earthworm. **Habits and habitat:** Rarely seen above ground except after heavy rain. No other snakes are recorded on San Salvador.

Bahamian Slender Blindsnake   *Typhlops biminiensis*
Family: Typhlopidae (Long-tailed Blindsnakes)
Range and status: Endemic. Great Bahama Bank (Bimini Islands, Andros, Ragged Islands, New Providence); Cay Sal Bank (Elbow Cay).

Description: TL: 370 mm (14.6 in). Long, slender species with two color morphs. Dorsum (upperside) can be brown or pink; pink variant is only described from the Bimini Islands. Brown variant can appear pale or transparent in the region of the head. Venter (underside) is yellowish-white. Rostral scale is rounded and does not extend as far as the (black) eyes. **Similar species:** The brown variant is difficult to distinguish from *T. lumbricalis* (see below). Both species co-exist through much

Tropidophis greenwayi lanthanus eating Silver Key Anole (*Anolis scriptus*)

Tropidophis greenwayi lanthanus, light color

Tropidophis greenwayi lanthanus, dark color

Epictia columbi

Typhlops biminiensis, pink morph

of their respective ranges, but *T. biminiensis* is typically longer and thicker and has a more rounded rostral scale. The two species are found in different habitats: *T. lumbricalis* is associated more with moist soils with surface rocks in areas of dense vegetation, whereas *T. biminiensis* occurs in dry, rock-free, sandy areas.

Bahamian Brown Blindsnake   *Typhlops lumbricalis*
Family: Typhlopidae (Long-tailed Blindsnakes)

Range and status: Endemic. Little and Great Bahama Banks including Abaco, Grand Bahama, Andros, New Providence, Great Exuma, Little Exuma, Eleuthera, Cat Island, Long Island, Bimini Islands, Berry Islands, Ragged Islands.

Description: TL: 257 mm (10.1 in). Medium-sized species with a brown dorsum and pale-colored venter (white or cream). Difficult to distinguish from brown variant of *T. biminiensis* but is typically smaller with a less-rounded rostral scale and found in different habitat (see above).

Caicos Blindsnake   *Typhlops platycephalus*
Family: Typhlopidae (Long-tailed Blindsnakes)

Range and status: Native. Caicos Bank. Also Puerto Rico.

Description: TL: c. 172 mm (c. 6.8 in). Medium-sized species with a variable pinkish to dark reddish-brown dorsum and a pale whitish venter. Some individuals have a pale tip to the tail, the same color as the venter, but this is variable among individuals and many specimens on the Caicos Bank appear to lack this characteristic. Also known as Puerto Rican White-tailed Blindsnake and Flat-headed Blindsnake.

Brahminy Blindsnake   *Ramphotyphlops braminus*
Family: Typhlopidae (Long-tailed Blindsnakes)

Range and status: Introduced. New Providence, Grand Turk, Providenciales. Native to Africa and Asia; introduced worldwide.

Description: TL: 65–165 mm (2.5–6.5 in). Small, thin species. Coloration varies from shiny silver gray, charcoal gray, to blackish-purple. Eyes are not completely covered. **Habits and habitat:** Found underground beneath logs, moist leaves, humus in wet coppice, and gardens. Can reproduce both parthenogenetically (asexual reproduction resulting in offspring that are genetic clones of the mother) and sexually. Sometimes referred to as "flowerpot snake" because of its accidental introduction by way of potted plants.

Typhlops lumbricalis

Ramphotyphlops braminus

Typhlops platycephalus

Birds

More than 360 bird species have been found in The Bahamas, including 109 breeding species, and more than 220 species are recorded from the TCI, including 58 breeding species. The resident avifauna of the archipelago has many species typical of the Caribbean region. Many North American species winter or migrate through the archipelago; about 50% of the bird species found in the northern Bahamas during winter are species that breed in North America. Species are classified as:

1) *Endemic*—species found only within the archipelago;
2) *Permanent residents*—species found year-round in the archipelago (occasionally includes non-breeding species);
3) *Summer resident*—species found in the archipelago April to October (breed);
4) *Winter resident*—species seen in the archipelago from September to April (non-breeding); and
5) *Transients*—regularly seen fall or spring migrant species but are not permanent summer or winter residents.

Parts of the archipelago are Important Bird Areas (IBAs), internationally recognized sites important for conservation of permanent summer and winter resident species as well as transient species. Approximately thirty IBAs are recognized in The Bahamas and ten in the TCI. The archipelago is also noted as an Endemic Bird Area (EBA), areas with significant numbers of endemic or near-endemic species. Six endemic species, twenty-four endemic subspecies, seven range-restricted species (i.e., with a geographically restricted or small area of distribution), and ten globally threatened species occur in all or parts of the archipelago. Post-Columbus avian extirpations in the archipelago include Brace's Hummingbird (*Chlorostilbon bracei*, recorded only from The Bahamas) and the Giant Kingbird (*Tyrannus cubensis*, extirpated in The Bahamas but persists on Cuba).

All wild birds in The Bahamas, except designated game birds (and then only for specified dates and under license), are protected by law. All birds are protected in wild bird reserves or national parks.

We use the following terminology. Underparts refers to the underside of the bird's body and includes the throat, chest, abdomen, flanks, and undertail coverts. Upperparts include the crown, nape, back, wings, and rump (see illustration). We describe the majority of permanent and summer resident species and many of the commonly seen winter resident species. Conservation status is given for endangered and threatened species. Names and sequence of species follow the most recent American Ornithological Society Check-list of North American Birds (Chesser et al. 2018).

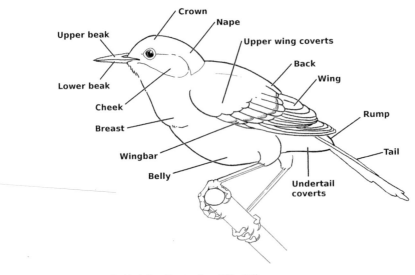

Anatomy of a bird, line illustration. Mike Hill.

Family Anatidae (Ducks)

Ducks, including geese and swans, have webbed feet and waterproof feathers, both of which are adaptations for their aquatic life. In most species, males are more brightly colored than are females. Most species are territorial during the breeding season, but in winter they are often seen in flocks. Diet includes plants, invertebrates, and small fish by diving, dabbling (up-ending), and grazing on vegetation. Waterfowl molt all their flight feathers at once and cannot fly for a period of c. four weeks in midsummer. Males often resemble females during this time (termed "eclipse plumage"). After the flight feathers have regrown, males molt again and regain their breeding plumage. Of the twenty-six species recorded in the archipelago, only three are permanent residents; the rest are winter resident or transient species. All ducks can be legally hunted *except* White-cheeked Pintail (*Anas bahamensis*), West Indian Whistling Duck (*Dendrocygna arborea*), and the Ruddy Duck (*Oxyura jamaicensis*), which are protected.

West Indian Whistling Duck   *Dendrocygna arborea*

Range and status: Permanent resident. Common on Long Island, Andros, Great Inagua; rarer on other islands; also the Greater Antilles.

Description: 56 cm (22 in). Large, brown, grazing duck with a distinctive upright (goose-like) posture; a long neck; and long, thick, gray legs and feet. Males and females are similar in appearance: brown collar with speckled black and white belly and gray upper wing coverts. **Habits and habitat:** Mostly nocturnal, roosting during the day, frequently in high dense vegetation close to water. Most often seen flying at dusk or dawn, sometimes in flocks, between feeding and roosting areas. Feeds on fruit, seeds, and shoots. **Vocalization:** Short, high-pitched squeaks

Spatula discors, female and male

Dendrocygna arborea

and harsh whistles; often given at night during flight. **Conservation status:** Hunting, predation by alien species, and the loss of feeding and roosting habitats have resulted in a marked decline in numbers. It is illegal to hunt this species. Listed as Vulnerable.

Blue-winged Teal   *Spatula discors*

Range and status: Common winter resident throughout the archipelago; introduced permanent resident on New Providence. Winters throughout Caribbean and from southern U.S. to South America. Breeds in North America.

Description: 39 cm (15.5 in). Small-sized, sexually dimorphic, dabbling duck. Males have a gray-blue head with a conspicuous white vertical band between the bill and the eye, a light brown body with a white patch near the rear, and a black tail. By contrast, the female is mottled brown and has a whitish area at base of the bill. Both sexes have a distinctive blue band on the wing, best seen in flight, for which this species is named. **Habits and habitat:** Diet is primarily plants but sometimes small insects and crustaceans taken from salt and freshwater marshes.

Northern Shoveler   *Spatula clypeata*

Range and status: Uncommon winter resident, rare in the southern archipelago. In the Caribbean region, winters primarily in the Greater Antilles; also winters from southern U.S. to Colombia. Breeds in western North America and Eurasia.

Description: 48 cm (19 in). Unmistakable, medium-sized, sexually dimorphic, dabbling duck with a large spatulate bill. Adult males have an iridescent dark green head, white breast, and chestnut belly and flanks; female is a drab mottled brown. **Habits and habitat:** Feeds by moving its bill from side to side, sifting invertebrates and particulate matter from the water and fine mud.

Mallard   *Anas platyrhynchos*

Range and status: Rare in northern Bahamas; unclear if individuals are winter residents or introduced. Breeds in the Northern Hemisphere. Rare in Cuba, vagrant elsewhere in Caribbean region. Migratory in northern parts of its breeding range; North America birds winter as far south as Central America and the Caribbean.

Description: 50–65 cm (20–26 in). Sexually dimorphic, dabbling duck. Males have a bottle-green head, a yellowish orange-red bill tipped with black, gray-brown wings, a pale gray belly, and a black tail with white borders. A white collar demarcates the head from the purple-tinged brown breast. Female is a mottled light brown with a black and orange bill. Both sexes have a purple patch on the wing bordered by white. **Habits and habitat:** Diet includes invertebrates, small vertebrates, and plants.

White-cheeked Pintail or Bahama Pintail   *Anas bahamensis*

Range and status: Permanent resident. Endemic subspecies *A. b. bahamensis* throughout the archipelago; species also in Greater and Lesser Antilles.

Description: 43 cm (17 in). Reddish-brown dabbling duck with a distinctive white cheek and throat, which contrasts with the mottled brown crown, nape, and neck. Upperparts are brown with black feathers bordered with pale buff on back; underparts are warm brown, spotted with black on breast and belly, with a pointed white tail. Dark bill has a pinkish-red base. **Habits and habitat:** Quiet and shy. Occurs on many water bodies throughout the archipelago. Diet is aquatic plants and invertebrates.

Northern Pintail   *Anas acuta*

Range and status: Rare winter resident and transient. Breeds in northern North America and Eurasia.

Description: Male 59–76 cm (23–30 in); female 51–64 cm (20–25 in). Sexually dimorphic, dabbling duck with long neck and wings, and a relatively small head. Males have a chocolate-colored head, white throat and belly with a white stripe extending up the side of the neck, gray flanks and back, brown and black patterning on the back and sides, and two long central black tail feathers. Females have a brown head and mottled brown body. **Habits and habitat:** Feeds on plants and occasionally invertebrates.

Spatula clypeata, female and male

Anas platyrhynchos, female

Anas platyrhynchos, male

Anas acuta, male

Anas acuta, female

Anas bahamensis

Ring-necked Duck   *Aythya collaris*

Range and status: Winter resident, most common in northern islands. Rare winter resident elsewhere. Also winters on Cuba, rare elsewhere in Caribbean. Breeds in northern North America. Winters primarily in the southern U.S.

Description: 43 cm (17 in). Small-sized, sexually dimorphic, diving duck. Males have a shiny purple-black head, black back, black breast bordered posteriorly by a contrasting white band, gray-white flanks, and a distinctive yellow eye; females have a pale brown head and body with a dark brown back, and brown eyes. Both sexes have a gray bill with a white band bordering the black tip. The cinnamon neck ring is rarely visible. **Habits and habitat:** Feeds on aquatic plants, invertebrates (mollusks and insects), and small fish.

Lesser Scaup   *Aythya affinis*

Range and status: Winter resident; uncommon to common in northern islands, less common in the southern archipelago. Mostly winters in southern U.S., but also in Mexico and Greater Antilles (rare elsewhere in Caribbean), south through Central America. Breeds in northern North America.

Description: 42 cm (16.5 in). Small-sized, sexually dimorphic, diving duck. Adult males have a light-blue bill, black head and breast (with an iridescent sheen), a finely barred black and white back, white flanks and a black tail. Females are dark brown and have a white band at the base of the bill. Immature birds resemble the adult females but are duller and have hardly any white at the bill base. **Habits and habitat:** Diet includes mussels, clams, seeds, and aquatic plants.

Hooded Merganser   *Lophodytes cucullatus*

Range and status: Rare winter resident and transient; rare elsewhere in the Caribbean. Mostly winters in southern U.S. Breeds in northern North America.

Description: 40–49 cm (15.7–19.3 in). Medium-sized, sexually dimorphic, diving duck with a thin bill. Both sexes have a crest at the back of the head that can be expanded. Males have a black head and back, a large white patch on the crest, and reddish-brown flanks; females have a shaggy, reddish crest and a gray-brown head and body. First-winter birds have a gray-brown neck and upperparts. All females are dark-eyed whereas in first-winter males, a pale eye develops during the winter. **Habits and habitat:** Diet includes small fish, crustaceans, and aquatic insects.

Red-breasted Merganser   *Mergus serrator*

Range and status: Rare winter resident, perhaps transient, in The Bahamas and elsewhere in Caribbean. Not recorded in the TCI. Mostly winters along coastal U.S. and Canada. Breeds in northern North America and Eurasia.

Description: 51–62 cm (20–24 in). Streamlined, sexually dimorphic, diving duck with a long neck, "shaggy" crest, and a long thin red bill with serrated edges. Males have a dark head with a green sheen, a white neck with a rusty breast, a black back, gray flank and sides, and white underparts. Females have a rusty head and a grayish body. Juveniles resemble females but lack the white collar and have a smaller white wing patch. **Habits and habitat:** Diet includes aquatic invertebrates and small vertebrates. Usually in nearshore waters of Atlantic Ocean.

Aythya collaris, male

Aythya collaris, female

Aythya affinis, male

Aythya affinis, female

Mergus serrator, female and male

Lophodytes cucullatus, male and female

Ruddy Duck *Oxyura jamaicensis*

Range and status: Permanent resident. West Indian subspecies (*O. j. jamaicensis*) is common on New Providence, but rare on Abaco, Andros, Eleuthera, Great Exuma; the Caicos Islands; also local permanent resident in Greater Antilles. Influx of migrant birds to the archipelago during the winter (rare winter resident). Species breeds mostly in western North America and winters in southern North America.

Description: 38 cm (15 in). Small-sized, sexually dimorphic, diving duck with distinctive rigid tail feathers, which are erected when the bird is at rest. Males have a white face, blue bill, black crown and neck, and chestnut body; females are overall gray-brown with a grayish face with a darker bill, cap, and cheek stripe. **Habits and habitat:** Diet includes plants and invertebrates.

Family Odontophoridae (New World Quails)

New World Quails are typically chicken-sized, ground-dwelling, brown birds with contrasting patterns in black, white, buff, or rufous. Males are typically brighter than females. Frequently reluctant to fly and will first run or hide from threats, flying only as a last resort. One introduced species occurs in the archipelago.

Northern Bobwhite *Colinus virginianus*

Range and status: Introduced, locally common, permanent resident on Abaco, Andros, New Providence. Not recorded on other islands.

Description: 25 cm (10 in). Small-sized, shy, elusive, chicken-like, ground-dwelling species, native to North America, and probably introduced to The Bahamas in the 19th century. Males have a white throat and supercilium bordered by black. Overall rufous plumage has gray mottling on the wings and a gray tail. Flanks show white scalloped stripes, while whitish underparts have black scallops. Females are similar but are duller overall and have a buff throat and brow without the black border. Both sexes have pale legs and feet. **Habits and habitat:** When threatened, will crouch and freeze, relying on camouflage to stay undetected, but will flush into low flight if closely disturbed. Associated with grassland and agricultural areas. Can be legally hunted. **Vocalization:** Common name derives from its characteristic whistling call, a clear whistle that sounds like *bob-WHITE* or *bob-bob-WHITE*.

Family Phasianidae (Pheasants and Grouse)

Family of large-sized, plump, ground-dwelling species with broad, short wings and strong legs (often with spurs). Males are larger and often more colorful with more ornate plumage and facial adornments than the females. Several species are important game birds and have been widely introduced for hunting; the red junglefowl (*Gallus gallus*, of southeast Asia) is the wild ancestor of the domesticated chicken, the most important bird in agriculture. One introduced species occurs in the archipelago.

Oxyura jamaicensis jamaicensis, male

Oxyura jamaicensis jamaicensis, female

Colinus virginianus, female and male

Phasianus colchicus, male

Ring-necked Pheasant   *Phasianus colchicus*

Range and status: Introduced on Eleuthera (Hatchet Bay), probably during the 19th century; and on North Andros during the late 1990s. Not recorded on other islands.

Description: Male 60–89 cm (24–35 in), tail c. 50 cm (20 in); female 50–63 cm (20–25 in), tail c. 20 cm (8 in). Large-sized, sexually dimorphic, chicken-like bird with a characteristic long tail. Males are distinctively colored with a bright gold and brown body with green, purple, and white markings. Head is bottle green with a small crest and distinctive red wattle and an incomplete white neck ring. Females (and juveniles) have mottled brown plumage and a shorter tail. **Habits and habitat:** Feeds on insects, seeds, grains, and fruits. **Conservation status:** Native to Asia but widely introduced elsewhere and is one of the world's most important game birds. Can be legally hunted.

Family Phoenicopteridae (Flamingos)

Flamingos are large-sized, long-legged, ungainly, gregarious wading birds with long necks and a uniquely shaped hooked bill. While feeding, flamingos hold their heads so they look backward between their legs. The bill acts like a pump, taking in mud and water and filtering out food items, such as small shrimp, seeds, blue-green algae, microscopic organisms, and mollusks. Their distinctive red or pink plumage is derived from a carotenoid pigment (canthaxanthin) in their diet. They nest in large colonies. The distinct mud mound nests are constructed in shallow, saline water bodies. One species occurs in the archipelago.

American Flamingo *Phoenicopterus ruber*

Range and status: Common, permanent resident on Great Inagua, Crooked and Acklins islands, Mayaguana, the Caicos islands. Small population on Andros. Rare elsewhere. Also the Caribbean, Central and northern South America.

Description: 117 cm (46 in). The national bird of the Bahamas is unmistakable, being uniformly pink with a long neck and long pink legs with webbed feet. Wing coverts are red, and primary and secondary flight feathers are black. **Vocalization:** Goose-like honking. **Conservation status:** By the 1950s, hunting for the pink feathers, as well as for meat, had devastated the Bahamas population and only about 5000 remained on Great Inagua. Only sixty years later, through conservation efforts by the National Audubon Society and The Bahamas National Trust involving the creation of the Inagua National Park and a ban on hunting, the flamingo was brought back from the brink of extirpation in The Bahamas. Lake Windsor (Rosa) in the Inaguan National Park is now home to c. 50,000 flamingos—the world's largest breeding colony. The flamingo is now listed as Least Concern because it continues to increase and expand its range.

Family Podicipedidae (Grebes)

Grebes are widely distributed, freshwater diving birds with short wings and tails. At first glance, they appear to lack tails and have a typically low profile in the water. Characteristically their legs are located far back on the body, making walking on land difficult. Their feet have lobed or rounded projections on their toes. They dive underwater to avoid danger and feed primarily on small fish, crustacean, amphibians (frogs), and other aquatic invertebrates. Some species winter in saline water bodies.

Least Grebe *Tachybaptus dominicus*

Range and status: Permanent resident. Greater Antillean subspecies *T. d. dominicus* is common in the northern islands, but less common in the southern archipelago. Species breeds from the southwestern U.S. and the Caribbean, south through Central and South America.

Description: 23 cm (9 in). Small-sized, dark-colored grebe with brown-gray upperparts, a brown crest and breast, pale brown underparts, a yellow eye, and a thin dark bill. **Habits and habitat:** Preferred habitat is freshwater with extensive vegetative cover. Diet is small fish and aquatic invertebrates.

Pied-billed Grebe *Podilymbus podiceps*

Range and status: Permanent resident and uncommon winter resident; less common in southern archipelago. Found throughout most of Western Hemisphere.

Phoenicopterus ruber, flock

Tachybaptus dominicus dominicus

Podilymbus podiceps

Phoenicopterus ruber

Description: 33 cm (13 in). Small-sized brown grebe with a dark eye and pale eye-ring. Breeding adults have a black throat and forehead, and a distinctive whitish bill with a black ring; the latter distinguishes it from other grebes in the region. Non-breeding birds have a dull yellow bill and lack the distinctive black facial characteristics and the ring on the bill. **Habits and habitat:** Found on both fresh and saltwater.

Family Columbidae (Pigeons and Doves)

Pigeons and doves are stout-bodied birds with short necks and short slender bills with a fleshy waxy structure at their base called a "cere." They bob their head while walking. Sexes are similar in appearance. Nests are simple stick structures. Unlike most birds, a special liquid "crop milk" is produced by both parents to feed their young. The adult diet comprises seeds and fruit. Songs typically consist of a series of low cooing notes. Eleven species are recorded in the archipelago including nine breeding species. Some resident species can be hunted outside the breeding season under license.

Rock Pigeon   *Columba livia*

Range and status: Introduced, common, permanent resident in residential areas of Grand Bahama, Abaco, New Providence, Eleuthera. Occurs as far south as Crooked Island but are rare or unrecorded on the southern islands. **Description:** 32 cm (12.5 in). Wild (type) Rock Pigeons are pale gray with a dark head and neck, black tips to wing and black tail. Two black bars on each gray wing and a white lower back. Domestic and feral pigeons *C. l. domestica*, which are often seen, are more variable in color and pattern. **Habits and habitat:** Rarely seen away from urban areas or cliffs.

White-crowned Pigeon   *Patagioenas leucocephala*

Range and status: Common, permanent resident in the north and central islands, less common in the southern archipelago. Also common on the Greater Antilles, locally found on Florida Keys, on the Lesser Antilles, islands off Mexico and Belize. **Description:** 34 cm (13.5 in). Unmistakable dark gray pigeon with a brilliant white forehead and crown, green and white bars on the nape, a white iris, and a pale-tipped red bill. Juveniles are a lighter shade of gray, lack the nape pattern and white iris, and show only a few pale feathers on the crown. **Habits and habitat:** Premier game bird in The Bahamas. Shy and difficult to approach. Requires two distinct habitats: breeds in coastal Red Mangroves but forages on berries and fruits, especially Poison Wood, in mature coppice. **Vocalization:** Call is a deep *coo, coo-aa-coo* or a *whoo-ca-coo*.

Eurasian Collared Dove   *Streptopelia decaocto*

Range and status: Introduced, permanent resident; common in residential areas in the north and central islands. Rare or unrecorded in southern islands. Native to Eurasia. Introduced to The Bahamas in 1970s, subsequently spreading to North America. **Description:** 33 cm (13 in). Slim pale-gray dove with a long square tail and a distinctive, incomplete black collar with a white border on the back of the neck. Short legs are red and the bill is black. Dark eye is surrounded by a small area of bare skin, which is either white or yellow. Juveniles have a poorly developed collar. **Habits and habitat:** Rarely seen away from urban or residential areas and usually seen in pairs. Can be legally hunted in The Bahamas. **Vocalization:** Call is a repeated *coo-COO-coo*.

Columba livia domestica

Streptopelia decaocto

Columbina passerina bahamensis

Patagioenas leucocephala

Common Ground Dove   *Columbina passerina*

Range and status: Common, permanent resident. Resident subspecies *C. p. bahamensis* is found throughout the archipelago except Great Inagua; *C. p. exigua* found on Great Inagua and the TCI. Species also widely distributed throughout subtropical and tropical Americas and the Caribbean.

Description: 16.5 cm (6.5 in). Small-sized, gray-brown dove with scaly appearance on the head and neck, characteristic dark spots on the wings. When approached, often quite closely, they rapidly walk away before flying, revealing distinctive rufous underwings and black corners on their short tails. **Habits and habitat:** Common in open urban and suburban areas where they feed on vegetation, seeds, and fruits, almost exclusively on the ground. Locally known as Tobacco Dove. **Vocalization:** Call is a series of repeated soft whoops that increase in pitch.

Key West Quail-dove   *Geotrygon chrysia*

Range and status: Permanent resident; uncommon in northern islands, rare in central islands, absent from the southern islands except on North Caicos and Pine Cay (TCI); also Greater Antilles.

Description: 30 cm (12 in). Large-sized, stocky, ground-dwelling dove, with rust-colored back and wings (upperparts), gray underparts, an iridescent green crown and nape, and a distinctive white stripe below the eye. **Habits and habitat:** Dry coppice and scrub habitats. Forages on the ground for seeds, berries, and fallen fruit, including Poison Wood. Shy and difficult to see. Frequently runs away rather than flying when disturbed. **Vocalization:** Call is a mournful monotone similar to someone blowing across the top of a bottle.

Caribbean Dove   *Leptotila jamaicensis*

Range and status: Introduced. Permanent resident, locally found on New Providence. Not found on other islands. Native to Jamaica, the Cayman Islands, San Andrés, Mexico.

Description: 30 cm (12 in). Stocky, ground-dwelling dove with a gray crown, white face, white-gray underparts, brown upperparts and tail, and a red-brown neck with traces of iridescence. **Habits and habitat:** Forages on the ground. Reluctant to fly when disturbed.

White-winged Dove   *Zenaida asiatica*

Range and status: Common, permanent resident in central and southern islands; rare in the northern archipelago; also southern U.S. south to South America.

Description: 30 cm (12 in). Brown-gray pigeon with a distinctive long white bar on the leading edge of the wing (appearing as a solid broad band in flight), which contrasts with the black wing tips and the gray-brown back. Adults have a patch of blue, featherless skin around each eye and a dark mark below the eye. **Habits and habitat:** Scrub, woodlands, urban, and cultivated areas. Feeds on seeds and fruits on the ground. **Vocalization:** Call is a hoarse *woo-koo woo woo.*

Zenaida Dove   *Zenaida aurita*

Range and status: Permanent resident; relatively common in the central and southern islands; less common in the northern islands. Found only in the Caribbean and the Yucatan, Mexico.

Description: 28 cm (11 in). Stocky, rich-brown dove, similar to Mourning Dove (MD, see below), but smaller and darker colored with a short, rounded tail and gray outer tail feathers. An iridescent violet patch occurs on the neck. Distinguished from the MD by a white spot on the wings that is visible at rest and in flight, and white outer tail feathers (most noticeable in flight). **Habits and habitat:** Open

Zenaida macroura

Geotrygon chrysia

Leptotila jamaicensis

Zenaida asiatica

Zenaida aurita

and semi-open habitats. Feeds primarily on seeds on the ground. Can be legally hunted in The Bahamas. **Vocalization:** Call is a mournful *cooOOoo-coo-coo* similar to that of a MD, but faster pace.

Mourning Dove   *Zenaida macroura*

Range and status: Common, permanent resident throughout the archipelago; also Greater Antilles, North and Central America.

Description: 30 cm (12 in). Medium-sized, slender, gray-brown dove with a long, tapered tail. Characteristic white outer tail feathers, which contrast with the black inner feathers, are most conspicuous in flight. Wings have black spotting. Below the eye is a distinctive crescent-shaped area of dark feathers. Adult male has bright, iridescent purple-pink patches on the sides of the neck, with light pink coloring reaching the breast and a distinctly bluish-gray crown. Females are similar, but with more brown coloring overall with reduced or absent iridescent feather patches. **Habits and habitat:** Open and semi-open habitats including urban areas, farms, grassland, and lightly wooded areas frequently close to human habitation. Can be legally hunted in The Bahamas. **Vocalization:** Call is a plaintive *cooOOoo-woo-woo-wooo*.

Family Cuculidae (Cuckoos)

Species in this family have slender bodies, long tails, and strong short legs but vary in size and bill shape. Unlike Old World cuckoos, North American cuckoos are not typically brood parasites; they build nests and raise their young. Five species are recorded in the archipelago, including three breeding species.

Mangrove Cuckoo   *Coccyzus minor*

Range and status: Permanent resident; rare in the northern islands; more common in the central and southern archipelago. Possible influx of migrants to the north and central islands in the fall. Also breeds in Florida and throughout the Caribbean.

Description: 30 cm (12 in). Adults have a distinctive black mask and gray crown, a long tail, gray-brown upperparts, white underparts with a buffy wash, and a black, curved bill with yellow on the lower mandible. **Habits and habitat:** Despite the common name, often seen in dense scrub and coppice habitats. Diet includes invertebrates, small vertebrates, and fruit. **Vocalization:** Most commonly heard call is a guttural *gawk gawk gawk gawk*. **Similar species:** Yellow-billed Cuckoo (*C. americanus*) (30 cm [12 in]) is a rare winter resident and transient. Resembles the Mangrove Cuckoo but lacks the black mask, has white underparts, a rufous-colored patch in the wing, and the lower mandible and base of upper mandible is yellow.

Great Lizard-cuckoo   *Coccyzus merlini*

Range and status: Permanent resident. Endemic *Coccyzus m. bahamensis* is a permanent resident on just three Bahamian islands: Andros, Eleuthera, and New Providence. The New Providence population may be extirpated. Species also in Cuba.

Description: 54 cm (21 in). Unmistakable large-sized cuckoo with an olive-brown back, wings, and crown; white throat and breast; and chestnut belly with a long tail, which is barred on the underside. Eye has a patch of bare red skin around it, and the dark bill is long. **Habits and habitat:** Agile on the ground and able to run rapidly through the undergrowth. Often surprisingly tame and curious. Diet includes lizards and invertebrates. Frequents broadleaved scrubs, coppice habitats, and occasionally pine forest. **Vocalization:** Call is a loud and distinctive *kek-kek-kek-kek*.

Coccyzus minor   *Crotophaga ani*   *Coccyzus merlini bahamensis*

Smooth-billed Ani   *Crotophaga ani*

Range and status: Common permanent resident throughout the archipelago. Breeds in south Florida, elsewhere in the Caribbean, Central and northern South America.

Description: 30–36 cm (12–14 in). Unmistakable all-black bird with a long, floppy tail and a large, high-ridged, conically shaped black bill. **Habits and habitat:** Commonly seen in open and semi-open country and cultivated areas, often in noisy groups. Weak and wobbly flight. Often feeds on the ground for invertebrates (termites and large insects) and small vertebrates (lizards and frogs). Has a complex cooperative breeding system with several females laying up to seven eggs each in one nest. The young of earlier broods help feed chicks in later nests. Locally known as Long-tail Crow or Cemetery Bird. **Vocalization:** Call is a whining *ooo-leeek*.

Family Caprimulgidae (Nighthawks and Nightjars)

These medium-sized, nocturnal birds have long, streamlined wings; short legs; small feet; large eyes; and short, broad, flattened bills. During the day they roost on trees or on the ground. Their plumage is camouflaged to resemble bark or leaves. Most active in the late afternoon, at dusk, and during the night, when they catch insects on the wing. Many species nest on the ground. Three species are recorded in archipelago, two of which breed.

Antillean Nighthawk   *Chordeiles gundlachii*

Range and status: Common summer resident throughout the archipelago; also Greater Antilles.

Description: 22 cm (8.5 in). Slender bird with a small head and bill, mottled dark brown upperparts, and lighter colored barred underparts. Sexes are similar in plumage, but males have a more distinctive white bar across the throat than females have. The long, pointed wings have a white band near the tip, which is very conspicuous in flight and also visible as a white square on the wing at rest. **Habits and habitat:** Roosts during the day on the ground. Feeds by catching flying insects on the wing, frequently at dusk or on cloudy days over open areas or above trees. They have a distinctive streamlined profile when flying, and their gliding flight is often punctuated by steep dives while chasing prey. Nests on open sandy ground including beaches. **Vocalization:** Males can be heard calling on the wing—a buzzy, insect-like *pitty-pit-pit* in spring and early summer. The nighthawk is known locally as Piramidik after its vocalization. **Similar species:** The Chuck-will's-widow (*Antrostomus carolinensis*) (30 cm [12 in]) breeds on Grand Bahamas and Abaco and is an uncommon winter resident in north and central Bahamas. It is also nocturnal, roosting during the day, and is similarly colored to the Antillean nighthawk but is larger and more heavy set, lacks the white wingbars, and has a distinctive whistled song *chup-whill-wheeeow*. The Common Nighthawk (*Chordeiles minor*) occasionally migrates through the archipelago. Similar to Antillean nighthawk and is most easily distinguished by its distinctive call, a nasal *peent*.

Family Trochilidae (Hummingbirds)

These small-sized, delicate birds are found only in the New World, and the family includes the smallest bird in the world. They feed primarily on nectar, using their thin bills to probe flowers, but they also consume insects and spiders. Males of most species sport iridescent plumage. Their aeronautical skills are unparalleled among birds; they can hover in midair to feed and are the only birds that can fly backward. Four species are recorded in the archipelago including three breeding species. The Ruby-throated Hummingbird, which is well-known in many areas, is not described here as it rarely occurs in the archipelago.

Bahama Woodstar   *Calliphlox evelynae*

Range and status: Endemic. Common permanent resident throughout archipelago except Great and Little Inagua.

Description: 9.5 cm (3.5 in). Males have metallic green upperparts with an iridescent purple-red throat, rufous belly and flanks, and a long, forked tail. A white band separates the throat from the dusky green chest. Females and immatures have metallic green upperparts, a white throat, rufous underparts, and the tail is more rounded compared to males. **Habits and habitat:** Common in coppice and gardens but less common and more locally distributed on islands with Cuban emerald. Breeds year-round, depending on food availability. **Vocalization:** Individuals are often detected by their high-pitched *chip* call notes.

Chordeiles gundlachii

Calliphlox lyrura, male

Calliphlox evelynae, male

Calliphlox evelynae, male in flight

Calliphlox evelynae, female

Calliphlox evelynae, female in flight

Inagua Woodstar   *Calliphlox lyrura*

Range and status: Endemic. Permanent resident. Great and Little Inagua.

Description: 9.5 cm (3.5 in). Very similar to Bahama Woodstar (*Calliphlox evelynae*) (see above), except the purple-red iridescence of males extends to the forehead and the tail is longer. Females are virtually identical to female woodstars (see above). **Vocalization:** Song characterized as *wet, squeaky shoes.*

Cuban Emerald *Chlorostilbon ricordii*

Range and status: Locally common, permanent resident on Grand Bahama, Abaco, Andros. Not found on other islands. Also Cuba.

Description: 9.5–10.5 cm (3.7–4.2 in). Sexually dimorphic species. Adult males have a uniformly iridescent, metallic green plumage with a white spot behind the eye and a deeply forked tail. Females have metallic green upperparts and a light gray throat and underparts. **Habits and habitat:** Common in pine and coppice habitats.

Family Rallidae (Rails, Gallinules, and Coots)

Rails are a large family of small- to medium-sized birds that tend to have short, rounded wings. They are weak fliers but have strong legs and long toes adapted to walking in densely vegetated wetlands. They are secretive birds and are much more often heard than seen. Their diet includes plant material (leaves, seeds, fruit), invertebrates (insects, snails, spiders), and small vertebrates (fish and frogs). Prey is taken while walking and pecking the substrate or while swimming (diving). Short-billed rails typically feed from the substrate surface, whereas species with longer bills can probe the substrate. Gallinules and coots are aquatic, often seen on ponds, and are more conspicuous than rails. Nine species are recorded in the archipelago. In addition to species described below, other transient rails recorded in the archipelago are the Yellow Rail (*Coturnicops noveboracensis*), the Virginia Rail (*Rallus limicola*), and the Black Rail (*Laterallus jamaicensis*).

Clapper Rail *Rallus longirostris*

Range and status: Permanent resident. Endemic subspecies *R. l. coryi* is common throughout most of the archipelago, less common in the TCI. Species is widely distributed along the east and west coasts of North America, the Caribbean, South America.

Description: 36 cm (14 in). Large (chicken-sized), gray-brown rail with a chestnut-colored breast, black barring on the flanks and belly, a distinctive white rump, and a long, thick, slightly down-curved bill. **Habits and habitat:** Typically frequents mangrove swamps but can occur in other wetlands. **Vocalization:** Difficult to see, but the distinctive *kek-kek-kek* call is frequently heard at dawn and dusk.

Sora *Porzana carolina*

Range and status: Uncommon winter resident and transient. Breeds in marshes throughout North America and overwinters in the southern U.S. and northern South America.

Description: 21 cm (8.5 in). Secretive, small-sized rail with a distinctive short, thick, yellow bill; blue-gray head with black patch at base of the bill and on the throat; blue-gray underparts; a brown back; and distinctive black and white barring along its flanks. **Vocalization:** A slow, whistled *ker-whee*, or a descending whinny.

Chlorostilbon ricordii, male

Chlorostilbon ricordii, female

Porzana carolina

Rallus longirostris coryi

Porphyrio martinicus

Purple Gallinule   *Porphyrio martinicus*

Range and status: Uncommon winter resident and transient in the northern and central islands; rare in the southern islands. Resident in southern Florida and tropical America. Also breeds in swamps and marshes in the southeast U.S. where they are migratory, wintering southward in the Americas.

Description: 33 cm (13 in). Distinctive purple-blue rail with a green back, red and yellow bill, and large yellow feet. Pale blue forehead shield and white undertail coverts. Juveniles are brown with a brownish-olive back.

Common Gallinule　*Gallinula galeata*

Range and status: Common permanent resident in the north and central islands. Less common in the southern archipelago. Breeds in eastern North America south to South America.

Description: 35 cm (14 in). Commonly seen rail with a slate-gray head and body, slate-gray wings and back, white flank stripe, dark brown tail, white under tail coverts, and large, yellow legs. Adults in breeding plumage have a distinctive red forehead shield and a red bill with a yellow tip. Juveniles are browner and lack the red shield. **Habits and habitat:** Forage beside or in water, by swimming and sometimes upending to feed.

American Coot　*Fulica americana*

Range and status: Common winter resident and uncommon permanent resident in the northern islands. Common to uncommon winter resident in the southern islands. North America south through Central America.

Description: 40 cm (15.5 in). Slate-gray rail with a black neck and head and a conspicuous short, white bill and white forehead shield with a red-brown spot at its top. Legs and lobed toes are pale olive. **Habits and habitat:** Territorial during the breeding season but often forms large flocks on ponds during the winter. **Vocalization:** High-pitched squeaking honk. The similarly sized Caribbean form of the American Coot is also recorded in The Bahamas, but it lacks the red-brown spots at the top of the frontal shield.

Family Aramidae (Limpkins)

Limpkins are ungainly heron or ibis-like birds. Only one species in this family.

Limpkin　*Aramus guarauna*

Range and status: Permanent resident on Abaco, New Providence, Andros, Eleuthera. Rare visitor to other islands. Also southern U.S., Central and South America.

Description: 66 cm (26 in). Large-sized, dark-brown, long-legged heron-like bird, with white streaking on the head, neck, back, and wings. Yellowish bill is long, heavy, and slightly down-curved, with a dark tip. **Habits and habitat:** Common name derives from its limp-like walk. Broad diet includes snails, insects, crustaceans, frogs, and lizards. Although widely distributed throughout wetlands in subtropical and tropical Americas, in The Bahamas it is also found in pine and coppice habitats. **Vocalization:** Easier to hear than see; produces a distinctive eerie, loud wail or scream *kwEEEeeer* or *klAAAar*, which is most often given during the night and at dawn and dusk, when most active.

Gallinula galeata

Fulica americana

Aramus guarauna

Himantopus mexicanus

Family Recurvirostridae (Stilts and Avocets)

Stilts have extremely long legs and long, thin, straight bills, whereas avocets have long legs and long up-curved bills. Two species are recorded in the archipelago, including one breeding species. The American Avocet (*Recurvirostra americana*) is a rare visitor to the archipelago.

Black-necked Stilt   *Himantopus mexicanus*

Range and status: Uncommon permanent or summer resident in the northern islands; common permanent resident in the central and southern archipelago.

Description: 36 cm (14 in). Distinctive slender, black-and-white shorebird with very long, pink legs and a long, thin, slightly up-curved black bill. White underparts contrast with the black cap, wings, and back. A small white spot above the eye. **Habits and habitat:** Diet includes a variety of small aquatic invertebrates and vertebrates by probing and gleaning primarily in mudflats and lakeshores. Breeding occurs from late April to August. **Vocalization:** A sharp *yip* or *kek* given repeatedly when alarmed or disturbed.

Family Haematopodidae (Oystercatchers)

Oystercatchers are large-sized, heavy-bodied, noisy waders, with strong, long bills used for smashing or prying open mollusks. One species is recorded in the archipelago.

American Oystercatcher   *Haematopus palliatus*

Range and status: Uncommon, locally distributed, permanent resident throughout the archipelago. Locally distributed in the Caribbean and from east coast of Canada south to South America.

Description: 46 cm (18 in). Large-sized shore bird, immediately identifiable by its conspicuous black-and-white body and long, heavy, thick, orange bill. **Habits and habitat:** Inhabits coastal habitats, typically rocky shorelines, where they feed and breed. **Vocalization:** Loud, whistled *wheep.*

Family Charadriidae (Plovers)

Plovers are small- to medium-sized birds with compact bodies; short, thick necks; and long, usually pointed wings. They are found in open country, mostly coastal habitats, and some in grasslands. Non-breeding plovers often form flocks that forage in marshland and coastal habitats. They have short bills and pick invertebrates from the surface using a run-and-pause technique. Nests are scrapes on the ground. Eight species are known from the archipelago, including three breeding species. Winter residents are mostly commonly seen in winter plumage (washed-out version of summer plumage). During spring, some individuals have partially molted into their breeding (summer) plumage before migrating.

Black-bellied or Gray Plover   *Pluvialis squatarola*

Range and status: Common winter resident throughout the archipelago. Breeds throughout the Arctic. Winters worldwide south of breeding range along coasts.

Description: 29 cm (11.5 in). Large-sized, gray plover typically seen in the dull winter plumage (September–April), similar to juvenile plumage, with mottled gray upperparts, a gray-speckled breast, and a white belly. Distinctive black "armpits" under the wings, only visible during flight, distinguish it from other plovers in the region. Conspicuous breeding plumage develops from April onward. Adults are spotted black and white on the back and wings; face, neck, and breast are black. **Habits and habitat:** Rocky and sand beaches. **Vocalization:** Two-syllable, high-pitched, clear whistle.

Snowy Plover   *Charadrius nivosus*

Range and status: Uncommon permanent resident and common summer resident in the central and southern islands.

Description: 16 cm (6.3 in). Small-sized, pale plover with a thin black bill, white forehead, and gray legs. Upperparts are brown-gray and underparts are white. Noticeable black marks on the forehead and behind the eye, with an incomplete black (breeding) or brown (non-breeding) breast band, which usually appears as

Charadrius nivosus

Charadrius wilsonia

Haematopus palliatus

Pluvialis squatarola, winter plumage

Pluvialis squatarola, partial breeding plumage

dark lateral patches on the sides of the breast. In flight, the flight feathers are blackish with a distinct white stripe running from wing tip to body on the upper wing surface. **Habits and habitat:** Largely restricted to sandy beaches. **Vocalization:** Flight call is a sharp *bip*.

Wilson's Plover   *Charadrius wilsonia*

Range and status: Permanent or summer resident; common summer resident in the central and southern islands, less so in the northern archipelago. Some overwintering birds may be permanent residents, e.g., on Abaco.

Description: 20 cm (8 in). Medium-sized, brown-backed plover with a distinct thick, heavy, black bill. Underparts are pale (white) with a noticeable dark breast band. Legs are pink. Breeding males have a black breast band, some black on forecrown and lores, white stripe above the eye (supercilliary stripe), and the back is dark brown. Females and non-breeding males have a similar plumage, and both appear as a faded version of the breeding male. Non-breeders have a grayer tint to the head and breast band. **Habits and habitat:** Prefers sandy beaches. **Vocalization:** High, weak whistle.

Semipalmated Plover   *Charadrius semipalmatus*

Range and status: Winter resident and transient. Common in southern archipelago, but less common in central and north. Breeds in northern Canada and Alaska and overwinters in the southern U.S., Caribbean, much of South America.

Description: 14–20 cm (5.5–7.9 in). Small-sized plover with a gray-brown back and wings, a white breast and belly, and one black neckband. A brown cap, a white forehead, a black mask around the eyes, and a short, orange bill with a black tip. **Habits and habitat:** Sandy beaches and mud flats.

Piping Plover   *Charadrius melodus*

Range and status: Winter resident in northern archipelago, rare in central and southern islands. Also winters in Cuba and coastlines of southeast U.S. and Mexico. Breeds in northern Great Plains of U.S. and Canada, Great Lakes shore-lines, and Atlantic Coast from Canada south to central U.S. Most wintering birds in archipelago are from the Atlantic Coast population, but some from the Great Lakes area.

Description: 15–19 cm (5.9–7.5 in). Small-sized, stout, sand-colored plover with a large, rounded head; a short, thick neck; and a stubby, orange bill with a black tip. Breeding adults have yellow-orange legs, a black band across the forehead from eye to eye, and a narrow, usually incomplete, black ring around the neck. During the non-breeding season, the bands become faint or brown, and the bill is black and thicker than that of the Snowy Plover (*C. nivosus*) (see above). **Habits and habitat:** Sandy beaches and coastal mudflats. **Vocalization:** Clear, rich whistles. **Conservation status:** Globally threatened species; historically hunted for its feathers, resulting in a population crash in the early 20th century. After hunting was outlawed, the population continued to decrease due to loss and disturbance of breeding habitat. Since the 1990s, populations have increased because of active nest protection on beaches. It is listed as Near Threatened.

Killdeer   *Charadrius vociferus*

Range and status: Permanent resident. Common in central and southern islands; less common in the northern archipelago.

Description: 27 cm (10.5 in). Large-sized, brown-backed plover with a red eye-ring. Face and cap are brown with a black band connecting the eyes above a white fore-head. Underparts are white, bisected by two black bands; one is a collar around the neck, the other is on the breast. Long tail and a distinctive tawny-orange rump. **Habits and habitat:** Restricted less to coastal habitats than other plovers in the archipelago and can even be found in grassland and bare ground habitats. **Vocalization:** Common name derives from their frequently heard call: a loud *kill-deer kill-deer*.

Numenius phaeopus

Charadrius vociferus

Charadrius melodus

Charadrius semipalmatus

Family Scolopacidae (Sandpipers)

This family is large and diverse with small- to large-sized shorebirds that include sandpipers, curlews, godwits, shanks, tattlers, woodcocks, snipes, and dowitchers. Sandpipers have long legs, narrow wings, and narrow bills of a variable length. Diet is largely invertebrates acquired by pecking (short-billed species) or by probing of substrates (longer-billed species). Most are long-distance migrants that breed in the Arctic and overwinter, frequently in groups or large flocks, in temperate and tropical areas of the Americas. Twenty-four species have been recorded in the archipelago, but only one breeding species. Most species in the archipelago are seen in their dull, winter plumage, which makes identification difficult, especially for the smaller species collectively known as "peeps" in North America or "stints" in Europe.

Whimbrel *Numenius phaeopus*

Range and status: Uncommon winter resident and transient; rarer in the northern islands. Breeds throughout subarctic North America, Europe, and Asia. Overwinters in southern Europe and U.S., Africa, South America, Asia, Australasia.

Description: 37–47 cm (15–19 in). Large-sized, brown shorebird with a long, down-curved bill and distinct striping on the head with a distinctive central crown stripe and strong supercilium. **Habits and habitat:** Probes in soft mud and sand for small invertebrates, especially crabs.

Ruddy Turnstone   *Arenaria interpres*

Range and status: Permanent non-breeding resident and common winter resident and transient throughout the archipelago. Breeds throughout the Arctic with a worldwide coastal distribution during the winter.

Description: 24 cm (9.5 in). Unmistakable small-sized, stocky shorebird with a harlequin-like plumage pattern of black and white. Breeding birds have reddish-brown upperparts with black markings. Head is mainly white with black streaks on the crown and a black pattern on the face. Breast is mainly black apart from a white patch on the sides. Rest of the underparts are white. Non-breeding adults are duller than are breeding birds and have dark gray-brown upperparts with black mottling and a dark head with little white. Juvenile birds have a pale brown head and pale fringes to the upperpart feathers creating a scaly impression. **Habits and habitat:** Rocky or stony shores and quays where they commonly push or turn over debris for invertebrate prey.

Sanderling   *Calidris alba*

Range and status: Common winter resident and transient throughout the archipelago. Breeds throughout the Arctic with worldwide coastal distribution during the winter.

Description: 20 cm (8 in). Small-sized, compact, pale-colored sandpiper with a short, straight, black bill and black legs. Winter and juvenile plumages are almost white, apart from a dark shoulder patch. Its pale color, especially prominent white wing stripe, and the absence of a hind toe distinguishes it from other small waders in the region. During the breeding season, face and throat become brick-red. **Habits and habitat:** Typically seen in small flocks, running along sandy beaches near the edge of the water and stopping only to feed. **Vocalization:** A simple *kip*.

Least Sandpiper   *Calidris minutilla*

Range and status: Uncommon winter resident and transient in north and central islands; common in the southern archipelago. Breeds in North American Arctic and overwinters in southern U.S., the Caribbean, Central and South America.

Description: 15 cm (6 in). Smallest sandpiper with brown-gray upperparts, white underparts with streaked and brown wash on the neck and upper breast, greenish-yellow legs, and a short, thin, dark, slightly down-curved bill. **Vocalization:** Flight call is a *kreeet*.

Semipalmated Sandpiper   *Calidris pusilla*

Range and status: Transient. Breeds in North American Arctic and overwinters in the southern U.S., Caribbean, Central and South America.

Arenaria interpres, winter plumage

Calidris alba

Arenaria interpres, partial breeding plumage

Calidris minutilla

Calidris pusilla

Description: 16 cm (6.3 in). Small-sized sandpiper with brown-gray upperparts, white underparts, black legs, and a short, stout, straight, dark bill. Has toes that are webbed for part of their length (hence common name). Similar to the Least Sandpiper (*Calidris minutilla*) (see above). **Habits and habitat:** Seen alone or in small groups on mudflats. **Vocalization:** Various calls, but most common is single *cheh* or *cherk*.

Short-billed Dowitcher   *Limnodromus griseus*

Range and status: Winter resident and transient. Common in southern islands, less common in the northern archipelago. Breeds in Alaska and Canada and overwinters in the Americas from the southern U.S. south to Brazil.

Description: 28 cm (11 in). Medium-sized, dark-colored, stocky, long-billed shorebird with dark brown upperparts, reddish underparts, and barred flanks. Tail has a black-and-white barred pattern. Legs are a yellowish color. **Habits and habitat:** Feeds on invertebrates, often by rapidly probing the bill into mud in a sewing machine fashion. **Similar species:** Long-billed Dowitcher (*L. scolopaceus*) (28–30 cm [11–12 in]). Rare winter transient in northern archipelago. Similar in coloration to *L. griseus* but has a longer, straighter bill; a longer, more cocked tail; a slightly steeper forehead; longer legs; and is associated more with freshwater habitats.

Spotted Sandpiper   *Actitis macularius*

Range and status: Uncommon winter resident and transient throughout the archipelago. Breeds in Canada and the U.S. and overwinters in southern U.S. south to South America.

Description: 19 cm (7.5 in). Small-sized, compact wader with gray-brown upperparts and white upperparts. In spring, individuals develop distinctive black spots on their underparts and a white supercilium. **Habits and habitat:** Frequently bobs its rear end up and down while walking and has a distinctive stiff-winged flight low over water. **Vocalization:** Call is a high whistled *peep-peep*.

Lesser Yellowlegs   *Tringa flavipes*

Range and status: Common to uncommon winter resident throughout the archipelago. Breeds in subarctic Canada and Alaska and overwinters along the Gulf Coast of the U.S. south to South America.

Description: 27 cm (10.5 in). Similar to, but smaller than, the Greater Yellowlegs (GYL; see below), with a shorter, straighter bill. **Vocalization:** Call softer than GYL, a two-noted, short, whistled *tu-tu*.

Willet   *Tringa semipalmatus*

Range and status: Eastern Willet (*T. s. semipalmatus*) is an uncommon summer and locally common permanent resident in the northern archipelago and a locally common summer resident in central and southern islands. Breeds in coastal eastern U.S. and the Caribbean and winters south to the Atlantic coast of South America. Western Willet (*T. s. inornatus*) is a rare winter resident or transient.

Description: 38 cm (15 in). Large-sized wader with gray legs and a long, straight, dark, stout bill. Non-breeders are uniformly gray with white underparts, whereas breeding individuals develop a mottled gray back and wings, and gray streaking on the head and neck. Distinctive bold black-and-white pattern on the wings conspicuous in flight. The two subspecies (eastern and western) are difficult to distinguish in the field. **Habits and habitat:** Nests on the ground, often in colonies in coastal salt marshes. **Vocalization:** Named after their distinctive song, which is a loud *pee wee willet*.

Tringa flavipes

Limnodromus griseus

Tringa semipalmatus semipalmatus in flight

Tringa semipalmatus semipalmatus, winter plumage

Actitis macularius, winter plumage

Actitis macularius, spring plumage

Greater Yellowlegs   *Tringa melanoleuca*

Range and status: Common to uncommon winter resident throughout the
archipelago. Breeds in subarctic Canada and Alaska and overwinters on Atlantic
and Pacific coasts of the U.S. south to South America.

Description: 36 cm (14 in). Large-sized, gray. long-legged shorebird with yellow legs;
a long, thin, dark bill with a slight upward curve; mottled gray-brown upperparts;
and white underparts. Neck and breast are streaked with dark brown, and the
rump is white. **Habits and habitat:** Forages in shallow water, sometimes using
the bill to stir up the water. **Vocalization:** 3–4 notes; a loud, ringing *dew-dew-dew*
or *twe-twe-twe*.

Family Laridae (Gulls, Terns, and Skimmers)

A large family of waterbirds with global distribution, most common along ocean coastlines. Certain species have adapted to and flourish in human-modified environments.

SUBFAMILY LARINAE (Gulls)

Gulls are medium- to large-sized seabirds with white or gray body plumage with
darker wings that often have black markings near the wing tips or on the head; thick,
longish bills; and webbed feet. Immature gulls are difficult to identify. All are omnivores, feeding on live prey, for example, fish; a variety of invertebrates including crustaceans, worms, and mollusks; plant material (seeds and berries); as well as offal and
carrion. Gulls have unhinging jaws, which allow them to consume surprisingly large
prey. Although typically associated with coastal and wetland habitats, many species
concentrate near harbors and garbage dumps. Eight species are recorded in the archipelago, but only one breeding species. In addition to species described below, other
transient gulls recorded in the archipelago are Bonaparte's Gull (*Chroicocephalus
philadelphia*) and Black-legged Kittiwake (*Rissa tridactyla*).

Laughing Gull   *Leucophaeus atricilla*

Range and status: Common permanent resident in The Bahamas and a summer
resident in the TCI. Widespread in southern North America, the Caribbean,
subtropical and tropical parts of Central and North America.

Description: 43 cm (16.5 in). Medium-sized gull with a distinctive black head during
the breeding season (February–September); white eye-ring; black to blackish-red
bill and black legs; and a gray back and wings with black wingtips. Tail and underparts are white. In winter, the head is white with a gray smudge behind the eye.
Habits and habitat: Breeds in colonies on offshore cays. Often seen in harbors,
beaches, and inland ponds. **Vocalization:** Common name is derived from its raucous *kee-agh* call, which sounds like a high-pitched laugh *ha . . . ha . . . ha*.

Tringa melanoleuca

Leucophaeus atricilla, breeding plumage

Leucophaeus artricilla, juvenile

Larus delawarensis, adults

Ring-billed Gull   *Larus delawarensis*

Range and status: Winter resident, common in northern and central islands, but rare in the south of the archipelago. Breeds in northern North America, wintering south to the Caribbean.

Description: 46 cm (18 in). Adults have a white head, gray back and wings, and white underparts with yellow legs. The short bill is yellow with a black ring near the tip. Eyes are yellow with red rims. Juvenile birds have a variable plumage; first year birds are mottled colored with a black tip to bill; second winter individuals resemble drabber versions of an adult and with a distinctive ring on their bill. **Habits and habitat:** Coastal.

Herring Gull   *Larus argentatus*

Range and status: Winter resident. Uncommon in the northern islands; rare in the south of the archipelago. Breeds in temperate Northern Hemisphere. Winters south to the Caribbean and Central America.

Description: 64 cm (25 in). Adults have a gray back and upperwings and a white head, tail, and underparts. Wingtips are black with white spots. Yellow bill has a red spot near the tip. A ring of bare yellow skin surrounds the pale eye. Legs are normally pink. Immature birds are mainly brown with darker streaks and have a dark bill and eyes. Third year birds have a whiter head and underparts with less streaking, and the back is gray. **Habits and habitat:** Coastal.

Lesser Black-backed Gull   *Larus fuscus*

Range and status: Winter resident. Uncommon in northern islands; rare or not recorded in the southern archipelago. Uncommon winter resident in North America. Breeds in northern Eurasia. Increasing in eastern North America.

Description: 53 cm (21 in). Adults have a dark gray back, white head and underparts, yellow bill with red tip, and yellow legs and feet. Juveniles are very similar to juvenile herring gulls. Dark gray-black back feathers are developed by their second winter. Similar to Great Black-backed Gull (see below), but with a slimmer build, yellow rather than pinkish legs, and smaller white "mirrors" at the wing tips.

Great Black-backed Gull   *Larus marinus*

Range and status: Winter resident. Uncommon in the northern islands; rare or unrecorded in the southern archipelago. Rarely winters in the Caribbean. Breeds in northern coastal regions of North America and Eurasia.

Description: 76 cm (30 in). Largest gull in the archipelago. Adults have black wings and back, with conspicuous white "mirrors" at the wing tips, white underparts, pink legs, and a yellow bill with a red spot at the end of the lower mandible. Young birds have a black to gray bill, whitish head, scaly brown upperparts, and black-and-white mottled wings.

SUBFAMILY STERNINAE (Terns)

Terns are small- to large-sized, slender, white or gray sea-birds, often with a black cap and nape, and pointed bills; more elegant and streamlined than most gulls. They dive for fish and squid or pick prey items from the surface. Some individuals live more than thirty years. Commonly found along shorelines, but some feed far out at sea when not breeding (e.g., Brown Noddy, Sooty and Bridled Terns). Most are colonial ground-nesting species with nests being no more than a shallow scrape on the ground on isolated beaches or small, isolated cays. Fourteen species are recorded in the archipelago, including eight breeding species. In addition to species described below, other transient terns recorded in the archipelago are the Common Tern (*Sterna hirundo*), Black Tern (*Chlidonias niger*), Forster's Tern (*Sterna forsteri*), and the Caspian Tern (*Hydroprogne caspia*). Also recorded is the Black Skimmer (*Rynchops niger*), a tern-like species, with black upperparts and white underparts, orange legs, black and white

Anous stolidus

Larus argentatus, adult

Larus fuscus, adult

Larus marinus, adult

Larus marinus, juvenile

wings, but with a distinctive heavy, orange bill with an elongated larger lower mandible; feeds by flying low over the water surface with the lower mandible just below the water (skimming) to catch small fish.

Brown Noddy   *Anous stolidus*

Range and status: Locally common summer resident archipelago-wide. Pantropical.

Description: 39 cm (15 in). Largest of the noddies, with dark brown upper- and underparts, a pale forehead and crown, a white crescent below the eye, and a long black bill. The distinctive wedge-shaped tail is visible in flight. **Habits and habitat:** Often feeds with other species and unlike many tern species will rest on water.

Sooty Tern   *Onychoprion fuscatus*

Range and status: Locally common summer resident archipelago-wide. Pantropical.

Description: 41 cm (16 in). Most common tern in the region with dark black upperparts and white underparts. Wings and deeply forked tail are long, and it has black legs and a black bill. Juveniles are scaly-gray above and below. Resembles the smaller Bridled Tern (*O. anaethetus*) (see below) but is darker-backed, has a broader white forehead, and most distinctively has no pale neck collar. **Vocalization:** A loud, piercing *ker-wack-a-wack* or *kvaark*.

Bridled Tern   *Onychoprion anaethetus*

Range and status: Locally common summer resident archipelago-wide. Pantropical except South America.

Description: 38 cm (15 in). Gray-black tern with dark gray upperparts, white underparts, long wings, a long, forked tail, black bill, and black legs. Forehead, eyebrows, and a characteristic collar on the hind neck are all white. Juveniles are scaly gray above and pale below. Similar to the Sooty Tern (see above), but smaller and with a grayer back, a narrower white forehead, and a pale neck collar.

Least Tern   *Sternula antillarum*

Range and status: Locally common summer resident archipelago-wide. Breeds primarily in North America locally south to South America and winters in Central America, the Caribbean, northern South America.

Description: 23 cm (9 in). Smallest tern in the region with gray upperparts and white underparts, a white head, black cap and lores, a small, white patch on the forehead, a distinctive thin, yellow bill with a black tip, and orange legs. **Habits and habitat:** While hunting, their characteristic fast, jerky wingbeats with the bill pointing slightly downward is distinctive. Nests individually or in loose colonies on sandy or rocky beaches. **Vocalization:** Call note is a distinctive *kip kip*, and a *chiti-whitit, chiti-whitit* call is heard near breeding colonies.

Gull-billed Tern   *Gelochelidon nilotica*

Range and status: Summer resident. Small, scattered populations common in the archipelago, especially in the south. Generally found in warmer portions of the world.

Description: 35 cm (14 in). Large-sized tern with a distinctive thick, gull-like, black bill. Breeding adults have a black cap and nape, black legs, gray back and wings, and a short tail. Non-breeding adults and juveniles have a white head and a dark patch behind the eye. **Habits and habitat:** Unlike other terns, does not normally dive for fish but instead snatches prey from water surface; also regularly feeds on insects and even lizards over fields and woodlands. **Vocalization:** Call is a *Ker-wik*.

Onychoprion anaethetus

Sternula antillarum

Onychoprion fuscatus

Gelochelidon nilotica, in flight

Gelochelidon nilotica

Roseate Tern   *Sterna dougallii*

Range and status: Uncommon summer resident archipelago-wide; also worldwide.

Description: 38 cm (15 in). Small- to medium-sized tern with pale gray upperparts and white underparts; a black cap and nape; long, slender, red bill with a black tip; and orange-red legs. Adults have a long, deeply forked tail. In summer, the underparts of adults take on the pinkish tinge, which gives this species its common name. Outside the breeding season the black on the head is restricted to the nape, and the bill is black. **Vocalization:** Call is a *Chuwit*.

Royal Tern   *Thalasseus maxima*

Range and status: Uncommon summer resident. Common winter resident in the northern islands but less common in the south of the archipelago.

Description: 51 cm (20 in). Large-sized, crested tern with yellow-orange bill and black legs, gray upperparts, and white underparts. In winter, the black cap becomes patchy. Juveniles have a scaly-backed appearance. **Habits and habitat:** Coastal, often perched on pilings.

Sandwich Tern   *Thalasseus sandvicensis*

Range and status: Common summer resident at some sites archipelago-wide; uncommon or rare winter resident in the southern archipelago. Breeds in North America south to South America and also Eurasia. Winter range includes Asia, Africa.

Description: 38 cm (15 in). Medium-sized, crested tern with pale gray upperparts and white underparts, a narrow black bill with a yellow tip, and black legs. The black crest becomes less noticeable in winter, and individuals develop a white forehead. Juveniles have dark tips to their tails and a scaly appearance on their back and wings. **Habits and habitat:** Nests in colonies on offshore cays. **Vocalization:** Vocal species with a loud, grating *kear-ik* or *kerr-ink*.

Family Phaethontidae (Tropicbirds)

Tropicbirds are large, white seabirds with long tail streamers and small weak legs. The feet are at the rear of the body, which makes walking impossible. They have large powerful bills and feed mostly on fish and squid caught by hovering followed by shallow plunge-diving. Nests are in cavities or crevices on bare ground. Pantropical distribution. One species is recorded in the archipelago.

White-tailed Tropicbird   *Phaethon lepturus*

Range and status: Summer resident; common in scattered sites throughout the archipelago. Pantropical distribution including rest of the Caribbean.

Description: 38 cm (15 in). Unmistakable white seabird with a red-orange bill, black eye-line and wing bars, and two long, thin, white tail streamers. **Habits and habitat:** Often near cliffs or limestone banks. Typically solitary, but breeds in loose colonies.

Phaethon lepturus

Phaethon lepturus, in flight

Sterna dougallii

Thalasseus sandvicensis, winter plumage

Thalasseus maxima, summer
plumage, in flight

Thalasseus maxima, winter plumage

Family Procellariidae (Shearwaters)

Shearwaters are medium-sized, long-winged, tube-nosed seabirds, most frequently found in temperate and cold waters. These long-lived birds spend most of their lives at sea, coming to land only to breed; they are rarely seen from shore. They nest colonially, typically on isolated islands or cays, free of land predators. Adults return to their breeding sites at night to minimize predation. They nest in burrows and frequently give eerie contact calls during these nighttime visits. They feed on fish and squid captured from both the surface and/or by diving. Five species have been recorded in the archipelago. In addition to the species described below, four other transient shearwater species occur in the archipelago: Cory's Shearwater (*Calonectris diomedea*), Greater Shearwater (*Ardenna gravis*), Sooty Shearwater (*Ardenna grisea*), and Black-capped Petrel (*Pterodroma hasitata*).

Audubon's Shearwater   *Puffinus lherminieri*

Range and status: Summer resident. West Indian subspecies *P. l. lherminieri* is a common summer resident in scattered sites in the central and southern islands. Species has a pantropical distribution, including rest of the Caribbean.

Description: 30 cm (12 in). Small-sized shearwater with dark brown-black upperparts and white underparts. **Habits and habitat:** It is a surface feeder, feeding in mixed flocks with other seabirds by shallow diving. During the breeding season, typically observed in offshore flocks at dusk prior to coming to land.

Family Fregatidae (Frigatebirds)

Frigatebirds are large seabirds with long, pointed wings and a forked tail giving a pterodactyl-like silhouette in flight. They soar on rising air currents and catch their food from the water's surface or by robbing other seabirds of their prey.

Magnificent Frigatebird   *Fregata magnificens*

Range and status: Common permanent resident in widely scattered sites archipelago-wide. Found throughout Caribbean and Mexico south to Galapagos and Brazil.

Description: 102 cm (40 in). Large-sized, slender, black, fork-tailed seabird with long, narrow wings. Males are glossy black with a red inflatable throat pouch (observed only in breeding season). Females are black with a white breast and no throat pouch. Juveniles have black head and white throat. **Habits and habitat:** Nests in colonies on uninhabited offshore cays, building stick nests on or near the ground.

Puffinus lherminieri lherminieri

Puffinus lherminieri lherminieri, in flight

Fregata magnificens, male in flight

Fregata magnificens, displaying males

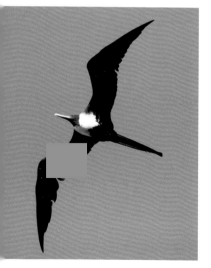

Fregata magnificens, female in flight

Family Sulidae (Boobies)

Boobies are large seabirds with distinctive pointed wings and tails and strong pointed bills. Fish are captured by plunge-diving. They are colonial breeders, typically nesting on uninhabited islands or cays. Four species are recorded in the archipelago. In addition to the species described below, transient booby species recorded in the archipelago are the Northern Gannet (*Morus bassanus*) and the Masked Booby (*Sula dactylatra*), which bred in the southern Bahamas in the 1800s.

Brown Booby   *Sula leucogaster*

Range and status: Permanent resident. West Indian subspecies *S. l. leucogaster* is common near widely scattered breeding islands, but rare elsewhere. Species has a pantropical distribution, including rest of the Caribbean.

Description: 76 cm (30 in). Large-sized brown-and-white seabird with dark brown upperparts and neck (often appearing black) sharply contrasting with white belly and undertail coverts. Large bill is yellow-gray in males and yellow-pink in females. Juveniles lack the white belly and are completely gray-brown in color.

Habits and habitat: Along coastal, open water of Atlantic Ocean and offshore.

Red-footed Booby   *Sula sula*

Range and status: West Indian subspecies *S. s. sula* is a rare, permanent resident on or near cays on San Salvador. Very rare or accidental elsewhere. Species has a pantropical distribution, including rest of the Caribbean.

Description: 71 cm (28 in). Smallest of the boobies with two regional color morphs; the pale morph is mostly white with black flight feathers, and the dark morph is mostly brown with a white tail and rump. Both morphs have red legs and a gray-blue bill.

Family Phalacrocoracidae (Cormorants)

Cormorants present in the archipelago are medium- to large-sized coastal seabirds with dark plumage, long necks, and thin bills. They sit low in the water and surface-dive to catch their prey, primarily fish, but also amphibians and crustaceans. The feathers lack a water-resistant coating and consequently they often perch with their wings spread open to dry their feathers. They are colonial breeders, usually nesting close to water. Two species occur in the archipelago.

Neotropic Cormorant   *Phalacrocorax brasilianus*

Range and status: Permanent resident. Common on Inagua, less so on New Providence, and rare on Eleuthera, Great Exuma, Cat Island, Long Island. Uncommon on the TCI. Very rare elsewhere. Found from extreme southern U.S. south through Argentina.

Description: 64 cm (25 in). Medium-sized species with primarily black plumage and a small, yellow-brown throat patch, which develops a white edge during the

Sula leucogaster leucogaster

Sula leucogaster leucogaster, in flight

Sula sula sula, white morph

Phalacrocorax brasilianus

Sula sula sula, dark morph

breeding season. Breeding birds have fine, white tufts on the sides of the head. It has a long tail and frequently holds its neck in an S-shape. Juveniles are brownish. Smaller and generally daintier compared to the Double-crested cormorant (see below). **Habits and habitat:** Most frequent on freshwater or brackish water close to tall coppice. Often seen with conspecifics while roosting and breeding. Rests and roosts in trees.

Double-crested Cormorant   *Phalacrocorax auritus*

Range and status: Winter resident in the northern Bahamas; permanent resident subspecies *P. a. auritus* on San Salvador and possibly Eleuthera. Rare elsewhere in the archipelago. Species is widely distributed across North America.

Description: 80 cm (32 in). Large, thickset, black cormorant with a long neck, medium-sized tail, and a yellowish-orange bare area of skin that joins the lower half of the beak to the neck. Breeding adults develop white and/or black crests located just above the eyes, a characteristic for which this species is named. Juveniles are dark gray or brownish with lighter underparts and a pale throat. **Habits and habitat:** Most frequent in coastal waters and saltwater bodies. Rests on water and trees and shrubs. Usually seen with conspecifics.

Family Pelecanidae (Pelicans)

Pelicans are large-sized water birds that possess a large, unique bill with an expandable pouch, which is used to capture prey. They capture fish by plunge-diving and/or fishing from the surface. After fishing, the pouch is drained before the prey are swallowed. In addition to fish, they may also consume amphibians, crustaceans, and even smaller birds. One species is recorded in the archipelago.

Brown Pelican   *Pelecanus occidentalis*

Range and status: Permanent and winter resident. West Indian subspecies *P. o. occidentalis* is a permanent resident on Great Inagua. *P. o. carolinensis*, an uncommon winter resident in northern islands, and *P. o. occidentalis* is a rare winter resident in the TCI. Species is found locally throughout the Caribbean and from southern Atlantic and Pacific coasts of U.S. south to South America.

Description: 122 cm (48 in). Smallest pelican species. Adults are gray-brown with a dark belly, and a white head and neck with yellow tinges around the face and head. Bill is pale colored. Juvenile is entirely brown with a pale belly. **Habits and habitat:** Flies with slow, laborious wing beats interspersed by short glides, often low over water. Hunts using steep dives and plunging head first into the water. Seen around sheltered coasts and harbors.

Phalacrocorax auritus auritus

Phalacrocorax auritus auritus,
juvenile

Pelecanus occidentalis carolinensis, in flight

Pelecanus occidentalis carolinensis,
juvenile

Pelecanus occidentalis carolinensis,
adult

Medium- to large-sized wading birds with long necks; long, sharp bills used to spear prey; short tails; and long, powerful, unfeathered legs. In contrast to other long-necked birds seen in the archipelago (e.g., storks, ibises, and spoonbills), herons fly with their neck retracted and legs and feet held backward. Their diet includes fish, reptiles, amphibians, crustaceans, mollusks, and aquatic insects, which they capture by standing motionless on the edge of, or in, shallow water, or by slowly stalking their prey. Sexes are usually similar in plumage coloration and size. Usually silent, calls are often a series of croaks. Many species develop feathery plumes on the head, neck, and back during the breeding season. They often nest in trees in colonies, sometimes with other species. Their habitat includes saltwater and freshwater swamps and even coppice. Eight of the twelve species recorded from the archipelago are known to breed.

Least Bittern   *Ixobrychus exilis*

Range and status: Permanent resident. Common on some scattered sites on New Providence; uncommon on Grand Bahama, Andros, Eleuthera, Abaco; winter resident on San Salvador, Pine Cay. Rare winter resident on other islands. Permanent resident in Greater Antilles. Breeds in North, Central, and South America.

Description: 33 cm (13 in). The world's smallest heron is primarily brown and bright buff with large, buff wing patches, buffy throat and white belly with brown streaking, and yellow eyes and bill. Crown and back are black in males and brown in females. **Habits and habitat:** Very secretive; more frequently heard than seen. **Vocalization:** Variety of cooing and clucking sounds, usually given in early morning or near dusk.

Great Blue Heron   *Ardea herodias*

Range and status: Winter resident; more common in the northern archipelago than in the central and southern islands. Breeds in North America south to Mexico and Cuba. Winters throughout the Caribbean south to extreme northern South America.

Description: 119 cm (47 in). Largest heron in the region with a big, heavy, yellow bill, gray legs, gray wings and body, and a black cap. Head is white with a black stripe above and behind the eye, and a pair of black plumes from just above the eye to the back of the head. Immature birds are duller in color and have no plumes. Occasional reports of an all-white form—Great White Heron (*A. h. occidentalis*). **Vocalization:** Call is a harsh croak.

Great Egret   *Ardea alba*

Range and status: Common winter resident (with sporadic breeding records) in the north of the archipelago; uncommon winter resident in south of archipelago. Found mostly in Greater Antilles. Worldwide distribution.

Description: 99 cm (39 in). Conspicuous, large, all-white egret, distinguished from other white egrets in the region by its large size, long neck, yellow bill, and black legs and feet. During the breeding season, long delicate ornamental feathers are

Ardea herodias

Egretta thula

Ardea alba

Ixobrychus exilis

on the back. Juveniles resemble non-breeding adults. **Habits and habitat:** Feeds by ambushing or stalking prey in shallow water but may also feed on lizards and large insects on land.

Snowy Egret   *Egretta thula*

Range and status: Common winter resident archipelago-wide, rarer in the southern islands; permanent resident on Great Inagua; also Greater Antilles, Virgin Islands, and locally elsewhere in the Caribbean, and from North America south to South America.

Description: 61 cm (24 in). Medium-sized, white heron with yellow lores; a thin, black bill; long, black legs; and distinctive bright yellow feet. The area of the upper bill, in front of the eyes, is yellow but turns red during the breeding season. Plumes on the back create a "shaggy" appearance. **Habits and habitat:** Feeds by ambushing or stalking prey in shallow water, running or shuffling their feet to flush prey, or they "dip-fish" by flying with their feet just over the water.

Little Blue Heron   *Egretta caerulea*

Range and status: Uncommon winter resident; permanent resident on Abaco, Andros, New Providence, Eleuthera; also throughout the Caribbean and from coastal southeast U.S. and Mexico south to South America.

Description: 61 cm (24 in). Medium-sized heron with two distinct plumages related to age. Adults are slate-blue with gray-green legs, and a long, gray-blue bill, with a distinctive black tip. Juveniles are entirely white, similar to Snowy Egret (see above), but have a two-toned bill and dull yellow-green legs. **Habits and habitat:** Stalk prey methodically in shallow water, often running.

Tricolored Heron   *Egretta tricolor*

Range and status: Common permanent resident especially in the central and southern archipelago; also Greater Antilles east to Virgin Islands and southeast U.S. south to South America.

Description: 66 cm (26 in). Medium-sized heron. Adults have a blue-gray head, neck, back and upper wings, with a white line running down the front of the throat and neck. Belly is white. Has a noticeably long, dark bill (with lower yellow mandible), long neck, and pale yellow legs in the non-breeding season, which turn reddish during the breeding season.

Reddish Egret   *Egretta rufescens*

Range and status: Permanent resident archipelago-wide; also found widely along coast in Greater Antilles, Florida, Mexico.

Description: 76 cm (30 in). Large-sized, active heron with two color morphs: the pale morph is completely white, and the dark morph has a slate-blue body and orangish gray-colored head and neck. Both morphs are easily identified by the long, pink bill with a black tip. Young birds have a brown body, head, and neck. Breeding adults have stringy plumes on rear of the crown, neck, and back. **Habits and habitat:** Saline water bodies. Very active while feeding, appearing to "dance," characterized by jumping, running, and rapid changes of direction.

Egretta caerulea, juvenile

Egretta caerulea, adult

Egretta rufescens, white morph

Egretta rufescens, dark morph

Egretta tricolor

Cattle Egret   *Bubulcus ibis*

Range and status: Permanent resident; more common in the northern islands; less common in the southern archipelago; widespread in the Caribbean; worldwide distribution.

Description: 51 cm (20 in). Short-necked, thickset, small, white heron with a short, yellow bill and dark legs and feet. Often exhibits a distinctive hunched posture. Breeding adults have rusty-colored plumes on the crown, back, breast, and legs. **Habits and habitat:** Unlike most herons in the archipelago, individuals feed predominantly on land, often near livestock and in garbage dumps. **Conservation status:** Widely distributed throughout the tropics, subtropics, and warm temperate zones, this egret has undergone one of the most rapid and extensive natural expansions of any bird species. Originally native to southwestern Europe, Africa, and Asia. First recorded as breeding in North America (Florida) in the 1950s; now commonly seen throughout the Americas.

Green Heron   *Butorides virescens*

Range and status: Permanent resident. Endemic subspecies *B. v. bahamensis* archipelago-wide. Species breeds from North America south throughout the Caribbean and Central America.

Description: 46 cm (18 in). This small-sized and dark heron is the most commonly observed in the archipelago. Adults have a dark (greenish-black) glossy cap, a dark back and wings with blue-green gloss, a long, dark bill, and short, yellow legs. Juveniles are similarly colored with brown streaks on underparts. Neck is often pulled in tight against the body. **Habits and habitat:** Usually solitary and often near water. Also in scrub and woodland habitats, where it often stands still and waits for prey species, such as large insects and lizards, to come within striking distance. Frequently crepuscular, most active at dusk and dawn. Unusual for most herons, they occasionally drop small items on the water's surface to attract fish. **Vocalization:** Call is a loud and sudden *kyow* but can also make a series of more subdued *kuk* calls.

Black-crowned Night-Heron   *Nycticorax nycticorax*

Range and status: Permanent resident on Grand Bahama, New Providence, Abaco, Great Inagua; winter resident elsewhere, rarer on southern islands; also Greater Antilles, Cayman Islands, and North America south to South America.

Description: 63 cm (25 in). Short-necked, short-legged heron. Adults have a black crown and back, gray wings, whitish underparts, a stout, black bill, and yellow legs. Breeding adults have two long head plumes. Juveniles are brown and heavily streaked. **Habits and habitat:** Primarily nocturnal, feeds in both fresh and saltwater, usually standing still at the water's edge to ambush prey. During the day, individuals frequently roost in trees or bushes.

Butorides virescens bahamensis, juvenile

Butorides virescens bahamensis, adult

Nycticorax nycticorax, juvenile

Nycticorax nycticorax, adult

Bubulcus ibis

Yellow-crowned Night-Heron   *Nyctanassa violacea*

Range and status: Common permanent resident archipelago-wide. Range includes southern U.S. south to coastal South America.

Description: 61 cm (24 in). Short-necked, short-legged heron. Adults have a distinctive black head with white crown and white stripe below the eye, a gray body and wings, a red eye, and a thick, heavy, black bill. Breeding adults have two long head plumes. Juveniles resemble young Black-crowned night-herons (see above) and are mottled gray-brown. **Habits and habitat:** Frequently feeds at night, mostly in saltwater, sometimes in uplands. Feeds heavily on crabs. Flushes from roads at night.

Family Threskiornithidae (Ibises and Spoonbills)

Ibises are a group of medium-sized, long-legged, wading birds, most with long, down-curved bills. The long bill is typically used to probe mud for invertebrates. Three species are recorded in the archipelago.

White Ibis   *Eudocimus albus*

Range and status: Winter resident; uncommon in northern and central islands, rare in the south of the archipelago; also in Cuba, Hispaniola, Jamaica, and from coastal southeast U.S. south to coastal northern South America.

Description: 53–70 cm (21–28 in). Adults are white with a bright red-orange down-curved bill, long red-orange legs, and black wing tips that are usually visible only in flight. Juveniles have a predominantly brown plumage and only the rump, under-wing, and underparts are white. **Habits and habitat:** Feeds primarily on small aquatic prey, such as insects, crabs, and small fishes, but sometimes observed foraging on grassy pastures. **Similar species:** Glossy Ibis (*Plegadis falcinellus*) (58 cm [23 in]) is an occasional winter resident. Adults of this species have reddish-brown bodies, shiny bottle-green wings, and a brown bill and legs.

Roseate Spoonbill   *Platalea ajaja*

Range and status: Uncommon permanent resident on Great Inagua and North Andros; rare elsewhere in the archipelago; occurs on Cuba, Hispaniola, and from Florida and Texas south to South America.

Description: 71–86 cm (28–34 in). Large, pink bird with long legs; long neck; long, spatulate bill; a greenish head ("golden buff" when breeding); white neck, back, and breast; and gray bill. Immature birds are typically paler, have white-feathered heads, and yellowish or pinkish bills. **Habits and habitat:** Feeds in shallow fresh or coastal waters by swinging its bill from side to side sifting prey as it walks. Diet is crustaceans, insects, and fish.

Nyctanassa violacea, adult

Platalea ajaja, adult

Eudocimus albus, adult

Cathartes aura

Family Cathartidae (New World Vultures)

These large, dark-colored birds have long, broad wings, featherless heads and necks, and they feed primarily on carrion. They lack a syrinx (a sound-producing structure in birds). Their vocalizations, rarely heard, are grunts or low hisses. They defecate on their legs to cool them. Two species are recorded, with one breeding species in The Bahamas.

Turkey Vulture   *Cathartes aura*

Range and status: Common permanent resident on Grand Bahama, Abaco, Andros; rare on other islands. Not recorded on the TCI.

Description: 66 cm (26 in). Large, uniformly brown-black bird with long, broad wings and a featherless, red head and upper neck. Head resembles that of a male

Wild Turkey (*Meleagris gallopavo*), hence its common name. **Habits and habitat:** Roosts in large groups. Carrion is found by smelling ethyl mercaptan, a gas produced by decaying carcasses, which is an unusual ability among birds. Frequently soars on thermals. **Similar species:** Black Vulture (*Coragyps atratus*) (56–68 cm [22–27 in]) with a 1.37–1.67 m (4.5–5.5 ft) wingspan with glossy black plumage. This is a rare transient species recorded in the northern archipelago. Head and neck are featherless, but in contrast to the Turkey Vulture, the skin is dark gray and wrinkled.

Family Pandionidae (Ospreys)

Large, distinctive black, brown, and white fish-eating raptors with a nearly worldwide distribution. Adaptations to a fish diet include reversible outer toes, sharp spicules on the underside of the toes, closable nostrils to keep out water during dives, and backward-facing scales on the talons, which act as barbs to help hold the catch.

Osprey   *Pandion haliaetus*

Range and status: Permanent resident subspecies *P. h. ridgwayi* archipelago-wide; most common in the central and southern islands. Winter residents from North American subspecies *P. h. carolinensis* more commonly seen in northern islands.

Description: 58 cm (23 in). Large raptor with brown upperparts, predominantly gray-white underparts, with a black eye patch and wings. Resident subspecies has a white head with a thin dark line behind eye, whereas winter residents have a streaked cap and thick, dark line through the eye. **Habits and habitat:** Frequently seen soaring over, hovering above, and/or perched close to inland lakes, blue holes, and coastlines. Its distinctive flight with arched wings and drooping "hands" gives a gull-like appearance. **Vocalization:** Series of sharp whistles, described as *cheep, cheep* or *yewk, yewk*.

Family Accipitridae (Hawks, Eagles, Kites, and Harriers)

Diurnal birds of prey with powerful hooked beaks for dismembering prey, strong legs, powerful talons, and excellent eyesight. Four species are recorded in the archipelago, including one breeding species. In addition to the species described below, other transient raptors recorded in the archipelago are the Swallow-tailed Kite (*Elanoides forficatus*, rare spring transient in the northern archipelago); Sharp-shinned Hawk (*Accipiter striatus*, rare transient archipelago-wide); and the Northern Harrier (*Circus hudsonius*, rare transient and winter resident archipelago-wide).

Red-tailed Hawk   *Buteo jamaicensis*

Range and status: Permanent resident subspecies *B. j. jamaicensis* found on Grand Bahama, Abaco, Andros; rare on other islands. Species is widely distributed throughout North and Central America and the Caribbean.

Pandion haliaetus ridgwayi, resident subspecies

Pandion haliaetus carolinensis, North American subspecies in flight

Buteo jamaicensis jamaicensis in flight

Buteo jamaicensis jamaicensis, adult

Description: 48 cm (19 in). Medium-sized raptor with dark brown upperparts; white underparts; a short, broad tail; and broad, rounded wings. Rusty-red tail, which gives this species its name, is uniformly brick-red in adults. Juveniles have a brown, finely barred tail. **Habits and habitat:** Frequents mixed forest habitats and fields with prominent perches (high bluffs or trees) from where individuals can locate prey. Diet mainly small mammals, but also birds and reptiles. **Vocalization:** Call is a short, hoarse, rasping scream, described as *kree-eee-ar*.

Owls are nocturnal birds of prey with comparatively large heads, forward-facing eyes, and the ability to turn their heads 180° to either side. This group comprises two families: the Tytonidae (barn owls) with heart-shaped faces and elongate bills; and the Strigidae, which have rounded heads and more hawk-like bills. Songs and calls include a range of distinct and often eerie vocalizations, the most recognizable of which is the *hoot*. One representative from each family is recorded in the archipelago, both of which breed.

Barn Owl   *Tyto alba*

Range and status: Uncommon permanent resident; found nearly worldwide.

Description: 41 cm (16 in). Nocturnal, large, light-colored owl, with orange-brown upperparts speckled with gray and white, and predominantly white underparts. The distinctive white heart-shaped face contrasts strikingly with the black eyes. **Habits and habitat:** Associated with open habitats. Nests in natural or man-made cavities. Specialized wing feathers allow silent flying. Diet includes small vertebrates, but also large invertebrates. **Vocalization:** Call is an eerie scream.

Burrowing Owl   *Athene cunicularia*

Range and status: Permanent resident subspecies *A. c. floridana* uncommon to rare, with a declining population. Recorded on at least Abaco, Cat Island, Great Inagua, Andros, New Providence, Eleuthera, Grand Bahama. Not recorded in the TCI. Species also in Florida, western North America south to Argentina.

Description: 24 cm (9.5 in). Small, brown, ground-dwelling owl with yellow eyes, prominent white eyebrows, and long, gray, bare legs. Head, back, and wings are brown with white spotting; the underparts are white with variable brown spotting or barring. **Habits and habitat:** Favors open dry habitats with low vegetation; nests in natural cavities, such as the sides of small sinkholes or man-made cavities. Often nests in or near human settlements. Unlike most owls, may be active during the day. Diet includes insects and small rodents, lizards, geckos, amphibians, birds, and occasionally fruits and seeds.

Kingfishers are medium-sized, frequently brightly colored birds with a large head, long, pointed bills, and short legs and tails. Individuals dive for fish from a perch or while hovering over the water. One species is recorded in the archipelago.

Belted Kingfisher   *Megaceryle alcyon*

Range and status: Uncommon winter resident archipelago-wide. Breeds in Canada and U.S.; winters in the Caribbean and southern U.S. to Central America.

Tyto alba

Megaceryle alcyon, female

Athene cunicularia floridana

Description: 33 cm (13 in). Unmistakable large, heavy-headed kingfisher with slate-blue upperparts and white underparts, a broad, white collar, and a prominent crest (which is often raised). Males have a slate-blue band across the chest, and females have both a blue and a rust-colored breast band, and rusty flanks. **Habits and habitat:** Solitary and shy. Perches on branches overhanging fresh and saltwater. **Vocalization:** Includes a mechanical-sounding rattle.

Woodpeckers are small- to medium-sized birds with chisel-like beaks, short legs, stiff tails, and long tongues. The strong bills are used to drill holes in the bark of trees to access invertebrates and to excavate nest cavities. Strong feet and toes enable individuals to grip tree trunks, and the stiff tail allows them to balance and brace themselves during hammering. They are often heard before seen. Four species occur in the archipelago, including two breeding species.

West Indian Woodpecker   *Melanerpes superciliaris*

Range and status: Permanent resident: two endemic subspecies *M. s. blakei* on Abaco and *M. s. nyeanus* on San Salvador and rarely on Grand Bahama. Not found on other islands. Species also in Cuba and the Cayman Islands.

Description: 25 cm (10 in). Largest woodpecker in the archipelago with bold black-and-white barring on the back, rump, and wings; creamy-gray underparts. Males have a red crown and nape, but only the nape is red on females. Males have a black mark above and behind the eye whereas the black extends to the crown in females. *M. s. nyeanus* is shown. **Habits and habitat:** Pine and coppice habitats and suburban areas, especially those with Coconut Palms. Diet includes invertebrates and fruits.

Yellow-bellied Sapsucker   *Sphyrapicus varius*

Range and status: Uncommon winter resident and transient. Breeds in northeastern North America; winters south through Mexico and Greater Antilles.

Description: 22 cm (8.5 in). Black-and-white woodpecker with a striped black-and-white face pattern and pale yellow belly. Adult males have a red forehead and throat; females have a red forehead and white throat. Both sexes have distinctive white patches on the wings, observed as a white line along the leading edge of the wing when perched. Juveniles lack the red head coloring, and their head, upper back, and throat are tinged brown. **Habits and habitat:** Drills shallow holes in tree trunks in horizontal rows from which they feed on sap.

Hairy Woodpecker   *Picoides villosus*

Range and status: Common permanent resident on northern Bahamas. Two endemic subspecies: *P. v. piger* on Grand Bahama and Abaco; *P. v. maynardi* on Andros and New Providence. Absent from other islands and elsewhere in the Caribbean. Species widely distributed in North America south to Panama.

Description: 20 cm (8 in). Black-and-white woodpecker. Distinguished from the Yellow-bellied sapsucker (see above) by the white, central panel on the black back, white spots on the black wing, and the uniform black tail with white outer tail feathers. Both sexes have a white stripe above and below the eye, while the male has a red spot at back of the head. **Habits and habitat:** Primarily in pine forests.

Melanerpes superciliaris nyeanus, female

Melanerpes superciliaris nyeanus, male

Picoides villosus maynardi, male

Sphyrapicus varius, female

Falcons are fast-flying, diurnal, highly maneuverable birds of prey with thin, tapered wings. Prey includes insects, small vertebrates (reptiles and mammals), and birds. Three species are recorded, including one breeding species.

American Kestrel *Falco sparverius*

Range and status: Permanent resident subspecies *F. s. sparveroides* has two color forms. Light color form is a common permanent resident on Abaco and in the central and southern islands. Dark color morph is an uncommon permanent resident on San Salvador, Rum Cay, Crooked and Acklins islands. Species breeds throughout most of North America. Migrants from North America are rare winter residents in the northern islands and are recorded throughout the Caribbean, Central and even South America.

Description: 25 cm (10 in). Small, sexually dimorphic falcon with a rufous (brown) back and tail, and a distinctive white face with black marks both below and behind the eye. Males have blue-gray wings with black spots and an unbarred tail; females have rufous-colored wings and a barred tail. There are two color morphs (or forms): the pale morph has a white breast, belly, and flanks with black spotting and is most common in the southern archipelago; the dark morph has a rufous chest with black spotting and occurs more frequently in the north. **Habits and habitat:** Hunts while hovering or from a perch. Diet consists of invertebrates and small vertebrates, including grasshoppers, lizards, mice, and small birds. Found in a wide variety of open, urban, and suburban habitats. Cavity nester; breeds in trees, cliffs, and buildings. **Vocalization:** Three basic calls: the *klee* or *killy*; the *whine*; and the *chitter*. The *klee, klee, klee, klee* call is often used when individuals are upset or excited.

Merlin *Falco columbarius*

Range and status: Uncommon winter resident archipelago-wide; northern North America populations typically overwinter in the southern U.S., Central America, the Caribbean, northern South America.

Description: 25 cm (10 in). Small, dark falcon. Males have a blue-gray back and streaked buffy-white underparts; females and juveniles are brownish-gray to dark brown above, with whitish-buff underparts (streaked with brown). **Habits and habitat:** Rely on speed and agility to hunt their prey and take most prey on the wing. During migration and on the wintering grounds, individuals can be found in most habitats.

Peregrine Falcon *Falco peregrinus*

Range and status: Uncommon winter resident, rarer on southern islands; nearly worldwide distribution.

Description: 46 cm (18 in). Largest falcon in the archipelago with blue-gray upperparts and white, finely barred underparts. Juveniles have dark brown upperparts

Falco sparverius sparveroides, light morph

Falco sparverius sparveroides, dark morph

Falco peregrinus

Falco columbarius

and heavily streaked underparts. Larger and more thickset than the Merlin (see above) with more distinct facial markings. Adults have a black crown and mask, with a white throat and breast. **Habits and habitat:** Frequently seen along the coast. Prey includes medium-sized birds taken on the wing, especially shorebirds. One of the fastest falcons and can reach speeds of 322 kmh (200 mph) during dives.

Family Psittacidae (Parrots)

Parrots are brightly colored, with short powerful bills and a characteristic down-curved larger upper mandible. They feed on fruits, seeds, and nuts. All parrots have four toes on each foot; two at the front and two back. Frequently in groups, the loud and raucous parrots are heard more than seen. One native species occurs in The Bahamas.

Cuban Parrot *Amazona leucocephala*

Range and status: Permanent resident. Endemic subspecies *A. l. bahamensis* is locally common only on southern Abaco and Great Inagua. Another subspecies occurs on Cuba and the Cayman Islands.

Description: 30 cm (12 in). Medium-sized, green parrot with a white head, green nape, rose-colored throat, and blue feathers in its wings. Juveniles have little or no red on the abdomen, less black edging on the green feathers, and some of the feathers on the top of the head may be pale yellow rather than white. **Habits and habitat:** Often seen in flocks. Diet includes fruits and seeds of palms and West Indian mahogany. **Conservation status:** There are two extant, disjunct populations: Abaco (mostly in Abaco National Park, c. 4000 individuals) and Great Inagua (c. 6500 individuals). As recently as the 1940s, it was recorded on Acklins and Crooked islands, and historical evidence also indicates that the parrot was on both New Providence and San Salvador. The Abaco population nests in natural cavities in sinkholes, whereas Inaguan birds, like all other Cuban Amazon subspecies, nest in tree cavities. Because of the small and potentially vulnerable, spatially separated populations, this parrot is listed as Near Threatened and is protected under international and Bahamian law. Cuban Amazons now occur in Nassau, presumably escaped from captivity.

Family Tyrannidae (Tyrant Flycatchers)

Occurring throughout the Americas, tyrant flycatchers are typically dull-colored birds with broad, flat, strong bills that have bristles at their base. They have a small repertoire of stereotypical vocalizations. Most are aerial insectivores supplemented by some fruit consumption; they characteristically fly from a perch to catch prey on the wing, often returning to the same perch, rather than search the foliage. Fourteen species occur in the archipelago; of the four that breed in the archipelago, three are permanent residents and one is a summer resident.

Cuban Pewee *Contopus caribaeus*

Range and status: Permanent resident. Endemic subspecies *C. c. bahamensis* found on Grand Bahama, Abaco, Andros, New Providence, Cat Island, Eleuthera; absent from other islands. Species also in Cuba.

Amazona leucocephala bahamensis   *Contopus caribaeus bahamensis*   *Myiarchus sagrae lucaysiensis*

Description: 15 cm (6 in). Smallest Bahamian flycatcher with dark gray upperparts, dull white underparts, a slight crest at the rear of the head, and a distinctive white crescent behind the eye. **Habits and habitat:** Occurs in both pine and coppice, but most common in pine forests. **Vocalization:** Call is a fast, soft *tit-tit-tit* and its song a high, whistled *DEE-dee-dee.*

La Sagra's Flycatcher   *Myiarchus sagrae*

Range and status: Permanent resident. Endemic subspecies *M. s. lucaysiensis* throughout The Bahamas; most common on Grand Bahama, Abaco, Andros, New Providence, Eleuthera. Not recorded in the TCI. Species also in Cuba and Grand Cayman.

Description: 19 cm (7.5 in). Medium-sized flycatcher with gray-brown upperparts, a pale whitish throat, a whitish-yellow belly, and a dark brown cap or crest. **Habits and habitat:** Frequently raises its crest. Found in most wooded habitats. **Vocalization:** Call is a distinctive rising, whistled *wheet* and a combination of softer *wheets* with buzzy or raspier notes. Dawn song is a short series of whistles.

Gray Kingbird   *Tyrannus dominicensis*

Range and status: Common summer resident throughout the archipelago.

Description: 23 cm (9 in). Very vocal flycatcher with gray upperparts, white under-parts, and a dark mask extending back from a broad bill. **Habits and habitat:** Commonly seen in urban and suburban areas utilizing prominent perches, frequently power lines. Spring arrival is mid-to-late April, and fall departure occurs in August–September. **Vocalization:** Call is a *pi-cheer*; the song is a longer variation of this.

Loggerhead Kingbird   *Tyrannus caudifasciatus*

Range and status: Permanent resident. Endemic subspecies *T. c. bahamensis* on Grand Bahama, Abaco, Andros, New Providence; absent from other islands. Species also in Greater Antilles.

Description: 23 cm (9 in). Medium-sized flycatcher with a distinctive dark head, which contrasts markedly with the white cheeks and underparts; dark upperparts, yellowish undertail coverts, and a large, heavy, black bill. **Habits and habitat:** Largely restricted to pine forest. **Vocalization:** Call is a loud, sharp, frequently repeated *tireet* and the song is a rolling *pirri-pirri-pirri*.

Family Vireonidae (Vireos)

Vireos are typically small, dull-colored green birds with short, stout bills. They are restricted to the New World. They feed opportunistically on invertebrates, fruit, and even anoles occasionally. Seven species are recorded in the archipelago, two of which breed, including one permanent resident and one summer resident. The Red-eyed Vireo (*Vireo olivaceus*), Philadelphia Vireo (*Vireo philadelphicus*), Warbling Vireo (*Vireo gilvus*), and Blue-headed Vireo (*Vireo solitarius*) occur rarely and are not described.

White-eyed Vireo   *Vireo griseus*

Range and status: Uncommon winter resident throughout the archipelago; also winters in Greater Antilles and from southern U.S. to Central America. Breeds in eastern U.S.

Description: 12.7 cm (5 in). Similar to, but daintier than, the Thick-billed Vireo (TBV; see below). Adults can be distinguished from the TBV by the distinctive white eye and yellow spectacles, the gray-green head and back, gray neck, and the marked contrast between the white underparts and yellow flanks. Juveniles lack the white eye. **Habits and habitat:** Coppice and secondary habitats where it commonly feeds on fruits of Gammalamme.

Thick-billed Vireo   *Vireo crassirostris*

Range and status: Permanent resident. Two endemic subspecies: *V. c. crassirostris* throughout most of The Bahamas; *V. c. stalagmium* is common on the Caicos Islands, but not found on the Turks Islands. Species also in Cayman Islands, and locally on islands off Nicaragua, Cuba, Hispaniola.

Tyrannus dominicensis

Tyrannus caudifasciatus bahamensis

Vireo crassirostris crassirostris

Vireo griseus

Vireo flavifrons

Description: 13.3 cm (5.3 in). Sexes are similar in appearance with a grayish-olive head and back, yellow-white underparts, a dark tail and wings, and two white wing bars. Lores are yellow with a black stripe and the eyes have a dark iris. **Habits and habitat:** The most common vireo, and perhaps the most common songbird, seen in the archipelago. Inhabits virtually all scrub and coppice habitats. **Vocalization:** Call is a combination of squeaky, buzzy, and nasal notes, while the song is a repeated *chick-wi-wea-chick* or *chik-didle-wer-chip*.

Yellow-throated Vireo   *Vireo flavifrons*

Range and status: Uncommon winter resident archipelago-wide; also winters in Cuba and locally elsewhere in the northern Caribbean and from Mexico to northern South America. Breeds in eastern North America.

Description: 14 cm (5.5 in). Distinctive, brightly colored, green-yellow vireo, with olive-green upperparts, a bright yellow throat, a white belly, dark eyes with yellow lores, and "spectacles." Tail and wings are dark with two white wing bars.

Black-whiskered Vireo   *Vireo altiloquus*

Range and status: Common summer resident but absent from San Salvador and Rum Cay. Also breeds from southern Florida south through the Caribbean to northern South America. Winters in South America.

Description: 16 cm (6.3 in). Both sexes have dull olive-green upperparts and white underparts, with yellow on the flanks and under the tail. Common name derives from the distinct facial markings—a dark line through the eyes, a white supercilium, and a distinctive black line (the "whisker") on either side of the neck. It has red eyes and a gray-brown crown with faint dusky edges, thick blue-gray legs, and a stout bill. Juvenile birds are similar but have brown-red eyes. **Habits and habitat:** Coppice and secondary habitats. Frequently sings throughout spring and early summer. Arrives in mid-to-late April and departs by August or September. **Vocalization:** Song is a *cheap-John Stirrup, sweet-Joe Claire*.

Family Corvidae (Crows and Jays)

Crows are found nearly worldwide and include the largest songbirds with a robust build, strong legs, and heavy bills, often with nostrils covered by bristle-like feathers. Many are strong flyers. Calls are characteristically short, harsh vocalizations. Temperate species often have black or blue (sometimes with iridescence) or pied (black-and-white) colored plumage, but some tropical species are brightly colored. Omnivorous and highly adaptable; some species have adapted well to co-existing with humans. The ability of some crow species to solve complex problems and to use tools is well-documented. One species is recorded in the archipelago.

Cuban Crow   *Corvus nasicus*

Range and status: Permanent resident. Common in certain sites on North and Middle Caicos, uncommon on Providenciales. Not found elsewhere in the archipelago. Extirpated from The Bahamas; fossils and sub-fossils recorded from New Providence, Little Exuma, Crooked Island; also Cuba.

Description: 46 cm (18 in). Large-sized, all-black crow with glossy plumage, a heavy, pointed bill, and short tail. **Habits and habitat:** Large, noisy flocks can be seen feeding in trees or on the ground. Omnivorous diet includes insects, lizards, fruit, and carrion. **Vocalization:** Produces un-crow-like calls including a combination of strange, liquid, bubbling notes and high, ringing sounds produced in various combinations and a thin screeched *aaaaauh*.

Vireo altiloquus

Corvus nasicus

Tachycineta cyaneoviridis in flight

Tachycineta cyaneoviridis

Family Hirundinidae (Swallows)

Swallows are agile, aerial feeders recognized by their pointed, slender wings, short bills with wide gape, and many species have forked tails. Ten species occur in the archipelago including two breeding species. Swallows that only migrate through the archipelago include the Purple Martin (*Progne subis*), Barn Swallow (*Hirundo rustica*), Tree Swallow (*Tachycineta bicolor*), Bank Swallow (*Riparia riparia*), and the Northern Rough-winged Swallow (*Stelgidopteryx serripennis*).

Bahama Swallow   *Tachycineta cyaneoviridis*

Range and status: Endemic. Permanent resident on Grand Bahama, Abaco, Andros; occasional on New Providence; rarely seen on other islands. Not recorded from the TCI.

Description: 14.6 cm (5.8 in). Green-blue swallow with a dark green crown, nape, and back; green-blue wings; black tail and wing tips; and bright white cheeks, throat, and underparts. Females are duller than the male. In flight, a marked contrast can be seen between the dark flight feathers and white underwing. The forked tail is longer than the wings when perched. **Habits and habitat:** Breeds in pine woodlands, nesting in old woodpecker holes (and sometimes openings in cell towers). Feeds over open areas, such as clearings in woodlands, marshes, fields, and along coastlines. **Conservation status:** The winter distribution, although poorly known, includes islands where they breed and rarely southern Florida and the southern Bahamas. The declining population is estimated at 1500–4000. It is listed as Endangered. **Similar species:** Only one other species of swallow breeds in the archipelago; a small population of Cave Swallows (*Petrochelidon fulva*) (14 cm [5.5 in]) on Andros Island (not found on other islands). This summer resident has a dark blue back and crown; a rufous collar; a chestnut rump, forehead, and cheeks; and dull white underparts with a rufous wash on the breast and flanks.

Family Sittidae (Nuthatches)

Small, thickset birds with big heads, short tails, and powerful bills and feet. This family is unique among birds, having the ability to climb down trunks and branches of trees head first. All other birds go upward. One breeding species is recorded in the archipelago.

Brown-headed Nuthatch   *Sitta pusilla*

Range and status: Rare permanent resident. Endemic subspecies *S. p. insularis* is only found on Grand Bahama. Species also in the southeast U.S.

Description: 10 cm (4 in). Small, woodpecker-like species found exclusively in the pine woodlands of Grand Bahama with gray upperparts, dull white underparts, and a brown cap. A dark line runs through the eye, and the cheek and the throat are white. The Bahamian subspecies has a longer bill, shorter wings, darker eye line, and gives a more rapid and higher pitched call than those found in the U.S. **Habits and habitat:** Locates insects by probing into bark; also feeds on seeds.

Family Polioptilidae (Gnatcatchers)

These small, dainty, active, insectivorous birds have long, thin bills and long tails (often held upright) and are typically blue-gray in color. One species occurs in the archipelago.

Blue-gray Gnatcatcher   *Polioptila caerulea*

Range and status: Common permanent resident on Grand Bahama, Abaco, Andros, Cat Island, Crooked Island, Acklins Island, Mayaguana, Great Inagua, and the TCI; less common on San Salvador, Long Island, Ragged Island; rare on other islands. An influx of birds to the archipelago from North America occurs during the winter. Breeds in temperate North America wintering south to Cuba and Central America.

Description: 11.4 cm (4.5 in). Small, dainty species with blue-gray upperparts, pale gray underparts, a long, slender bill, a distinctive white eye-ring, and a long, black tail with white outer tail feathers. Males are bluer than females and also have a

Polioptila caerulea

Turdus plumbeus plumbeus

Sitta pusilla insularis

black line above the eye. **Habits and habitat:** Individuals flit rapidly through vegetation as they feed on small invertebrates caught on the wing or gleaned from foliage. Observed in pine woodlands in the northern islands as well as in scrub and thickets in the southern islands. Often found in pairs or small groups. **Vocalization:** Call is a thin, whining *spee spee*, while the soft song is a mixture of thin warbles, whistles, chips, and buzzes.

Family Turdidae (Thrushes)

Thrushes are plump, upright, small- to medium-sized birds, often with brown backs, that are frequently found in wooded areas. The diet includes invertebrates and fruit. Seven species have been recorded in the archipelago, including one breeding species. In addition to the species described below, other thrushes recorded in the archipelago (all winter residents or transients) include Hermit Thrush (*Catharus guttatus*), Gray-cheeked Thrush (*Catharus minimus*), Bicknell's Thrush (*Catharus bicknelli*), Swainson's Thrush (*Catharus ustulatus*), American Robin (*Turdus migratorius*), and Eastern Bluebird (*Sialia sialis*).

Red-legged Thrush *Turdus plumbeus*

Range and status: Permanent resident. Endemic subspecies *T. p. plumbeus* common permanent resident on Grand Bahama, Abaco, Andros, New Providence; less common on the Berry islands, Eleuthera, Cat Island; not recorded on other islands. Outside The Bahamas, species found only in Greater Antilles, Cayman Islands, Dominica.

Description: 27 cm (10.5 in). Unmistakable slate-gray thrush with a black throat and a distinctive orange-red eye-ring and legs. **Habits and habitat:** Feeds on insects, fruit, small vertebrates (frogs and lizards), and even birds' eggs. Observed in pine woodlands, coppice habitats, and gardens.

Family Mimidae (Mockingbirds and Thrashers)

The mimids are typically dull-colored (gray-brown), thrush-like birds, with relatively long tails, and slightly to strongly down-curved bills. They are notable for their complex songs and vocalizations, especially their ability to mimic a wide variety of other bird species and background sounds. The sexes are typically similar in size and plumage coloration, but males are generally more vocal than females. They are omnivorous, feeding primarily on insects and fruit. Five species are recorded in the archipelago including three breeding species.

Gray Catbird   *Dumetella carolinensis*

Range and status: Common winter resident archipelago-wide. Breeds in much of the U.S. and southern Canada. Also winters in Florida, the Caribbean, Mexico, Central America.

Description: 22 cm (8.5 in). Shy, medium-sized, dark gray species with a black cap, rust-colored undertail coverts, slim, black bill, and black eyes, legs, and feet. **Habits and habitat:** Often heard rather than seen. Diet is primarily fruit. **Vocalization:** Common name derived from its mewing cat-like call.

Pearly-eyed Thrasher   *Margarops fuscatus*

Range and status: Permanent resident. Recorded on San Salvador and Crooked and Acklins islands, but rare on other islands; also Bonaire and Puerto Rico to the northern Lesser Antilles.

Description: 29 cm (11.5 in). Distinguished from the Bahama Mockingbird (see below) by its darker brown upperparts, heavily speckled pale underparts, pale yellow eyes, and a heavy, pale colored bill. **Habits and habitat:** Aggressive, opportunistic omnivore that feeds primarily on large insects, but also fruits and occasionally lizards, frogs, small crabs, and other bird's eggs and nestlings.

Bahama Mockingbird   *Mimus gundlachii*

Range and status: Common permanent resident throughout the archipelago; also Cuba, Jamaica, and as a stray in Florida.

Description: 28 cm (11 in). Large-sized, dull-colored, thrush-like species with streaked gray-brown upperparts and streaked pale gray underparts. **Habits and habitat:** Unlike the Northern Mockingbird (see below), is almost exclusively found in natural habitats (scrub, pine, and coppice). Feeds on insects, fruit, and small lizards. Sings incessantly and can be very tame. **Vocalization:** Call is a sharp *check*. Typi-

Dumetella carolinensis

Margarops fuscatus

Mimus gundlachii

Mimus polyglottos polyglottos

cal song consists of hoarse, loud phrases including *jer-key jer-key jer-key up jup jup*. Often jumps in the air while singing, returning to the same perch.

Northern Mockingbird   *Mimus polyglottos*

Range and status: Common permanent resident in residential areas archipelago-wide. Subspecies common to the southeast U.S. *M. p. polyglottos* found in northern Bahamas; Greater Antillean subspecies *M. p. orpheus* found in the southern Bahamas and the TCI. Species also found throughout much of the Caribbean and temperate North America south to Mexico.

Description: 25 cm (10 in). Very vocal, slender, thrush-like species with gray upperparts, pale gray underparts, and a distinctive white bar on wings. **Habits and habitat:** Most common in urban and suburban areas. Other birds and animals, including humans, can be chased from territories especially while breeding. Typically sings from a prominent perch. Often simultaneously sings and jumps into the air in a looping motion, with the wings outstretched to display the conspicuous white wing bars. **Vocalization:** Excellent mimic of many species. Sings at night.

Family Sturnidae (Starlings)

Medium-sized songbirds with strong legs and feet; they typically walk on the ground or forage in trees and shrubs where they use their slender straight bills to feed on a variety of invertebrates, small vertebrates, nectar, and fruits. Most are highly gregarious with swift and direct flight. Starling wing shapes are round or triangular, and tails are usually short and square. Plumage of many species may have a metallic iridescent sheen. Starlings are native to Eurasia, Africa, Oceania, and Malaysia south to northern Australia but have been introduced to other parts of the world.

European Starling *Sturnus vulgaris*

Range and status: Permanent and winter resident. Northern and central islands, absent from southern archipelago with the exception of Grand Turk. First recorded in The Bahamas in 1956. Common on some parts of Jamaica, scarce elsewhere in Caribbean. Native to Eurasia; introduced to North America where abundant throughout the U.S. and southern Canada.

Description: 19–23 cm (7.5–9.1 in). Stocky, dark-colored species with shiny, dark (purple to green-black), speckled plumage and a long bill, short tail, and triangular-shaped wings (visible in flight). Breeding adults have glossy, all-black plumage with spotting on their backs and a conspicuous yellow bill. Non-breeding birds have a dark bill and very heavy, white, spotted upper and underparts. Juveniles are gray-brown. **Habits and habitat:** Walks rather than hops. Primarily found in urban areas, often in small flocks seen feeding on grassy areas.

Family Motacillidae (Pipits)

These small, slender birds have thin bills, medium to long tails, and long legs, feet, and toes. They are ground-feeding insectivores of open country. When feeding on the ground they walk as opposed to hopping. One species is recorded in the archipelago.

American Pipit *Anthus rubescens*

Range and status: Rare winter resident in the northern Bahamas. Not recorded in southern archipelago; very rare winter resident in rest of Caribbean. Winters from southern U.S. to Mexico. Breeds on tundra and alpine mountaintops of U.S. and Canada.

Description: 17 cm (6.5 in). Drab, brown bird with brown-gray upperparts, streaked buffy-white underparts, a pale supercilium, and a dark bill and legs. White outer tail feathers conspicuous during flight. **Habits and habitat:** Usually seen in small flocks in open grassy areas, including golf courses and pastures. Often close to water. Numbers highly variable between years. **Vocalization:** Call is a high-pitched *speet* or *speet-tit*.

Anthus rubescens

Sturnus vulgaris, breeding plumage

Bombycilla cedrorum

Family Bombycillidae (Waxwings)

The three species of waxwings are found in northern parts of the Eastern and Western Hemispheres. They have soft, silky plumage and unique red tips to the wing feathers, which have a waxy appearance and give the group its common name. They are arboreal and feed on insects in summer and fruit in winter. One species is recorded in the archipelago.

Cedar Waxwing   *Bombycilla cedrorum*

Range and status: Rare winter resident archipelago-wide.

Description: 18 cm (7.3 in). This unmistakable bird has soft brown plumage on the head, back, and crest with a narrow, black mask outlined in white running through the eye. Rump is gray and wing tips are black; a red spot on the wing in adult birds. Belly is pale yellow, and the short square tail has a bright yellow tip. Short, wide bill is black. **Habits and habitat:** Usually observed in flocks around fruiting plants. **Vocalization:** Call is a faint *sree*.

Small, plump, dull-colored (brown or gray) birds with short tails and short, powerful beaks. Sparrows are seed-eaters but may also consume small insects. One species is recorded in the archipelago.

House Sparrow   *Passer domesticus*

Range and status: Introduced permanent resident common in the northern archipelago. Not recorded elsewhere in the archipelago except for an isolated population in Matthew Town, Great Inagua. Native to Eurasia and Africa; also Greater Antilles, North and Central America, especially in urban regions.

Description: 16 cm (6.3 in). Males have chestnut-brown upperparts with black markings on back and wings, and dull-gray underparts. Breeding males have a gray crown, a black bill and lores, and a black throat and breast, while non-breeders are duller with less black on breast. Females have dull-brown upperparts streaked with black, with a light brown crown and pale stripe behind the eye, and light gray underparts. **Habits and habitat:** Common around human habitations, gregarious and forages on ground for seeds, insects, and food scraps. **Vocalization:** Various chips, chirps, and chittering notes.

The Passerellids are a large New World family of seed-eating birds with distinctively short, cone-shaped bills that include North and South American sparrows. Many species largely feed on the ground. Males may be similar to or more brightly colored than females, and juveniles resemble females. At least seven species are recorded in the archipelago. The White-crowned Sparrow (*Zonotrichia leucophrys*), Lincoln's Sparrow (*Melospiza lincolnii*), Swamp Sparrow (*Melospiza georgiana*), and Chipping Sparrow (*Spizella passerina*) occur very rarely in The Bahamas.

Savannah Sparrow   *Passerculus sandwichensis*

Range and status: Uncommon winter resident in northern and central Bahamas; not recorded on southern islands. Breeds in northern and western North America to Mexico. Winters from southern U.S. south to Mexico and Cuba.

Description: 15 cm (5.8 in). Dull-colored bird forages on the ground, sometimes in small flocks on open grassland or agricultural areas. Sexes are similar in appearance with dark streaked upperparts, white underparts with heavy streaks on the breast and flanks, a pale bill, and pink legs and feet. Head is distinctively marked with a light brown supercilium, pale yellow lores, brown cheeks, and two dark crown stripes. **Similar species:** Clay-colored Sparrow (*Spizella pallida*) (11.4 cm [4.5 in]) and the Grasshopper Sparrow (*Ammodramus savannarum*) (10–14 cm [3.9–5.5 in]); both are rare winter residents/transient species.

Passer domesticus, male

Passerculus sandwichensis

Spindalis zena townsendi, male

Family Spindalidae (Spindalis)

Sexually dimorphic species restricted to The Bahamas Archipelago and Greater Antilles; colorful males typically orange, black, and white; females brown and gray. Previously named Stripe-headed Tanagers. Feeds heavily on fruit and insects; reasonably heavy bill. One resident species recorded in the archipelago; three other species in Greater Antilles.

Western Spindalis *Spindalis zena*

Range and status: Locally common permanent resident; most common in the northern islands. Two endemic subspecies: *S. z. townsendi* on Grand Bahama and Abaco, and *S. z. zena* elsewhere in the archipelago. Not recorded from the Turks Islands. Species also in Florida, the Cayman Islands, Cuba.

Description: 16.5 cm (6.5 in). Males of this brightly colored species have a black head with a prominent white supercilium and equally prominent white stripe below the eye. They have black wings with a white shoulder patch and stripes; an orange-yellow breast, neck, and rump; and pale belly and undertail coverts. Male *S. z. towsendi* have green backs, in contrast to the black backs of male *S. z. zena*. Females are dull-colored with brownish-olive upperparts and pale underparts, with a washed-out supercilium and more distinct whitish stripe under the eye. **Habits and habitat:** Most commonly observed in pine, coppice, and suburban habitats, often near fruiting plants. **Vocalization:** Call is a soft, thin *seet*.

The Icterids are a New World family of small- to medium-sized, often colorful species, including the grackles, cowbirds, New World blackbirds, and New World orioles. Most species are predominantly black, often enlivened by patches of yellow, orange, or red. Eleven species are recorded for the archipelago including three breeding species.

Bahama Oriole   *Icterus northropi*

Range and status: Endemic. Andros.

Description: 20 cm (8 in). Distinctive black-and-yellow oriole. Both sexes have black upperparts, a black bill, and bright yellow underparts. A striking yellow panel is on the shoulder. Immatures have an olive-gray back, olive head, dark wings and tail, black lores, and dull yellow underparts. **Habits and habitat:** Frequents pine, coppice, and residential areas. Nests in Coconut Palms often found close to human habitations as well as in native palms in pine woodlands and in pine trees. **Conservation status:** Formerly known as the Greater Antillean Oriole (*Icterus dominicensis northropi*); only recognized as a distinct species in 2010. Originally on both Abaco (until the 1990s) and Andros; now found only on Andros. Threats include habitat loss through development, the introduction of lethal yellow (a disease that kills Coconut Palms), and nest parasitism by the Shiny Cowbird (*Molothrus bonariensis*, a recent arrival to the archipelago). **Similar species:** Baltimore Oriole (*I. galbula*) (21.6 cm [8.5 in]), a rare winter resident or transient species found archipelago-wide. Males are black and orange and can be further distinguished from the Bahama Oriole by having a white wing bar.

Red-winged Blackbird   *Agelaius phoeniceus*

Range and status: Permanent and winter resident. Endemic subspecies *A. p. bryanti* common resident on Grand Bahama and Abaco; less common on Andros, The Berry Islands, New Providence. Not found elsewhere in the Caribbean. Species widely distributed throughout North and Central America. Northern populations are migratory while southern populations are resident.

Description: 22 cm (8.8 in). Male is all black with red shoulder epaulets (not always visible); the drab female has dark upperparts, pale, heavily streaked underparts, and a prominent white supercilium and throat. Young birds resemble the female. **Habits and habitat:** Freshwater and saltwater wetlands but can also occur in drier grassland habitats. Diet includes seeds, insects, and invertebrates. Colonial nesting species. One male can defend up to ten females.

Shiny Cowbird   *Molothrus bonariensis*

Range and status: South American species, which expanded north in past 100 years by island hopping through eastern Caribbean to Greater Antilles and southern U.S.; breeding recorded first in archipelago in 1994. Locally common on Andros. Sporadic records from other islands.

Description: 18 cm (7 in). Sexually dimorphic species with short, conical bill. Males are all black with an iridescent purple-blue gloss; the female is smaller than the

Agelaius phoeniceus, juvenile male

Agelaius phoeniceus bryanti, male

Icterus northropi

Molothrus bonariensis, female

Molothrus bonariensis, male

male, dark brown, and with a paler stripe above the eye. **Habits and habitat:** Occurs in open woodland, cultivated lands, and pastures often with cattle. Forages on seeds and insects on the ground, often in flocks. Obligate nest parasite (i.e., lays eggs in other birds' nests); parasitizes many species throughout its range. **Vocalization:** Male's song is a purr and whistle, *purr purr purrte-tseeeee* and its call is a sharp *tsee-tsee*; the female call is a harsh rattle. **Conservation status:** Range overlaps with the endangered and single island endemic Bahamas Oriole and is considered a threat to this endangered species. Has been controlled elsewhere in its range to protect other threatened bird species.

The New World wood warblers are small, typically brightly colored birds with thin bills; males are usually brighter than females. Most are arboreal insectivores on their breeding grounds, but on their wintering grounds some species also feed on fruit and nectar. They tend to move often while foraging. Most migrant species exhibit a drab and less distinctive fall–winter plumage compared to breeding (spring) plumage, which begins to appear from February–March. Vocalizations on the wintering grounds are normally limited to species-specific *chip* call notes, although some species can be heard singing partial or even full songs, just prior to spring migration. Many of the migrant warblers defend territories on the wintering grounds. Forty-two species are recorded in the archipelago, most of which are winter residents or transients. There are five breeding species, all of which are permanent residents.

Ovenbird *Seiurus aurocapilla*

Range and status: Common winter resident and transient throughout the archipelago. Breeds in Canada and eastern U.S. Also winters in Florida, the Caribbean, Mexico south through Central America.

Description: 14.6 cm (5.8 in). Common but inconspicuous species with olive-brown upperparts and white underparts heavily streaked with black. Distinctively marked head with a white eye-ring, a black stripe running below the white cheek, and a line of orange feathers with olive-green tips on the crown bordered by two black stripes. Feet and long legs are pinkish. Sexes are similar in appearance. **Habits and habitat:** Frequents dense coppice habitats where it forages on the ground for ants and other invertebrates by walking, often cocking its tail and head bobbing simultaneously. **Vocalization:** Call is a sharp *chup*.

Worm-eating Warbler *Helmitheros vermivorum*

Range and status: Uncommon winter resident and transient. Breeds in the eastern U.S. Winters mostly in Greater Antilles and from Mexico to Panama.

Description: 13.3 cm (5.3 in). Dull-colored warbler. Sexes are similar in appearance with olive-brown upperparts and light buffy-brown underparts, a large, pointed bill, and pink legs. Head is distinctively marked with black and light brown stripes, two distinct black stripes on the crown, and another through each eye. **Habits and habitat:** Frequents wooded areas where it characteristically forages in dead leaves, on the bark of trees and shrubs, or searching leaf litter on the forest floor.

Northern Waterthrush *Parkesia noveboracensis*

Range and status: Common winter resident and transient. Breeds in northern North America. Winters from the southern U.S. and Mexico south through the Caribbean to Central and northern South America.

Description: 14.6 cm (5.8 in). Commonly seen dull-colored warbler with brown upperparts, dingy white or yellowish heavily streaked underparts, a distinctive white supercilium, contrasting with a dark stripe through the eye, and a dark, pointed bill. Sexes are similar in appearance. **Habits and habitat:** Forages for

Seiurus aurocapilla

Helmitheros vermivorum

Parkesia noveboracensis

Vermivora cyanoptera, male

insects, mollusks, and crustaceans on the ground. Walks instead of hopping, bobbing their tail as they do so. Often seen in mangrove swamps and adjacent uplands during winter. Easily confused with the Louisiana Waterthrush (*P. motacilla*; 15.2 cm [6 in]), a rare winter resident and transient. Differences between the two species are subtle. Louisiana Waterthrush has pink-buff flanks as opposed to white or yellow, lacks streaking on the throat, and has brighter pink legs. **Vocalization:** Call is a hard *chink*.

Blue-winged Warbler   *Vermivora cyanoptera*

Range and status: Rare winter transient. Breeds in eastern North America and primarily winters in Mexico and northern Central America.

Description: 11.5 cm (4.5 in). During spring, males have greenish upperparts, bright yellow underparts, gray wings with two white wing bars, and a black eye stripe. Females, juveniles, and males in winter plumage are duller, but otherwise similar. **Vocalization:** Call is a sharp *chip*.

Black-and-white Warbler   *Mniotilta varia*

Range and status: Common winter resident and transient. Breeds throughout eastern North America and winters in the southern U.S., Mexico south to South America, the Caribbean.

Description: 12.7 cm (5 in). Distinctive black-and-white streaked warbler with a slightly down-curved bill. Males are more distinctly marked than the females, with a black cheek and throat; females have a white throat and cheek and a distinct black line behind the eye. **Habits and habitat:** Feed by probing tree trunks and branches as they creep along woody parts of trees. **Vocalization:** Hard *tick*. **Similar species:** Blackpoll Warbler (*Setophaga striata*) is black and white with a straight bill, and the male has a white cheek.

Prothonotary Warbler   *Protonotaria citrea*

Range and status: Uncommon transient. Breeds mostly in the southeast U.S. Winters from Mexico and the Caribbean region south to northern South America, especially in mangrove forests.

Description: 13.3 cm (5.3 in). Brightly colored yellow and green sexually dimorphic warbler with an olive back, blue-gray wings and tail, yellow underparts, a relatively long, pointed bill, and black legs. Adult males have a bright orange-yellow head; females and immature birds are duller and have a yellow head. Other than white spots on the tail, it lacks other conspicuous markings. **Habits and habitat:** Prefers wooded habitats near water but also occurs in coppice. **Vocalization:** Call is a sharp *tink*.

Tennessee Warbler   *Oreothlypis peregrina*

Range and status: Uncommon transient. Breeds in northern North America and winters in southern Central America and northern South America.

Description: 12 cm (4.8 in). Dull-colored warbler with greenish-brown upperparts and white (pale) underparts. Crown and nape are gray and has a white supercilium. Females are duller, showing less contrast in their color patterns. Non-breeding and young birds resemble the adult females. **Vocalization:** Call is a sharp *sit*.

Kentucky Warbler   *Geothlypis formosa*

Range and status: Rare winter transient. Breeds mostly in the southeast U.S. Winters in southern Mexico, northern Central America, rarely in the Caribbean.

Description: 13 cm (5.1 in). Heavyset warbler with olive-green upperparts, a brilliant yellow throat and belly, and a short tail. Some black flecks on their crown, black cheeks, and a black border to the yellow throat and supercilium that arcs behind the eye. Females have slightly less black on the sides of their head, and immature birds may have almost no black at all. **Habits and habitat:** Frequently observed near the ground.

Mniotilta varia, female

Protonotaria citrea, male

Oreothlypis peregrina, male

Geothlypis formosa, male

Geothlypis rostrata coryi, male

Geothlypis rostrata tanneri, female

Bahama Yellowthroat   *Geothlypis rostrata*

Range and status: Endemic. Permanent resident in the northern and central Bahamas. Three subspecies: *G. r. tanneri* on Grand Bahama and Abaco; *G. r. rostrata* on Andros and New Providence; *G. r. coryi* on Eleuthera and Cat Island. Not found on other islands.

Description: 14 cm (5.5 in). Yellow-green, sexually dimorphic warbler: Males have olive green upperparts, a bright yellow throat, pale yellow-green belly, and a prominent black mask; females are similar but lack the black mask and have a gray crown. The most obvious differences between the three subspecies are seen in the intensity of the males' yellow plumage and extent of the black mask; *G. r. coryi* is the yellowest of the subspecies. **Habits and habitat:** Observed in broadleaf understory under pine, low coppice, and thickets. Forages low in vegetation and feeds on invertebrates and occasionally *Anolis* lizards. **Similar species:** Common Yellowthroat (CYT; *G. trichas*; see below) is smaller, has a thinner bill, and is usually more active. In addition, compared to the CYT, male Bahama Yellowthroats (BYT) have more extensive yellow underparts and a larger face mask extending onto the nape; females have a gray wash on the head. BYTs frequent dense, low-stature woody thickets, whereas the CYT favors open grassy wetlands or old fields. **Vocalization:** Call is a distinctive *tuck*. Song is a loud *wichety wichety wichety wich*, with slower tempo than CYT song.

Common Yellowthroat   *Geothlypis trichas*

Range and status: Common winter resident and transient archipelago-wide. Breeds across North America. Northern populations winter in the southern parts of the breeding range and from Mexico south to Central America and the Caribbean.

Description: 12.7 cm (5 in). Commonly seen olive-green warbler with a yellow throat and chest and a white belly. Adult males have a black face mask, bordered above with white or gray. Females and immatures lack the black mask and have paler underparts. First-year males have a faint black mask, which darkens completely by spring. **Similar species:** Bahama Yellowthroat (*G. rostrata*; see above for differences). **Vocalization:** Call is a soft *jip*.

Hooded Warbler   *Setophaga citrina*

Range and status: Uncommon winter resident and transient. Breeds in eastern North America. Winters mostly from Mexico south through Central America, less commonly in the Caribbean east to the Virgin Islands.

Description: 13.3 cm (5.3 in). Yellow-green warbler with plain olive-green-brown upperparts and bright yellow underparts and white tail spots, best seen when the tail is fanned. Males are unmistakable with a black hood, which surrounds their yellow faces; the hoodless females and immature males have an olive-green cap. **Habits and habitat:** Actively forages for invertebrates in coppice understory; often fans its tail while foraging. **Vocalization:** Call is a loud, rich *chip*; heard more often than seen.

American Redstart   *Setophaga ruticilla*

Range and status: Common winter resident and transient. Breeds throughout much of North America and winters from Mexico south to northern South America and the Caribbean.

Description: 12.7 cm (5 in). Commonly seen, sexually dimorphic warbler. Adult males are unmistakable with a jet-black face, breast, and upperparts except for large orange-red patches on their wings, tails, and flanks, and a white belly. Females have yellow patches in the same pattern as the male but have dull-gray heads, olive backs, and dull-white underparts. Juveniles are similar to females with immature

Setophaga citrina, male

Setophaga ruticilla, male

Setophaga ruticilla, female

Geothlypis trichas, male

Geothlypis trichas, female

males developing a few black feathers on the head and breast. **Habits and habitat:** Frequently observed in coppice and disturbed habitats. Very active species, often fanning its tail to flush insects as it forages through foliage. **Vocalization:** Call is a soft *chip*.

Kirtland's Warbler   *Setophaga kirtlandii*

Range and status: Rare winter resident, most frequent in the central Bahamas. Almost the entire population of this threatened species breeds in young jack pine forest in the northeastern Lower Peninsula of Michigan and winters in the archipelago.

Description: 14 cm (5.5 in). Bluish-brown upperparts with dark streaks on the back, and the underparts are yellow with black, streaked flanks. Undertail feathers (coverts) are white. Both sexes have a distinctive broken white eye-ring. During spring, males have black in their lores and blue-gray upperparts; females and juveniles are generally duller with more brown on the back. **Habits and habitat:** Distinctively, they continually pump their tails. Only two other warbler species exhibit the same behavior: Palm warbler and Prairie warbler (see below), and both have different facial patterns and plumage. The blue-gray plumage, split white eye-ring (both sexes), and constant tail bobbing distinguish the Kirtland's warbler from other warblers seen in the archipelago. Shy and secretive on the wintering grounds, it is rarely seen. Typically associated with broadleaf scrub or secondary habitats, often with fruiting species (White Sage [*Lantana involucrata*], Snow-berry [*Chiococca alba*], Black Torch [*Erithalis fruticosa*]). Frequently seen on or close to the ground. **Vocalization:** Call is strong *tchit*.

Cape May Warbler   *Setophaga tigrina*

Range and status: Common winter resident and transient. Breeds in the boreal forests of Canada and far northern U.S. Winters almost exclusively in the Caribbean.

Description: 12 cm (4.8 in). During spring, males have an olive-green back streaked with black, a yellowish rump, and black crown, while the underparts are yellow, with black streaking. Throat is bright yellow, and the face is chestnut with a black eye stripe. Has a narrow white wing bar. Other plumages variable but may resemble washed-out versions of the male's spring plumage and lack the distinctive head pattern. **Habits and habitat:** Insectivorous on the breeding grounds but feeds extensively on fruit and nectar in winter. Unique among warblers, they have a tubular tongue to facilitate feeding on nectar. **Vocalization:** Call is a thin *sip*.

Northern Parula   *Setophaga americana*

Range and status: Common winter resident and transient archipelago-wide. Breeds throughout eastern North America. Winters in southern Florida, Central America, the Caribbean.

Description: 11 cm (4.3 in). Small, brightly colored warbler with blue-gray upperparts, a yellow breast, white belly, and a short tail. They have a noticeable green patch on the upper back, two white wing bars, a split, white eye-ring, and a black stripe through eye. Males in spring plumage have black lores and a rufous breast band. Females are duller and lack the breast band. **Vocalization:** Call is a soft *chip*.

*Setophaga
kirtlandii*,
spring male

Setophaga tigrina, spring
male

Setophaga tigrina, first fall/
winter plumage

Setophaga americana, female

Setophaga magnolia, fall male

Setophaga americana, fall male

Magnolia Warbler   *Setophaga magnolia*

Range and status: Uncommon winter resident and transient archipelago-wide.
  Breeds in northeastern North America. Winters primarily in Mexico, Central
  America.

Description: 12 cm (4.8 in). Small, brightly colored warbler with a very variable
  seasonal plumage. All plumages have bright yellow underparts, white undertail
  coverts, two white wing bars, a complete eye-ring, yellow rump, and distinct

black tail with a white patch in the middle of the tail. Breeding males have a white supercilium; white, gray, and black backs and yellow throat; a yellow and black-streaked belly; and white, gray, and black foreheads. Females are duller and have less black on head, throat, and belly. Winter plumage is duller gray upperparts and gray streaking on the yellow belly and flanks. **Vocalization:** Call is a distinctive hoarse, but squeaky *choof.*

Yellow Warbler   *Setophaga petechia*

Range and status: Permanent and winter resident. Endemic subspecies *S. p. flaviceps* is uncommon to locally common permanent resident in the northern islands and more common in the southern archipelago. Species is widely distributed throughout North America and northern South America. Some birds from North America breeding population may overwinter.

Description: 12 cm (4.7 in). Unmistakable yellow warbler with golden-green upper-parts, brilliant yellow underparts, rusty-red streaks on the breast and flanks, yellow tail spots, and a dark (black) iris and yellow eye-ring. Males are yellower than females. Resident subspecies is difficult to distinguish from migrant individuals from North American. **Habits and habitat:** Most commonly seen in mangroves, swamps, and coastal scrub. **Vocalization:** Call is a soft or harder *chip* or *ship*, and the song is a *sweet sweet sweet, I'm so sweet.*

Blackpoll Warbler   *Setophaga striata*

Range and status: Transient more commonly seen in the northern islands in spring and the southern archipelago in the fall. Breeds in northern North America and winters in South America.

Description: 13.3 cm (5.3 in). During spring, males are mostly black and white with black-streaked gray upperparts, a prominent black cap, white cheeks, two white wing bars, and pale heavily streaked white underparts. Spring females are duller with olive-gray-green streaked upperparts, pale lightly streaked underparts, and they lack the strong facial patterns of the male, instead having a gray face and crown. Winter plumage of adult and juveniles is similar: greenish heads, dark-streaked greenish upperparts, streaked yellowish breasts, and pale underparts. A distinctive contrast between the dark legs and the pale feet regardless of season. **Vocalization:** Weak *zeet* given in flight.

Black-throated Blue Warbler   *Setophaga caerulescens*

Range and status: Common winter resident and transient. Breeds in northeastern North America; mostly winters in Greater Antilles.

Description: 13 cm (5 in). Striking sexually dimorphic warbler; adult males are unmistakable and have dark blue upperparts and white underparts with a black throat, face, and flanks. Immature males are similarly colored but with more green in the upperparts. By contrast, females have olive-brown upperparts and olive underparts with darker wings and tail, a gray crown, and a brown patch on the cheek. All birds have a small, white patch on the wing and a thin, pointed bill. **Habits and habitat:** Forages for fruits, insects, and spiders in understory in relatively moist coppice sites and gardens. **Vocalization:** Call is a flat *ctuk.*

Setophaga striata, spring male

Setophaga petechia flaviceps, male

Setophaga palmarum palmarum,
winter plumage

Setophaga caerulescens, male

Setophaga caerulescens, female

Palm Warbler   *Setophaga palmarum*

Range and status: Common winter resident and transient. Western Palm Warbler (*S. p. palmarum*) archipelago-wide. Breeds in Canada and extreme northern U.S. Winters in the southern U.S., the Caribbean, locally in Central America.

Description: 12.7 cm (5 in). Winter plumage birds have dull-brown upperparts with a brown to rusty-brown crown, pale supercilium, and yellow-green rump; pale gray-brown streaked underparts; and bright yellow undertail coverts. Spring plumage birds have a more distinct chestnut crown, yellow throat and supercilium, yellow undertail coverts, and yellow-white underparts. **Habits and habitat:** Conspicuous. Often observed in monospecific small flocks on the ground in open areas, e.g., grassland, agricultural lands, dumps, towns, and other areas near human habitation. Constantly bob their tails. **Vocalization:** Call is a sharp *chick* but also a *seep* call in flight.

Olive-capped Warbler   *Setophaga pityophila*

Range and status: Permanent resident. Common on Grand Bahama and Abaco; absent from other islands. Outside the archipelago found only on Cuba.

Description: 13 cm (5 in). Sexually dimorphic warbler with gray upperparts and dingy white underparts. Males have a bright yellow throat and upper breast, an olive-yellow crown, and black barring on the flanks. Females have a duller-colored crown, throat, and breast. **Habits and habitat:** Found exclusively in pine woodlands, typically foraging in the highest part of the trees. **Vocalization:** Song is high, fast, whistled *swee swee swee* notes followed by a rapid *chu chu chu*.

Pine Warbler   *Setophaga pinus*

Range and status: Permanent resident. Endemic subspecies *S. p. achrustera* common on Grand Bahama, Abaco, north Andros, New Providence. Rare winter resident and transient elsewhere, but not recorded on the TCI. Species is also a permanent resident on Hispaniola, and partial migrant to permanent resident in eastern U.S.

Description: 14 cm (5.5 in). Yellow-green warbler with olive upperparts; white belly; two white wing bars; dark legs; thin, relatively long, pointed bills; and yellowish lines over their eyes. Males have bright yellow throats and breasts and split yellow eye-rings; those of females and juveniles are paler. **Habits and habitat:** Occurs only in pine forests. Forages primarily on tree trunks, branches, and pine cones, and less often on the ground. Diet includes insects, seeds, and berries. **Vocalization:** Call is a sharp, rich *chip*. Song is a musical trill.

Yellow-rumped Warbler   *Setophaga coronata*

Range and status: Locally common winter resident and transient of the subspecies *S. c. coronata* (Myrtle Warbler), which breeds in the northeastern U.S. and across boreal Canada. Winters from the central U.S. south to the northern Caribbean region and Mexico south along the Caribbean coast of Central America.

Description: 13 cm (5.3 in). Spring plumage is a mosaic of slate-gray-blue upperparts; pale streaked underparts, with conspicuous yellow patches on the crown, flank, and especially on the rump; a white throat; eye stripe; and a contrasting black cheek patch. Females are less bright than males but have conspicuous yellow rumps. Winter plumage is duller (more brown-gray) with less conspicuous yellow patches on the crown and flank. Numbers vary dramatically from year to year. **Habits and habitat:** Highly frugivorous during winter; occasionally found in large monospecific flocks. **Vocalization:** Call is a dry *check*.

Setophaga pityophila, male

Setophaga pinus achrustera, male

Setophaga coronata, winter plumage

Setophaga dominica, male

Yellow-throated Warbler   *Setophaga dominica*

Range and status: Winter resident and transient. More common in the northern and central Bahamas, less common in the south of the archipelago. Breeds in the southeast U.S. In the Caribbean, also winters in Greater Antilles.

Description: 13–14 cm (5–5.5 in). Strikingly colored warbler with gray upperparts and wings, a yellow throat and breast, and white belly. Two white wing bars and black streaking along the flanks. Head is strongly patterned in black and white, with a long, white supercilium, black cheeks, and a white crescent under the eye. Females, juveniles, and non-breeding males resemble washed-out versions of males, especially a less clearly defined head pattern. **Habits and habitat:** Found across a wide range of natural, urban, and suburban habitats. Often seen near buildings and in palms. **Vocalization:** Call is a high *see* or sharp *chip*.

Bahama Warbler   *Setophaga flavescens*

Range and status: Endemic. Permanent resident found only on Grand Bahama and Abaco where locally common in pine woodlands.

Description: 13–14 cm (5–5.5 in). Once considered a subspecies of the Yellow-throated warbler (YTW) (*S. dominica*) (see above). Differs from YTW by having a longer, slightly down-curved bill; the yellow on the breast extends into the black barring on the flanks; and lacks the white patch on the neck bordering the black cheek. **Habits and habitat:** Restricted to pine woodlands. Forages in the upper branches and on pine cones; often climbs on tree trunks. Can be confused with the Kirtland's Warbler (*S. kirtlandii*; see above). **Vocalization:** Song is a sweet, whistled *sweer, sweer, sweer* ending with *swee-up,* which can be heard year-round.

Prairie Warbler   *Setophaga discolor*

Range and status: Common winter resident and transient. Breeds in eastern North America. Primarily winters in the Caribbean.

Description: 11.5 cm (4.5 in). Commonly seen, small, brightly colored green-yellow warbler with yellow underparts, dark streaking on the flanks, olive upperparts with rusty streaks on the back, and pale wing bars. Individuals have a yellow supercilium, a dark line through the eye, and a yellow crescent under the eye bordered by black. Females and juveniles are duller than males and less distinctly marked around the face. **Habits and habitat:** One of the most common wintering warblers in the archipelago, they are found in many habitats, particularly along forest edges and secondary habitats. They typically bob their tails. **Vocalization:** Call is sharp *tchit.*

Black-throated Green Warbler   *Setophaga virens*

Range and status: Uncommon winter resident and transient. Breeds primarily in northern North America. Winters from Mexico south through Central America and the Caribbean, mostly western Greater Antilles.

Description: 12 cm (4.8 in). Both sexes have an olive-green crown, a yellow face with an olive patch behind the eye, a thin, pointed bill, white wing bars, an olive-green back, and pale underparts with black streaks on the flanks. Adult males have a black throat and upper breast; females have a pale throat and black markings on their breast. **Vocalization:** Call is a sharp *tsip.*

Family Cardinalidae (Grosbeaks, Buntings, and Cardinals)

New World family of mostly sexually dimorphic, seed-eating species with relatively heavy, cone-shaped bills or insect- and fruit-eating species with thinner bills. Six species are described for the archipelago; all are transients or winter residents. In addition to the species described below are two species commonly referred to as "tanagers" (and were formerly considered in the family Thraupidae; see below): the Summer Tanager (*Piranga rubra*) (17–18 cm [6.8–7.2 in], a rare migrant and winter resident) and Scarlet Tanager (*Piranga olivacea*) (16–19 cm [6.3–7.5 in], a rare migrant).

Setophaga virens, male

Setophaga discolor, male

Setophaga flavescens

Pheucticus ludovicianus, female

Pheucticus ludovicianus, spring male

Rose-breasted Grosbeak   *Pheucticus ludovicianus*

Range and status: Uncommon winter transient. Breeds in eastern North America. Most winter from Mexico south to northern South America.

Description: 20 cm (8 in). Large, thickset species with a heavy, cone-shaped bill. In spring, males have distinctive black-and-white plumage with a black head and upperparts, and a bright rose-red patch on the breast. Wings have two white patches and white underparts. By contrast, during winter males have mottled brown upperparts and white underparts and resemble females. Females and immatures have dark gray-brown upperparts, a white supercilium, a buff stripe along the top of the

381

head, and black-streaked underparts that are white tinged with buff. **Vocalization:** Call is a sharp squeaky *eek* or *peek*.

Blue Grosbeak   *Passerina caerulea*

Range and status: Uncommon winter transient. Breeds in the southern U.S. and northern Mexico. Winters in Central America south to northern South America; rarely in the Caribbean.

Description: 17 cm (6.8 in). Sexually dimorphic, medium-sized species with large, cone-shaped bill. In spring, males are uniform dark blue; females are brown. Both have two rufous wing bars, which distinguish this species from the smaller Indigo Bunting (*Passerina cyanea*) (see below), which lacks these wing bars and has a smaller bill. **Habits and habitat:** They feed primarily on insects but also eat snails, spiders, seeds, and fruits. This species is found in partly open habitat with scattered trees, scrub, coppice edges, and overgrown fields.

Indigo Bunting   *Passerina cyanea*

Range and status: Uncommon to locally common winter resident and transient. Breeds in eastern North America. Winters in Greater Antilles and from southern Florida and Mexico south through Central America.

Description: 14 cm (5.5 in). Spring plumage males are a uniform indigo color; in winter they are primarily brown, as are females, but males frequently retain some blue feathers. **Habits and habitat:** Feeds on seeds in the winter, on the ground; often in small flocks in brushy areas and grasslands.

Painted Bunting   *Passerina ciris*

Range and status: Uncommon winter resident and transient in the northern and central Bahamas. Rare elsewhere in the archipelago. Breeds in southeast U.S. and Mexico. Winters from Florida, occasionally Cuba and Mexico south to Central America.

Description: 14 cm (5.5 in). Sometimes described as the most beautiful bird in North America, the adult male is unmistakable, maintaining his colorful plumage year-round. It has a blue head, red eye-ring, red underparts, yellow-green back, and a red rump. Females and juveniles have olive-green upperparts and pale yellow-green underparts. Juvenile males exhibit a combination of adult and juvenile plumages in spring. **Habits and habitat:** Often shy, inhabiting thickets and woodland edges bordering open areas. This species is illegally trapped and kept as a cage bird in many parts of its wintering grounds.

Family Thraupidae (Tanagers and allies)

Tanagers are a diverse and large New World group of small- to medium-sized, brightly to dull-colored, forest and grassland dwelling birds found mainly in the tropics. They feed on fruit, nectar, insects, and seeds. Most have short, rounded wings. Four permanent resident species are recorded in the archipelago.

Passerina ciris, male

Passerina ciris, female

Passerina caerulea, male

Passerina caerulea, juvenile

Passerina cyanea, spring male

Coereba flaveola bahamensis

Bananaquit   *Coereba flaveola*

Range and status: Common permanent resident. Endemic subspecies *C. f. bahamensis* throughout the archipelago. Species also widely distributed throughout the Caribbean, and Mexico south through tropical South America.

Description: 11.4 cm (4.5 in). Small-sized species with a distinctive short, dark, slender, decurved bill; dark upperparts; a white throat and supercilium; yellow belly and rump; and a short, square tail. Juveniles have brownish upperparts and dull-green underparts with "blacked out" facial markings. **Habits and habitat:** Feeds primarily on nectar, but also insects and fruit. Unlike hummingbirds, individuals perch while feeding and may pierce the base of larger flowers to obtain nectar. Found in many habitats, including early and late successional coppice, pine forests, and suburban areas. **Vocalization:** Wide range of calls including squeaks, clicks, and buzzes.

Cuban Grassquit   *Tiaris canorus*

Range and status: Locally common permanent resident on New Providence; accidentally introduced in 1963. Also Cuba.

Description: 11 cm (4.5 in). Small-sized, dark-colored, and compact; short tail. Males have an olive-green back, dark gray underparts, and a black face, throat, and breast. The black face is circled by a yellow band. Females are similar but duller, and the coloring of the face is more red-brown than black. **Vocalization:** Call is a high, soft *seet*, and the song is a repeated high, raspy *chit chita-lee*.

Black-faced Grassquit   *Tiaris bicolor*

Range and status: Common permanent resident on most major islands. Not recorded on the Turk islands. Found throughout the Caribbean except Cuba.

Description: 11 cm (4.5 in). Small-sized, dull-colored bird with short tail. Adult males have olive-green upperparts, paler gray-olive underparts, with a black head and breast. Females and immatures have dull olive-gray upperparts and head, and paler gray underparts becoming whiter on the belly. **Habits and habitat:** Often observed in small groups in grassy and scrub habitats. Feeds on the ground, primarily on seeds and often flush from the ground when approached. **Vocalization:** High soft *sip*, and the song is a repeated *tink tink tink kzcheee*.

Greater Antillean Bullfinch   *Melopyrrha violacea*

Range and status: Permanent resident. Endemic subspecies *M. v. violacea* is commonly found on most of the larger islands in The Bahamas; *M. v. ofella* recorded on Middle and East Caicos, with occasional records from North Caicos. Species also in Hispaniola and Jamaica.

Description: 17 cm (6.5 in). Unmistakable, all-black bird with a heavy head and bill, a rusty-red eye stripe above the eye, and a rusty-red throat and undertail coverts. Females duller black than males. Juveniles are brown with the red markings of the adults; individuals molting into adult plumage have a mixture of brown and black feathers. **Habits and habitat:** Coppice. Diet is fruit, seeds, and insects. **Vocalization:** Call is a soft *seet seet* call, and the song is repeated *seet seet seet seet*. Locally known as the Police Bird.

Tiaris canorus, male

Tiaris canorus, female

Tiaris bicolor, male

Melopyrrha violacea violacea, male

Mammals

Fossil evidence indicates that the archipelago has supported few mammals. This is likely because the islands are geologically recent; have had no land connections with the mainland United States or other islands in the Caribbean; and except for bats, mammals are generally poor dispersers over water.

Currently twenty-three species of terrestrial mammals have been recorded in the archipelago; fourteen native and nine introduced species. All but one native mammal is a bat. Only two endemic mammals are in the archipelago: a bat and a rodent, and no endemic mammals are on the TCI. Several domestic species have established feral populations including goats (*Capra aegagurus hircus*), donkeys (*Equus asinus*, on Great and Little Inagua), and feral cats (*Felis catus*). Several marine mammals occur in the archipelago, primarily whales and dolphins and although not covered extensively in the text, a checklist of these species is provided in the Appendix.

Order Chiroptera (Bats)

Bats are the only mammals capable of sustained flight, but unlike birds they do not flap their entire forelimbs. Instead, they flap their spread-out long digits (fingers), which support a wing-like thin membrane (patagium) that is also attached to the body (see illustration). Of more than 1300 species (c. 20% of all mammal species), the majority are insectivorous (70%), feeding on prey that includes moths, beetles, dragonflies, flies, bugs, wasps, and even ants, most of which are typically taken on the wing in midair, or from vegetation and the ground. Most other bat species are fruit-eaters, especially in the subtropics and tropics, and a few species specialize in feeding on flower nectar, fish, frogs, small birds, and even blood. Bats play important ecological roles as insect predators, plant pollinators, and seed dispersers. Most species become active one to two hours after sunset.

There are two groups of bats: 1) suborder Pteropodiformes (alternatively termed Yinpterochiroptera), which are primarily large bats, with many species commonly referred to as flying foxes, relying on sight to fly and to find their food (nectar, pollen, fruit); and 2) suborder Vespertilioniformes (alternatively termed Yangochiroptera), which primarily are small, nocturnal species, roosting in natural or man-made cavities during the day and using echolocation (ultrasonic sounds) at night to fly and feed. All bats in the archipelago are in the suborder Vespertilioniformes. Although many species are dark brown or gray, some brightly colored and distinctly marked species also occur. Bats are often difficult to identify in the field, especially when in flight.

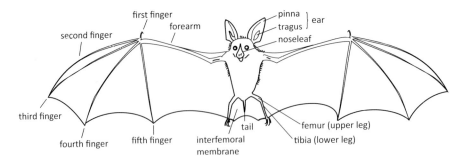

Anatomy of a bat, line illustration. D. Jean Lodge.

Like other mammals, bats can contract rabies and can transmit the disease through bites. Veterinarians and other individuals who handle wild animals, including bats, should be vaccinated against rabies. Even sick bats rarely bite unless handled. Thus, for anyone who simply does not handle bats, the odds of contracting a disease from one are extremely remote. Nonetheless, histoplasmosis, a respiratory disease (also known as cave disease, Ohio valley disease, and reticuloendotheliosis), can be contracted from exposure to dust containing spores of the fungus *Histoplasma capsulatum*, which occurs in decomposing feces of either bats or birds.

Thirteen extant species of bat have been recorded in the archipelago; ten are regularly recorded in archipelago. Most are Greater Antillean in origin and all are small insectivorous, frugivorous, or nectarivorous nocturnal species. The status of some species on some islands remains unclear and seasonal inter-island movements and even movements between the mainland and/or the other islands in the region and the archipelago is possible.

In addition to species described below, three accidental bat species have been recorded in the archipelago: 1) the Greater Bulldog or Fisherman Bat (*Noctilio leporinus* [Family Noctilionidae]), a large fish-eating species from Central and South America, recorded once on Great Inagua; 2) Silver-haired Bat (*Lasionycteris noctivagans* [Family Vespertilionidae]), a species of vesper bat recorded once on the TCI; and 3) Tomes' Sword-nosed Bat (*Lonchorhina aurita* [Family Phyllostomidae]), a species of leaf-nosed bat typically found in Central and South America with large ears and a very prominent sword-like noseleaf, which is as long as the head, recorded once on New Providence.

FAMILY PHYLLOSTOMIDAE (New World Leaf-nosed Bats)

New World Leaf-nosed bats are a common and diverse group that includes c. 140 species. Their most conspicuous identifying characteristic is a "noseleaf," a fleshy protuberance from the top of the nose used in echolocation. In most species this is a small, spear-shaped structure; it can range in size from being virtually nonexistent to nearly as long as the head itself. Most species are insectivorous although some species feed on fruit, nectar, pollen, bats, frogs, other small vertebrates, and even blood. Noseleaf appendages are reduced in size for non-insect-feeding species. This family includes the most common bats found in the archipelago; five species are recorded.

Macrotus waterhousii

Macrotus waterhousii

Monophyllus redmani

Waterhouse's Big-eared Bat   *Macrotus waterhousii*

Range and status: Found on most islands in the archipelago (not recorded on Little
   Abaco, Grand Bahama, or Mayaguana); also the southern U.S., Greater Antilles,
   Central America.

Description: Body length 50–69 mm (1.9–2.7 in), weight 12–18 g (0.4–0.6 oz). Unmistakable medium-sized species with distinctively long, rounded ears (25 mm [1 in])
   connected across the forehead by a flap of skin. They have brown upper parts and
   paler brown-gray underparts. The noseleaf is small and the tail extends slightly
   beyond the interfemoral membrane. **Habits and habitat:** Foraging flight is slow
   and maneuverable, often hunts close to the ground. Found in a variety of dry
   habitats. Diet includes fruit and insects. Unlike other bats, does not crawl across
   cave ceilings on feet and thumbs but can walk rapidly in an upside-down position.
   During the day, roosts in large caves or buildings in groups up to 500 individuals.

Leach's Long-tongued Bat   *Monophyllus redmani*

Range and status: Uncommon to locally common in the southern archipelago
   north to Crooked Island; also Greater Antilles.

Description: Body length 48–66 mm (1.9–2.6 in), weight 8–14 g (0.3–0.5 oz). Small,
   brown-gray colored bat with a long slender snout and a small noseleaf. Tail extends

approximately half its length beyond the interfemoral membrane. **Habits and habitat:** Primarily nectarivorous and has a long tongue that is adapted for the consumption of nectar but also consumes pollen, insects, and occasionally fruit. During the day, roosts in caves, often in large colonies.

Jamaican Fruit-eating Bat *Artibeus jamaicensis*

Range and status: Uncommon to locally common in the southern archipelago (Mayaguana, Little and Great Inagua, Providenciales); also Greater Antilles, Central and northern South America.

Description: Body length 70–90 mm (2.8–3.5 in), weight 36–48 g (1.3–1.7 oz). Medium-sized thickset bat, with short, velvety dark fur (often exhibiting a silvery tinge, especially on the neck and shoulders), and broad dark gray wings, but no tail. Head is large and typically has two pale stripes on the face each running from forehead to end of the snout and each passing near the inner side of the large eyes. Broad ears are triangular, pointed, ridged, and relatively widely separated. The short tragus has four or five relatively small serrations on the outer margin. Conspicuous noseleaf has a cluster of sebaceous glands, and the lower lip has a large central wart. It also has large canines in comparison to other members of its genus. **Habits and habitat:** Found in diverse habitats, depending on availability of suitable food, which is almost exclusively fruit, especially figs, but also includes pollen, nectar, flowers, leaves, and insects. In addition to more typical roosting sites, this species can also modify leaves to make a tent-like structure to provide a temporary daytime roost.

Cuban Fruit-eating Bat *Brachyphylla nana*

Range and status: Only known from one cave on Middle Caicos and from fossil records in The Bahamas; also Cuba, the Cayman Islands.

Description: Body length 70–85 mm (2.6–3.4 in), weight 28–32 g (0.9–1.1 lb). Large-sized species with a very small vestigial noseleaf and a V-shaped groove margined by small wart-like projections (tubercles) on the lower lip. Medium-sized ears with tragus form a slightly rounded V-shape at the top. Typically, light brown to ivory in color with reddish or yellowish tones. Neck, shoulders, and sides are usually paler in color (grayish-white) than the rest of the body, and the underside is brown. Has a well-developed interfemoral (tail) membrane with a concealed, vestigial tail. **Habits and habitat:** Diet includes fruit, pollen, nectar, and insects. Highly gregarious and large colonies are found in caves, sometimes with other species.

Buffy Flower Bat *Erophylla sezekorni*

Range and status: Widespread and common throughout the archipelago including Great Abaco, Eleuthera, Long Island, Cat Island, Great and Little Exuma, San Salvador, Crooked Island, Acklins Island, Great Inagua, Andros, Mayaguana, New Providence, East, West and North Caicos, The Providenciales; also Cuba, Jamaica, Cayman Islands.

Artibeus jamaicensis

Artibeus jamaicensis

Brachyphylla nana

Erophylla sezekorni

Description: Body length 60–75 mm (2.4–2.9 in), weight 14–18 g (0.5–0.6 oz). Little-known, medium-sized species with highly variable colored upper parts, ranging from cream to a pale gray-brown. Facial skin is pale, often white, while the underparts are a more uniform lighter brown. Individuals have a distinctive steep forehead and elongated nose, with a small notched or forked noseleaf and a long tongue to feed on nectar, and less frequently on insects and fruit. It has a narrow, pointed tragus; broad, naked wings; and a short tail that extends slightly beyond the tail membrane. **Habits and habitat:** Foraging typically begins later at night than for other bat species.

Funnel-eared bats are widely distributed in forests throughout Central and South America and the Caribbean. Their common name originates from their large, pale, funnel-shaped ears, with glandular papillae on the surface of the external ear and a short, typically thick triangular tragus. Body is small and slim with relatively long legs that can be longer than the head and body combined. Wings are long and narrow, and the slender tail is about as long as or longer than the legs and is completely enclosed in the tail membrane. An additional characteristic of this family is a unique structure on the face or muzzle of adult males, commonly known as the "natalid organ," which is a visible (glandular) swelling or projection made of sensory cells, the purpose of which is unknown.

Bahamian Funnel-eared Bat   *Chilonatalus tumidifrons*

Range and status: Endemic. Recorded only from Abaco, Andros, San Salvador.

Description: Body length 38–45 mm (1.5–1.8 in), weight 4–7 g (0.14–0.25 oz). Small-sized species with very wide ears shaped like a funnel (hence its common name), a noticeable moustache on the upper lip, and a distinctive ridge on the chin, which resembles a second lower lip. Fur is dense and long, and the upper parts are pale yellowish-brown with either a reddish or chestnut-brown tinge, and the underparts are uniformly pale yellowish-brown (buff). Male natalid organ on the face is very noticeable. **Habits and habitat:** Insectivorous and an agile flyer, able to forage among dense foliage. Populations may be limited by availability of suitable roosting caves. Listed as Near Threatened.

Gervais' Funnel-eared Bat   *Nyctiellus lepidus*

Range and status: Recorded throughout the central Bahamas (including Cat Island, Eleuthera, Little Exuma, Long Island); also Cuba.

Description: Body length 30–40 mm (1.2–1.6 in), weight 3–5 g (0.1–0.2 oz). Small-sized species (smallest bat in the archipelago) similar in appearance to *Chilonatalus tumidifrons* (see above) but has smaller ears and lacks the lip-like projection on the chin. **Habits and habitat:** Insectivorous and an agile flyer, able to forage among dense foliage. Typically roosts in large, single species colonies in caves.

FAMILY VESPERTILIONIDAE (Vesper or Evening Bats)

Vesper or evening bats have the most species—more than 300—and are the most well-known family of bats. Species often have long ears but lack the enlarged noses of other bats, hence another of their common names: plain-nosed bats. The tails are typically long and frequently extend beyond the edge of the tail membrane. Members of this diverse family of bats of varying sizes (4–50 g [0.14–1.76 oz]) are often dark-colored, and the family includes both colonial and solitary roosting species.

Big Brown Bat   *Eptesicus fuscus*

Range and status: Uncommon to locally common on the Great Bahama Bank (including New Providence, Andros, Great and Little Exuma, Long Island), San Salvador, the southern Bahamas (including Crooked Island, Acklins Island).

Chilonatalus tumidifrons

Chilonatalus tumidifrons

Eptesicus fuscus

Eptesicus fuscus

Nyctiellus lepidus

Not recorded on the TCI. Two endemic subspecies: *E. f. bahamensis* is found on New Providence and San Salvador (sometimes referred to as Bahamian brown bats); *E. f. dutertreus* elsewhere in the archipelago. Species also in most of temperate North America, Central and South America.

Description: Body length 50–65 mm (1.9–2.6 in), weight 11–23 g (0.4–0.8 oz). Medium-sized, heavy-bodied species with long, shiny, brown hair and dark-colored (brown or black) skin; upper parts are typically darker than underparts. Tail extends beyond the interfemoral (tail) membrane. **Habits and habitat:** Insectivorous, capturing prey in flight; hunts throughout the night. Occasionally audible sounds (clicks) are produced during flight. Frequently roosts, in sometimes large, single-species colonies, but may also roost with *Tadarida brasiliensis*.

Red Bat   *Lasiurus borealis**

Range and status: Recorded but scarce throughout the archipelago (including Andros, Cat Island, Grand Bahama, Great Inagua, Long Island, Mayaguana, New Providence, Middle Caicos, The Providenciales); also temperate North America, Greater Antilles. *Considered by some authorities as an endemic subspecies *L. borealis minor* or even a separate species known as Minor Red Bat (*L. minor*).

Description: Body length 47–57 mm (1.9–2.2 in), weight 6–11 g (0.2–0.4 oz). Medium-sized, sexually dimorphic species with dense fur and small rounded ears. Males have red fur; females are chestnut-colored with whitish frosting on the tips of the hairs. Underparts are typically paler in color than the upper parts. Upper side of tail is furry but does extend beyond the tail membrane. **Habits and habitat:** Insectivorous, typically foraging over open spaces. Individuals often use solitary roosts in tree foliage. Listed as Vulnerable.

FAMILY MOLOSSIDAE (Free-tailed Bats)

Free-tailed bats are named for their thick tail that extends beyond the tail membrane, with their tail making up almost half of their length. Muzzles of all Molossids are generally short and wide and often have wide, fleshy lips that may have folds or creases. Also, many have a distinctive pad over their noses, which may have small horn-like projections. Their ears are relatively short and thick, often joined across the forehead and point directly forward. The eyes are relatively small, while the lips are large. All species have long, narrow, thick wings and, as with the tail, are covered in a leathery membrane. Molossids have short, strong legs and broad feet. On the outer toes of each foot are curved bristles, which are used for grooming their fur. Species generally have short, velvety fur that is initially black but discolors over time giving the impression of reddish and brownish hues. Molossids are swift fliers and generally obtain insects above forest canopies.

Lasiurus borealis

Tadarida brasiliensis

Brazilian or Mexican Free-tailed Bat   *Tadarida brasiliensis*

Range and status: Locally common from Little Abaco to Crooked-Acklins (including Acklins, Crooked Island, Eleuthera, Great and Little Abaco, Great and Little Exuma, Long Island), Middle Caicos; also southern U.S., the Caribbean, Central and South America.

Description: Body length 50–60 mm (1.9–2.4 in), weight 10–12 g (0.35–0.42 oz). Medium-sized bat with a short snout and a very distinctive upper lip with a series of deep, vertical grooves, which distinguish it from other bats in the archipelago. Fur color varies from dark brown to gray. Tail is very long, extending well beyond the tail membrane and making up almost half the bat's length. Wings are long, narrow, and pointed, suitable for rapid, direct flight. **Habits and habitat:** Foraging is most frequent at dusk and dawn, corresponding with peak prey activity; insects are normally taken on the wing above the canopy or in open areas. Feeds on a wide variety of insects, including large numbers of crop pests. Frequently associated with dry coppice. One of the most widespread and abundant native mammals in the Western Hemisphere; daytime roosts can contain millions of individuals (the record is 20 million in Texas, U.S.). Has the highest recorded flight altitude among bats (c. 3300 m [10,826 ft]).

Order Rodentia (Rodents)

Rodents possess two continuously growing, typically prominent incisor teeth in both the upper and lower jaws. Most species eat seeds or plants, though some have more varied diets. One endemic and three introduced species of rodent occur in the archipelago.

FAMILY CAPROMYIDAE (Hutias)

Hutias are moderately large, guinea pig-like rodents that are endemic to and widely distributed throughout the Caribbean. At least twenty species have been described, ranging in size from 20–60 cm (7.9–24 in), and are typically herbivorous and nest in trees or rock crevices. Many species are now extinct through a combination of habitat loss, introduction of alien predators, and (historic) overhunting.

Bahamian or Ingraham's Hutia *Geocapromys ingrahami*

Range and status: Endemic. East Plana Cay (southeast Bahamas); introduced on several cays in the Exumas.

Description: Body length 308 mm (12 in); total length 378–395 mm (14.9–15.5 in). Medium-sized, gray-brown, herbivorous rodent with a stout body, large head, and short tail. **Habits and habitat:** Nocturnal, typically remaining underground during the day, emerging at night to feed on fruits, leaves, bark, and possibly seaweed. Feeds on ground and lower tree branches. Formerly widespread in The Bahamas but reduced to a single population on East Plana Cay from which individuals were successfully translocated to two cays in the Exumas. Hutias are listed as Vulnerable and are protected in The Bahamas. Locally known as Hooties.

FAMILY MURIDAE (Rats, Mice, and allies)

Three species have been introduced to the archipelago; two species of rat and one species of mouse. All three species originated in Asia but now have a worldwide distribution except in Antarctica, through accidental introductions by man. These species are highly prolific breeders, largely nocturnal, often associated with cities and settlements, and often invade houses.

Brown or Norway Rat *Rattus norvegicus*

Range and status: Introduced. Archipelago-wide; also worldwide.

Description: Body length 200–270 mm (7.9–10.6 in); total length c. 500 mm (c. 20 in). Large, strongly built rodent with coarse hair; typically brownish-gray with a paler underside. Tail is shorter than the body and is nearly always darker on top. Snout is blunt, and the ears are small and furry. **Habits and habitat:** Usually nocturnal, good swimmers (but poor climbers), and excellent diggers, often constructing extensive burrow systems. Omnivorous, but seeds and cereals form a substantial part of their diet. Dominant introduced rat in Europe and much of North America, most common in urban areas or settlements. Often a threat as a predator on native species on small islands or cays.

Black Rat *Rattus rattus*

Range and status: Introduced. Archipelago-wide and worldwide.

Description: Body length 150–220 mm (5.9–8.7 in); total length 278–403 mm (10.9–15.9 in). More sleek and graceful than the brown rat, with a typically dark (brown or black) body, pale or white underside, a more pointed snout, and large, translucent furless ears. Tail is thin and longer than the body. **Habits and habitat:** Omnivorous; diet includes seeds, fruits, stems, leaves, fungi, and a variety of invertebrates and vertebrates. Nocturnal. Excellent climbers. Exploit a wide range of suburban and natural habitats. Most common in suburban areas, having been replaced elsewhere by the brown rat (*R. norvegicus*). Because of its broad diet and climbing abilities, *R. rattus* is often a serious threat to native island species.

Mus musculus domesticus

Geocapromys ingrahami

Geocapromys ingrahami

Rattus norvegicus

Rattus rattus

House Mouse   *Mus musculus domesticus*

Range and status: Introduced. Archipelago-wide and worldwide.

Description: Body length 75–100 mm (2.9–3.9 in); total length 125–200 mm (4.9–7.9 in). Small, dainty, short-haired, brown-gray rodent with a pale belly and a thin tail approximately the same length as the body, a pointed nose, and small, finely haired ears. **Habits and habitat:** Nocturnal. Excellent jumper and climber. Omnivorous, but primarily a vegetarian, feeding mostly on seeds. Can be a threat as a predator on some native species on small islands.

Carnivores are a diverse group of mammals with adaptations including specialized teeth and claws for catching and eating other animals. Carnivores occur worldwide and are terrestrial and aquatic. Dogs, cats, weasels, mongooses, raccoons, bears, and seals are familiar representatives of this order. All terrestrial carnivores recorded in the archipelago have been introduced.

FAMILY PROCYONIDAE (Raccoons and allies)

Procyonids are a group of small- to medium-sized, highly dexterous, tree-climbing carnivores and/or omnivores found throughout the New World, often with distinct facial markings and banded tails.

Raccoon *Procyon lotor*

Range and status: Introduced. New Providence, Grand Bahama, Eleuthera, Abaco; also North and Central America.

Description: Body length 40–70 cm (16–28 in); total length 60–110 cm (23.6–43.3 in). Medium-sized carnivore with a thick, grayish coat and unmistakable facial mask: an area of black fur around the eyes, which contrasts sharply with the surrounding white face. Fur-covered tail is relatively long with dark rings. Raccoons on The Bahamas often have a slight ocher tint on the neck and shoulders, and the mask is interrupted by a distinct gap between the eyes. **Habits and habitat:** Intelligent, typically nocturnal mammal. Omnivorous; diet includes plant material (nuts and fruits), invertebrates, and vertebrates (including fish, amphibians, sea turtle eggs and hatchlings, birds and small mammals). Front paws are used to find, manipulate, and rapidly process food items. **Conservation status:** Recent analyses, including molecular genetics studies, indicate raccoons were introduced from North America in previous centuries. As alien predators they are recognized as a threat to native wildlife, including the ground-nesting Cuban Parrot on Abaco.

Artiodactyls are hooved, typically herbivorous mammals whose weight is borne equally by the third and fourth toes, and many important domestic species including sheep, pigs, and cattle are in this order.

FAMILY SUIDAE (Pigs, Hogs, or Boars)

Small- to medium-sized mammals with large heads, short necks, relatively small eyes, prominent ears (they have excellent hearing), and a distinctive snout (ending in a disc-shaped nose). They have a highly developed sense of smell and are frequently omnivorous. Wild Suids (swine) typically have short, bristly fur and a short tail ending in a tassle. Communication is by squeals and grunts. Males are larger than females. All Suids are native to Africa, Asia, and Europe, and because of their intelligence,

Sus scrofa, female

Procyon lotor

Sus scrofa

adaptability, and omnivorous diet their introduction to the New World has frequently had detrimental effects on the native species. One introduced species is recorded in the archipelago.

Wild Hog   *Sus scrofa*

Range and status: Introduced. Abaco, Andros, the Exumas, Inagua.

Description: Body length 120–180 cm (47–70 in), weight 50–100 kg (110–200 lb). Feral Bahamian hogs resemble their wild boar ancestors with a compact body, large head, and relatively short legs. Dark-colored fur consists of a mixture of stiff bristles and finer hair. Males are larger and have more prominent tusks than females. **Habits and habitat:** Omnivorous; diet comprises mostly vegetable matter including fruits, seeds, and roots but will also eat a variety of other food including earthworms, mollusks, crabs, other arthropods, fish, and turtle eggs. Poses a threat to native species including rock iguanas. Gregarious; normally most active in the early morning and late afternoon. Populations of feral domestic pigs (*S. s. domesticus*), which lack the long hair and tusks of wild-type hogs, also occur on several islands. Source of populations is unknown, but likely multiple introductions from Europe starting with Columbus.

AQUATIC MAMMALS

Order Sirenia (Sirenians or Sea Cows)

Sirenians, including dugongs, sea cows, and manatees, are large, rotund but stream-lined, nearly hairless, completely aquatic mammals with fluke-like tails. They are widespread in tropical coastal waters in the Americas, Africa, Indo-Australia region, and the western Pacific and Indian Oceans. One species occurs in the archipelago.

FAMILY TRICHECHIDAE (Manatees)

The family Trichechidae includes manatees, of which there are three species, one each in the Caribbean region, Amazon and Orinoco river basins, and West Africa. Characteristics of the family are illustrated by those of the West Indian manatee as described below.

West Indian Manatee   *Trichechus manatus*

Range and status: Rare, more commonly seen in north and central Bahamas. Traditionally considered as having two subspecies: Antillean or Caribbean Manatee subspecies (*T. m. manatus*) recorded throughout the Caribbean and northern South America; and Florida Manatee (*T. m. latirostris*) throughout coastal Florida, as well as along the Atlantic Coast. Recent genetic studies now indicate three distinct populations: 1) Florida and Greater Antilles; 2) Mexico, Central America, and northern South America; and 3) northeastern South America.

Description: Length 3 m (10 ft), weight 400–600 kg (800–1200 lb). Large, aquatic, gray-brown, seal-like mammal with a streamlined body, two front paddle-shaped flippers, and a powerful flat-rounded horizontal tail with no hind limbs. Bulbous face with small, wide-set eyes and stiff whiskers. Nostrils, located on the dorsal surface of the snout, can be closed when submerged, and the eyes have inner membranes that can be drawn across the eyeballs for protection. Small, lobeless ear openings are located just behind the eyes. **Habits and habitat:** Manatees are primarily vegetarians feeding on wide range of plant species and consuming 30–45 kg (60–100 lb) per day using a flexible upper lip that draws food into the mouth. Found in freshwater rivers, estuaries, and in the coastal waters of the Gulf of Mexico and the Atlantic Ocean, preferring waters above 20 °C (68 °F). Solitary or in small groups. **Conservation status:** Conservation efforts have resulted in the population increasing since the 1970s, but it is still considered at risk and is currently listed as Vulnerable.

Infraorder Cetacea (Whales, Dolphins, and Porpoises)

Cetaceans are aquatic, finned, carnivorous mammals consisting of two groups: toothed whales (Order Odontoceti), which feed on fish, invertebrates, and other aquatic mammals; and baleen whales (Order Mysticeti), which have a filter feeding

Tursiops truncatus

Trichechus manatus, mother and calf

Trichechus manatus

mechanism comprising bristle-like projections on a series of plates in their mouths to filter out food, such as small invertebrates (e.g., krill), from the water. At least twenty-two species representing both groups have been recorded in the archipelago (see the Appendix). They occur from relatively shallow, nearshore waters to deep, offshore waters. Most species are seen offshore.

Bottlenose Dolphin   *Tursiops truncatus*

Range and status: Widely distributed throughout the archipelago; also globally throughout most tropical and temperate seas.

Description: Length 2–4 m (6.6–13.1 ft), weight 150–650 kg (330–1430 lb). Medium-sized, gray-colored species and most commonly seen cetacean in the archipelago. Lacks distinctive markings but is typically darker gray along the back, lighter gray along the side of the body, and pale gray to almost white on the belly. **Habitat and habitat:** Two ecotypes described: a shallow water or coastal ecotype and an offshore ecotype. Their ranges overlap, and although genetically distinct they are not considered to be different species. Consumes small fish, crustaceans, and squid.

Acknowledgments

This book benefited from the input and suggestions of many people from The Bahamas, the Turks and Caicos Islands, and elsewhere. Our Bahamian and Turks and Caicos colleagues and friends have been exceedingly generous in sharing their knowledge of the region's geology and biota, in providing constructive reviews of various drafts of the book, and in facilitating our work in many other ways. We thank the participants of the Bahamas Natural History Conferences for helping us gain a broader perspective of the ecology and natural history of the archipelago along with the conservation challenges and successes within the region. We also acknowledge the scientists, various organizations, and naturalists whose books, scientific publications, and informative websites have increased our knowledge of the archipelago's flora and fauna. We gratefully recognize all their collective contributions to the book, while acknowledging that any errors in this book are the responsibility of the authors; we encourage readers to let us know of any errors.

We specifically thank Sandra Buckner, Paul De Luca (University of the Bahamas), Katherine Sullivan Sealey (University of Miami), and an anonymous referee for providing constructive reviews of the initial and a later draft of the book. Their collective comments were invaluable in shaping the content and style of the book. Neil E. Sealey (Media Enterprises) provided valuable critical comments on the original book proposal and his detailed suggestions for the Introduction, especially the sections on geology, were most helpful. We also appreciate Neil's permission to reproduce a figure from his geology text in our Introduction (Figure 1). Others provided invaluable suggestions for various sections of the field guide including Graham Reynolds (www.rgrahamreynolds.info); Sandra Buckner; Jessie Glaeser (USDA Forest Service); Beatriz Ortiz-Santana (USDA Forest Service); Merlin Tuttle (www.merlintuttle.com); Jaboury Ghazoul (ETH, Zurich); Janie Currie; Mike Hill; and Bret Ratcliffe (University of Nebraska). We thank Keith Salvesen (https://rollingharbour.com) for his enthusiasm reviewing drafts of the text, and his invaluable help in sourcing photos. Eric Carey, Lynn Gape, and the management and staff of The Bahamas National Trust (BNT, https://bnt.bs), the staff of The Leon Levy Plant Native Plant Preserve (www.levypreserve.org), and Eleanor Philips of The Nature Conservancy provided assistance and support throughout the development of the book. The Bahamas Marine Mammals Research Organisation (www.bahamaswhales.org) provided data on cetacean records. The International Union for Conservation of Nature and Natural Resources (IUCN) allowed use of their data. We also acknowledge the late Anthony "Tony" White for his interest and support of this book during its development, especially regarding birds, history, and conservation in the archipelago. DC thanks Dominica, Emma, Ian, GOB, DOM, and Esther for their support and encouragement during the writing of this book.

We thank Olga Ramos (USDA Forest Service) for preparing the archipelago map and modifying Figure 1 of the Introduction, Mike Hill for the bird and provisional bat artwork, Laura Sloan Crosby for the plant artwork, and Katrien Van Put for help with the images.

We thank the numerous individuals who generously donated their photos (see list of photo credits), without whose help it would have been impossible to illustrate the diversity of species from the archipelago. We also thank Matthew A. Young for facilitating access to photos in the Macaulay Library of the Cornell Lab of Ornithology (www.macaulaylibrary.org) and Lynn Gape (BNT) in helping to source photos in The Bahamas.

In addition to support from the authors' various institutions during their field work in the archipelago, funds for travel were provided by International Programs of the United States Department of Agriculture (USDA) Forest Service, the James Bond Fund of the Smithsonian Institution, and The Nature Conservancy. We thank the USDA Forest Service Northern Research Station for supporting Dr. Lodge, and the Athens, Georgia, office of USDA FS, Southern Research Station for providing an office to Drs. Lodge and Wunderle after hurricanes Irma and Maria made communication from their offices in Puerto Rico almost impossible. We thank the International Institute of Tropical Forestry, United States Department of Agriculture (USDA), Forest Service (FS) and the Leon Levy Native Plant Preserve for their financial contribution to the publication of this book. Finally, we are very grateful to Kitty Liu, Meagan Dermody, and Susan Specter (Cornell University Press), Candace Akins (independent editor), and Kendra H. Millis (professional indexer) for their help, advice, and patience in bringing the manuscript to fruition. We also extend our thanks to the production team, without whom there would be no book.

Appendix

Dolphin and Whale Species Reported in the Archipelago

Order follows standard taxonomic grouping of scientific nomenclature for Cetaceans.

1. Common Minke Whale (*Balaenoptera acutorostrata*)[1]
2. Sei Whale (*Balaenoptera borealis*)[2]
3. Bryde's Whale (*Balaenoptera brydei*)[1]
4. Blue Whale (*Balaenoptera musculus*)[2]
5. Fin Whale (*Balaenoptera physalus*)[1]
6. Humpback Whale (*Megaptera novaeangliae*)[1]
7. Short-beaked Common Dolphin (*Delphinus delphis*)[2]
8. Pygmy Killer Whale (*Feresa attenuata*)[1]
9. Short-finned Pilot Whale (*Globicephala macrorhyncus*)[1]
10. Fraser's Dolphin (*Lagenodelphis hosei*)[1]
11. Risso's Dolphin (*Grampus griseus*)[1]
12. Killer Whale (*Orcinus orca*)[1]
13. Melon-headed Whale (*Peponocephala electra*)[1]
14. False killer Whale (*Pseudorca crassidens*)[1]
15. Pantropical Spotted Dolphin (*Stenella attenuata*)[1]
16. Clymene Dolphin (*Stenella clymene*)[1]
17. Striped Dolphin (*Stenella coeruleoalba*)[1]
18. Atlantic Spotted Dolphin (*Stenella frontalis*)[1]
19. Spinner Dolphin (*Stenella longirostris*)[2]
20. Rough-toothed Dolphin (*Steno bredanensis*)[1]
21. Common Bottlenose Dolphin (*Tursiops truncatus*)[1]
22. Sperm Whale (*Physeter macrocephalus*)[1]
23. Pygmy Sperm Whale (*Kogia breviceps*)[1]
24. Dwarf Sperm Whale (*Kogia sima*)[1]
25. Gervais' Beaked Whale (*Mesoplodon europaeus*)[1]
26. Blainville's Beaked Whale (*Mesoplodon densirostris*)[1]
27. True's Beaked Whale (*Mesoplodon mirus*)[2]
28. Cuvier's Beaked Whale (*Ziphius cavirostris*)[1]

Species list courtesy of Dr. Charlotte Dunn and Dr. Diane Claridge, Bahamas Marine Mammal Research Organisation (BMMRO, www.bahamaswhales.org).

[1] Confirmed by BMMRO.   [2] Possible but unconfirmed.

Glossary

Achene Dry, single-seeded, indehiscent fruit that does not split to release the seed.

Acuminate Gradually tapering to a point.

Adventitious Roots that arise from locations other than primary roots, often from aerial parts of a stem.

Aggregate Fruit from multiple carpels in a single flower.

Allopatric Species whose geographic ranges do not overlap.

Alternate Leaf arrangement of a single leaf per node, so that leaves grow alternately on either side of the stem.

Ambient Medium (air, water, soil) surrounding an organism.

Annulus A ring or skirt-like structure on the stalk of a basidiomycete fungus fruiting body that is the remnant of a partial veil.

Antenna (of animals) A pair of long, thin sensory appendages on the heads of insects and arthropods. Plural: Antennae.

Anterior Near the head or front end of an animal.

Anther Part of a stamen that produces and contains pollen, usually borne on a stalk or filament.

Aposeomatic Conspicuous color and/or pattern associated with distastefulness or toxicity; a warning signal to predators.

Arboreal Lives in trees.

Areoles Small bumps on cacti from which a cluster of spines or hairs grow.

Aril(s) Special seed covering that commonly develops from the seed stalk and is often colored to attract avian seed dispersers.

Axil Angle between the leaf and the stem.

Axillary Situated in or growing from an axil (*see* Axil), and the axis from which it arises.

Basal Arising from the base of a stem.

Basal areas Areas where the wing of a butterfly is attached to the thorax.

Basal rosette Leaves arising from a single location on a stemless plant.

Batesian mimicry Resemblance of an innocuous species to another species that is protected from predators by unpalatability or other characteristics.

Berry Fleshy, multi-seeded fruit, e.g., grape.

Bilateral Having two opposite sides.

Binately Growing in pairs or couples.

Biological control Reduction in numbers or elimination of pest organisms by interference with their ecology, e.g., by the introduction of parasites or diseases.

Bipinnate Having pinnae that are further divided pinnately into leaflets, making the whole leaf fern or feather-like, i.e., a twice pinnately divided leaf.

Brackish Somewhat salty.

Bract(s) Modified or specialized leaf, especially one associated with a reproductive structure, e.g., flowers, but often different in appearance from both foliage leaves and the parts of a flower (sepals and petals).

Buttresses Broadened base of a tree trunk or a thickened vertical part of it.

Capsule Dry fruit that opens along three or more suture lines.

Carapace Protective, shell-like covering of a turtle or crustacean.

Carpel Female reproductive organ of a flower.

Caryopsis Dry, indehiscent fruit in the grass family (Poaceae).

Cellulose Fibrous compound made of chains of sugar that forms the main constituent of cell walls of most green plants.

Cephalium Elongated stem.

Cephalothorax Fused head and thorax in arachnids (spiders).

Chelicerae "Jaws" of spiders and relatives; may contain venom glands.

Chip calls Short duration or single note vocalizations.

Chitin Tough, semitransparent compound that is the main component of the exoskeletons of arthropods and cell walls of fungi.

CITES Convention on the International Trade in Endangered Species.

Compound Having two or more parts in a single structure.

Compound leaf A leaf in which the blade is divided to the midrib, forming two or more distinct blades or leaflets on a common axis.

Conspecific Member of the same species.

Coppice Dry, broadleaf evergreen forest, woodland, or shrubland.

Cord A rubbery structure produced by some fungi that is used to move nutrients and water and to colonize new pieces of wood, litter, or host plants.

Cordate Heart-shaped.

Corona (in flowers) A ring or crown of appendages arising from the petals (or edges of the stamens).

Corymb Flat-topped arrangement of flowers with outer flowers having longer stalks.

Costal margin Leading edge of the forewing of a butterfly.

Costal vein Nearest vein to the leading edge of the forewing in insects.

Crepuscular Most active around dawn or dusk.

Cyathium An inflorescence consisting of a cuplike involucre of fused bracts with the flowers arising from its base, typical of spurges (Euphorbiaceae).

Cyme Broad, flat flower cluster in which the first flower is terminal and subsequent flowers arise in the flower axil.

Dehiscent Spontaneous opening of a plant structure at maturity, e.g., a fruit or sporangium, to release its contents.

Dentate Having tooth-like or pointed conical projections.

Detritivore Organisms that feed on and break down dead organic matter, especially plant detritus.

Detritus Recently dead or decomposing organic matter, e.g., leaf litter, dead animals.

Dewlap Fold of loose skin hanging from the neck or throat of many vertebrates, e.g., dogs, cows and reptiles; may be brightly colored and expanded down and forward during displays in some lizards.

Dichotomous Dividing into two parts.

Dichromatic Two color forms of a species.

Dimorphic Occurring in two distinct forms.

Dioecious Having separate male and female plants.

Dorsal Upper side or surface of an animal or plant.

Drupe Usually a fleshy fruit with a stony layer around the seed, e.g., cherry.

Effused-reflexed Form of some polypore fungi that have an extensive flat surface on the lower side of a piece of wood and a small cap at the uppermost edge.

Elliptical Oval shape and slightly pointed at both ends.

Endemic An organism that is native only to a particular locality or region.

Entire Without teeth or projections, refers to leaf margins.

Epaulet A shoulder ornament or patch.

Epilithic Growing on rock substrate.

Epiphyte Plant that derives its moisture and nutrients from the air and rain and usually grows on another plant.

Eusocial (in insects) An advanced level of social organization in which a single female or caste produces the offspring, and the non-reproductive individuals cooperate in caring for the young.

Fertile frond Fern frond with reproductive structures (*see* Sporangia).

Filaments Thin elongate stalks, often referring to the stalk of an anther.

Fleshy Having a juicy or pulpy texture.

Follicle Dry, dehiscent, one-celled, many-seeded fruit formed from a single carpel that opens along one suture.

Forcipules In centipedes, first appendages or legs modified as venomous claws.

Frond A leaf or compound of a fern or a large compound leaf of a palm.

Frontal scales Noticeable (large) scales of reptile species between eyes and mouth.

Funnelform Gradual widening to the mouth of the flower (trumpet- or funnel-shaped).

Glabrous Hairless (smooth).

Gleba Spore-bearing mass found as a green, sticky, odiferous coating on the arms of stinkhorn fungi or a powdery mass enclosed in a ball-shaped structure called a "peridium" of earthstars, earthballs, and puffballs.

Glochids Sharp, hair-like, barbed spines found on *Opuntia* cacti.

Haustoria Attachments of fungi or parasitic plant for absorbing nutrients and water from a host.

Heartwood Older, harder, nonliving, central wood of trees that is usually darker, denser, less permeable, and more durable than the surrounding sapwood.

Herbaceous Non-woody.

Hermaphroditic Having both male and female reproductive organs in a single individual (or flower).

Holocene Current geological epoch from 12,000 years ago to present; Cenozoic era, Quaternary Period; time of human dominance.

Hymenium Underside of cap where spores are produced.

Indehiscent Remaining closed at maturity.

Inflorescence Complete flower head, including stems, bracts, and flowers; a flower cluster.

Insectivores Organism that feeds on insects.

Intromittent organ External male copulatory organ of an animal specialized to deliver sperm during copulation.

Invertebrate Animals lacking a backbone, e.g., insects and crustacea.

Involucre Bracts forming a structure around or just below the base of a flower head.

Lamellae A thin plate, scale, membrane, or layer of tissue, e.g., a spore-bearing gill on the underside of the cap of a fungus.

Lanceolate Narrow, oval shape tapering to a point at each end, like a lance head.

Lateral line Series of sensory pores along the head and sides of fish by which water currents, vibrations, and pressure changes are detected.

Latex Milky sap, sometimes irritating.

Leaflet Leaf-like structure that is a division of a compound leaf.

Legume Fruit that opens along two sutures (pod).

Lenticel One of the many raised pores on the stem of a woody plant that allow gas exchange.

Ligule A membranous scale on the adaxial portion of a leaf, e.g., grass blade.

Lithophyte A plant that grows on rock.

Macrofungi Fungi that produce fruiting bodies that are visible without magnification.

Margin Outside limit and adjoining surface of something (edge).

Median fins Unpaired fin located on the median (middle) vertical plane of the body, specifically the dorsal, anal, and caudal fins.

Meristem Undifferentiated plant tissue consisting of actively dividing cells forming new tissue, e.g., at the growing tip of a stem or root.

Mesic (of an environment or habitat) Contains a moderate amount of moisture.

Mesopleuron Side plate of the middle segment of the thorax, the region between the head and abdomen ("the waist").

Mesosoma Middle segment of the abdomen of arthropods with three-parted bodies; legs and wings attach to this segment.

Midvein Central vein of a leaf.

Molt Periodic replacement (usually annual) of old, worn feathers with new, fresh feathers.

Monoecious Having both male and female reproductive organs in the same individual. In plants, having separate male and female flowers on the same individual plant.

Morph Different forms of a species.

Mucronate Having a short, narrow point, e.g., a leaf.

Multiple Having, relating to, or consisting of more than one individual, element, part, or other component.

Mycelium Vegetative part of a fungus, consisting of a network of fine, branching, thread-like filaments (hyphae).

Mycorrhiza Symbiotic (usually mutually beneficial) association of the mycelium of a fungus with the roots of a vascular plant.

Native Of a species occurring naturally within a region or ecosystem having evolved there (endemic) or having been established without human intervention (indigenous).

Nectaries Nectar-secreting structure in a flower, leaf, or stem.

Node A point on the stem where a leaf or leaves grow.

Nominate Of a taxonomic grouping, bearing the same name as its immediately superior taxon and is its type.

Oblanceolate Elongate (leaf) with the wider portion to the leaf tip and tapering to the base.

Oblong Longer than wide.

Obovate Egg-shaped and flat, with the narrow end at the base.

Ocelli Simple eyes, common in invertebrates, e.g., insects.

Offal Discarded material, often organic material, such as garbage, remains of plants and animals.

Operculum A structure that covers or closes an opening or aperture, e.g., horny or calcareous plate attached to the foot of gastropods, used to close the aperture when the animal retracts into its shell.

Opposite Placed or located directly across from something else or from each other, e.g., opposite leaves arise from the stem on either side at the same level (at the same node).

Orbicular Spherical, circular, rounded.

Osmotic stress Physiological dysfunction caused by a sudden change in the solute concentration around a cell, which causes a rapid change in the movement of water across its cell membrane, e.g., saline conditions that result in water loss.

Oval, Ovate Egg-shaped.

Oviparous Lays eggs and young develop outside female.

Ovipositors Projection at the end of an insect (tip of the abdomen) used for egg-laying or stinging.

Ovoviviparous Females retain and hatch their eggs internally, giving birth to live young.

Ovule Part of the ovary of a seed plant that gives rise to and contains the female reproductive cells; after fertilization it becomes the seed.

Palmate Resembling the shape of a hand with the fingers spread.

Panicle Multi-branched cluster of flowers.

Pantropical Found in tropical areas around the world.

Papilla Small, nipple-like projection.

Parallel Being an equal distance apart everywhere.

Parasitoid An insect whose larvae live as parasites and eventually kill their host.

Parthenogenesis Reproduction from an unfertilized egg.

Parthenogenetically Reproducing without fertilization.

Pedicel Stalk of a single flower.

Pedipalpal patella Projections on the second pair of appendages to the cephalothorax (pedipalps) in arachnids.

Pedipalps Second pair of appendages near the mouth of chelicerate arthropods, e.g., spider or scorpion, used for reproductive, predatory, or sensory functions.

Pellucid Transparent or translucent.

Pendulant Suspended, so as to swing freely.

Pendulous Hanging.

Perennial Living for more than two years and usually flowering each year.

Petiole Stalk that attaches the leaf to the stem.

Phyllary (pl. phyllaries) Bracts or modified leaves that form a collar around the head of a flower in the Family Asteraceae.

Phyllode, phyllodia A flattened leaf stalk or stem that resembles and functions as a leaf.

Pileus Spore-bearing cap of a basidiomycete fungus attached to the top of the stem with gills or pores on the underside.

Pinna (pl. pinnae) A leaflet or primary division of a pinnate leaf or frond.

Pinnate Having pinnae (leaflets) arranged on either side of leaf stalk (petiole). An evenly pinnate leaf has an even number of pinnae and no terminal leaflet; and odd pinnate leaf has a terminal leaflet. Resembles a feather where paired barbs emerge opposite each other from the central shaft (rachis).

Pistil Female, ovule-bearing organ of a flower, including the stigma, style, and ovary.

Pleistocene Time frame 2.6 million years ago to 12,000 years ago; Cenozoic era, Quaternary Period; ice age epoch-extensive glaciation.

Pliocene Time frame 2.6 to 5.3 million years ago; Cenozoic era, Tertiary Period; cool and dry period.

Pneumatophore Specialized respiratory root in certain aquatic plants that protrudes above the water or mud into the air and functions especially in the intake of oxygen from the atmosphere.

Pollen Fine, powder-like, often yellow material with individual grains that contain the male reproductive cells of seed plants; produced in the anther of angiosperms and in the male cone in gymnosperms.

Polymorphic Occurring in two or more forms, such as color or shape, of a feature of an animal or plant.

Pores Minute opening in a surface through which gases, liquids, or microscopic particles may pass, e.g., those on the lower surfaces of shelf fungi and boletes through which spores are released.

Post basal spots Spots just to the side of the basal area (see definition) of a butterfly wing.

Posterior Behind or at the back of, e.g., near the hind end of an animal.

Prop root Aerial roots growing from the stem or trunk that penetrate the soil and help support the plant, e.g., mangroves.

Pseudobulbs A solid, bulbous enlargement of the (base of the) stem for storage as found in many epiphytes, e.g., orchids.

Pterodactyl Extinct flying reptile having membranous pointed wings supported on an elongated fourth digit.

Pubescent Covering of short, dense soft hairs (down or fuzz).

Raceme Unbranched conical cluster of flowers on stalks with the youngest (with shortest stalks) at apex.

Radial Having a central axis to which all parts are attached.

Recurved Bent backward.

Reflexed Bent or curved backward or downward.

Reticulate Resembling a net or network; especially having veins, fibers, or lines crossing.

Retuse Rounded leaf tip with a small notch.

Revolute Rolled backward or downward.

Rhizome Underground creeping stem that produces shoots and roots.

Rhombic Quadrilateral with all four sides equal in length.

Saline Salty; brackish ponds or lakes with salt concentrations greater than in freshwater.

Samara A dry, indehiscent, winged, one-seeded or two-seeded fruit as of an ash, elm, or maple.

Sapwood Recently created lighter-colored, outer layers of a woody tree that moves vascular fluids.

Scabrous Rough, covered in scales.

Scales (plants) Small, leaf-like structures covering buds or immature flowers.

Sebaceous glands Glands in the surface of the skin of mammals, usually opening into the hair follicles, and secreting an oily or greasy material (sebum) that lubricates and waterproofs the hair and skin.

Secondary compounds Compounds not essential to plant development, growth, or reproduction, but may deter herbivory (plant consumption) or otherwise increase survival.

Seed Fertilized mature ovule of a flowering plant containing an embryo and capable of germination to produce a new plant.

Sepal One of the usually green, leaf-like structures composing the outermost part of a flower, often enclosing and protecting the bud and may remain after the fruit forms.

Serrate Sharp-toothed structure.

Sessile Attached directly without a stalk.

Sheath Enveloping leaf-like structure protecting the stem.

Simple Leaf Undivided leaf.

Slash and burn Agriculture involving the cutting and burning of forests to grow crops or raise livestock until the soil is depleted and the site abandoned.

Solitary Single or existing separately.

Sori Reproductive structures of a fern; found on the underside of fertile fronds and contains spores.

Spatulate Spoon-shaped.

Spike Unbranched cluster of stalkless flowers.

Spinnerets Small projections at tip of abdomen in spiders; ejects silk from silk glands to build webs.

Sporangia Reproductive structures of lower vascular plants that produce spores.

Spores Primitive, unicellular, environmentally resistant reproductive body produced by lower plants, fungi, and some microorganisms; capable of development into a new individual either directly or after fusion with another spore.

Squamules Small scales, e.g., those on the cap of a fungus.

Stamen Male organ of a flower, typically having a filament with a pollen-bearing anther at its tip.

Sterile frond Fern frond lacking sporangia; shape often differs from fertile fronds.

Stigma Receptive surface of the pistil to which pollen grains adhere and germinate.

Stipule Small, leaf-like structures at base of leaf stalk.

Style part of the female part of a flower (often long and slender) that connects the stigma to the ovary.

Subcostal vein Vein in the fore and hindwings of butterflies just below the costal vein, which runs along the top of each wing.

Subleaflets Divisions of a compound leaflet.

Submarginal band Contrasting colored or patterned band inside outer edge of wing or other structure.

Substrate Medium on which a mycelium grows, e.g., wood or soil.

Subtended Structure that arises below another to support or protect.

Subtropics Regions immediately adjacent to the tropics referring to latitudes 23.5–35°, both north and south of the Equator.

Succulent Fleshy and non-woody, adaptation for water storage in arid environments.

Summer resident (birds) Species found in the archipelago April–October (breeding).

Supercilium A line or stripe immediately above a bird's eye, but below the crown, extending from forehead to nape of head.

Suture Line marking the point of dehiscence (splitting at maturity) to release the contents, e.g., pod or capsule to release seeds.

Symbiosis Close and long-term association between organisms: can be mutually beneficial, beneficial to one organism with little effect on the other, or parasitic in which one individual benefits at the expense of the other.

Sympatric Species whose geographic ranges at least partially overlap.

Synconium A type of inflorescence borne by figs, with multiple ovaries on the inside surface of an enlarged, fleshy, hollow receptacle.

Tendrils Slender, twining structure on vines attaching them to other plants and structures.

Tergum Dorsal or upper surface of the abdominal segment in insects.

Terminal In plants, growing at the end of a branch or stem, e.g., terminal panicles, terminal spike.

Tragus A prominent feature on the inner side of the external ear, in front of and partly closing the passage to the organs of hearing, common on bat ears.

Transient Passing through, e.g., bird species, mostly migrants, regularly seen in the archipelago, but not permanent, summer or winter residents.

Transverse bars Bars at right angles to the long axis (head–tail) of the body.

Trifoliate Having three leaflets.

Tropics Region of the earth surrounding the equator and bounded by the Tropic of Cancer (23°26'N) to the north and the Tropic of Capricorn (23°26'S) to the south.

Tube-nosed Tube-shaped structure on tops of bills of shearwaters and albatrosses from which saltwater is ejected.

Tubercles Rounded projection or bump.

Tympanum Eardrum; thin membrane that transmits sound waves to ear (vertebrates) or sensory organs (insects). Plural: Tympana.

Umbel Umbrella-shaped cluster of flowers from which all stalks arise from a central point and are of equal lengths.

Underparts (birds) Ventral parts of body including the throat, breast, belly, and undertail coverts.

Undulating Wavy.

Upperparts (birds) Dorsal parts of the main body including the nape, the back, and the rump, but excluding the wings.

Veil, partial A layer that protects the spore-producing surface (gills, lamellae, or pores) of a basidiomycete fungus fruiting body (mushroom or toadstool), extending from the edge of the cap to the stalk.

Veil, universal A layer that completely envelopes a basidiomycete fungus fruiting body (mushroom or toadstool) when young, often leaving remnants on the cap and stipe (stalk) base as the fungus expands.

Venter Belly or underside.

Ventral Lower side or surface of an animal or plant.

Vestigial Retention, usually reduced and modified, of an earlier stage of development or ancestral trait.

Viviparous Young develop inside female, born live.

Volva A cup-like, or sometimes ring-like structure at the base of the stalk of some basidiomycete fungi. It is the remnants of a universal veil.

Warts (bats) Epidermal growths of viral or nonviral origin.

Warts (over fungi) Removable patches on the caps of basidiomycete fungi that are the remnants of a universal veil.

Wing cell Distinct unveined area at the base of both a butterfly's fore and hindwings.

Winter resident (birds) Species seen in the archipelago from October–April (nonbreeding).

Xerophilic Of organisms that can grow and reproduce in conditions with a low availability of water.

Selected References and Additional Reading

Abbott, J.C. n.d. OdonataCentral: An online resource for the distribution and identification of Odonata. https://www.odonatacentral.org (accessed 1 January 2018). Tuscaloosa, AL: The University of Alabama.

Abbott, R.T., and P.A. Morris. 2001. A field guide to shells: Atlantic and Gulf coasts and the West Indies. New York, NY: Houghton Mifflin Harcourt.

Alden, P., R.B. Cech, R. Keen, A. Leventer, G. Nelson, and W.B. Zomlefer. 1998. National Audubon Society field guide to Florida. New York, NY: Alfred A. Knopf.

Allen, J.A., and D.P. Ingraham. 1891. Description of a new species of *Capromys* from the Plana Keys, Bahamas. Bulletin of the American Museum of Natural History 3: Article 23.

Allen, P.E. 1996. Breeding biology and natural history of the Bahama Swallow. Wilson Bulletin, 108(3): 480–495.

American Museum of Natural History. n.d. Digital depository, http://digitallibrary.amnh.org (accessed 1 January 2018). New York, NY: American Museum of Natural History.

AmphibiaWeb. n.d. https://amphibiaweb.org (accessed 1 January 2018). Berkeley, CA: University of California.

Andersen, K.W. 1994. On the status of bats on Andros and San Salvador island, Bahamas. *In* Proceedings of the 5th Symposium on the Natural History of the Bahamas, pp. 1–4. Bahamian Field Station, San Salvador, Bahamas.

Anonymous. 2011. The fourth national biodiversity report of The Bahamas to the UNCBD. Ministry of the Environment, Nassau.

Atkinson, T.H., P.G. Koehler, and R.S. Patterson. 1990. Annotated checklist of the cockroaches of Florida (Dictyoptera: Blattaria: Blattidae, Polyphagidae, Blattellidae, Blaberidae). Florida Entomologist, 732: 303–327.

Audubon. n.d. https://www.audubon.org (accessed 1 January 2018). New York, NY: National Audubon Society.

Austin Bug Collection. n.d. http://www.austinbug.com (accessed 1 January 2018). Austin, Texas.

Bahamas Marine Mammal Research Organisation. n.d. http://www.bahamaswhales.org (accessed 28 January 2018). Abaco, Bahamas.

The Bahamas National Trust. n.d. Protecting Bahamian Biodiversity. https://bnt.bs (accessed 1 January 2018). Nassau, The Bahamas.

Bainton, A.M., and A.W. White. 2006. A bibliography of birds, ornithology and birding in The Bahamas and Turks & Caicos Islands, 1492–2006. Media Enterprises Ltd., Nassau.

Baker, R.J., and H.H. Genoways. 1978. Zoogeography of Antillean bats. Zoogeography in the Caribbean, Academy of Natural Sciences of Philadelphia, Special Publication, 13: 53–97.

Banks, N. 1906. Arachnida from The Bahamas. Bulletin of the American Museum of Natural History 22: 185–189.

Baranowski, R.M., and J.A. Slater. 1986. Coreidae of Florida (Hemiptera: Heteroptera). Gainesville, FL: Florida Department of Agriculture and Consumer Services.

Barber, H.G., C.M. Breder, M.A. Cazier, W.J. Gertsch, F.H. Rindge, C. Vaurie, and J.A. Oliver. 1954. A report on the Hemiptera heteroptera from the Bimini Islands, Bahamas, British West Indies. American Museum Novitates, no. 1682.

Baroni, T.J. 2017. Mushrooms of the northeastern United States and Eastern Canada. Portland, OR: Timber Press.

Beccaloni, G., D.C. Eades, and H. Hopkins. n.d. Cockroach species file online. http://cockroach.speciesfile.org/HomePage/Cockroach/HomePage.aspx (accessed 1 January 2018). London: Wallace Correspondence Project.

BEST Commission. 2003. The national invasive species strategy for The Bahamas. The Bahamas Environment, Science & Technology Commission, Nassau, The Bahamas, 40.

Beuttell, K., and J.B. Losos. 1999. Ecological morphology of Caribbean anoles. Herpetological Monographs, 13: 1–28.

BirdLife International. 2016. *Charadrius melodus*. The IUCN red list of threatened species 2016: e.T22693811A93424518. Downloaded on 29 January 2018.

BirdLife International. 2016. *Dendrocygna arborea*. The IUCN red list of threatened species 2016: e.T22679770A84497213. Downloaded on 29 January 2018.

BirdLife International. 2016. *Icterus northropi*. The IUCN red list of threatened species 2016: e.T22736369A95132045. Downloaded on 29 January 2018.

BirdLife International. 2016. *Phoenicopterus ruber*. The IUCN red list of threatened species 2016: e.T22729706A95020920. Downloaded on 29 January 2018.

BirdLife International. 2016. *Tachycineta cyaneoviridis*. The IUCN red list of threatened species 2016: e.T22712080A94318203. Downloaded on 29 January 2018.

BirdLife International. 2017. *Amazona leucocephala* (amended version of 2016 assessment). The IUCN red list of threatened species 2017: e.T22686201A110628050. Downloaded on 29 January 2018.

Boardman, M.R., A.C. Neumann, and K.A. Rasmussen. 1989. Holocene sea level in The Bahamas. *In* 4th Symposium on the Geology of The Bahamas, Proceedings, pp. 45–52. Bahamian Field Station, San Salvador, Bahamas.

Bonato, L., A. Chagas Jr., G.D. Edgecombe, J.G.E. Lewis, A. Minelli, L.A. Pereira, R.M. Shelley, P. Stoev, and M. Zapparoli. 2016. ChiloBase 2.0. A world catalogue of centipedes (Chilopoda) (online). http://chilobase.biologia.unipd.it (accessed 10 October 2018).

Borror, D.J., and R.E. White. 1970. A field guide to insects of America north of Mexico. New York, NY: Houghton Mifflin.

Brock, P.D., T. Büscher, and E. Baker. n.d. Phasmida species file online. Version 5.0/5.0. http://Phasmida.SpeciesFile.org (accessed 12 May 2018).

Browne, D.J. 1992. Phrynidae (Amblypygi) from Andros Island, Bahamas, with notes on distribution patterns, recent origin and allometry. Journal of Arachnology, 20(1): 18–24.

Browne, D.J., S.B. Peck, and M.A. Ivie. 1993. The longhorn beetles (Coleoptera Cerambycidae) of the Bahama Islands with an analysis of species-area relationships, distribution patterns, origin of the fauna and an annotated species list. Tropical Zoology, 6(1): 27–53.

Buckner, S.D., R. Franz, and R.G. Reynolds. 2012. Bahama Islands and Turks and Caicos

Islands. Island lists of West Indian amphibians and reptiles. *In* Island Lists of West Indian amphibians and reptiles, eds. R. Powell, and R.W. Henderson, pp. 93–110. Florida Museum of Natural History, University of Florida.

Buden, D.W. 1986. Distribution of mammals of The Bahamas. Florida Field Naturalist, 14(3): 53–63.

Buden, D.W. 1987. A guide to the identification of the bats of The Bahamas. Caribbean Journal of Science, 23(3–4): 362–367.

BugGuide. n.d. http://bugguide.net (accessed 20 May 2018). Ames, IA: Iowa State University, Department of Entomology.

Butterflies of Cuba. n.d. http://www.butterfliesofcuba.com (accessed 1 January 2018).

Byrne, R. 1980. Man and the variable vulnerability of island life: A study of recent vegetation change in the Bahamas. Atoll Research Bulletin Vol. 240. Washington DC: Smithsonian Institute.

Cambridge, F.P. 1901. On a collection of spiders from the Bahama islands made by J.L. Bonhote, Esq.; with characters of a new genus and species of Mygalomorphæ. Journal of Natural History, 7(40): 322–332.

Campbell, D.G. 1978. The ephemeral islands: A natural history of The Bahamas. London, UK: Macmillan.

Carey, E.S., S.D. Buckner, A.C. Alberts, R.D. Hudson, and D. Lee. 2001. Protected areas management strategy for Bahamian terrestrial vertebrates: Iguanas and seabirds. Apple Valley, MN: IUCN/SSC Conservation Breeding Specialists Group Report.

Carey, E., L. Gape, B.N. Naqqi Manco, D. Hepburn, R.L. Smith, L. Knowles, D. Knowles, M. Daniels, M.A. Vincent, E. Greid, B. Jestrow, M.P. Griffith, M. Calonje, A.W. Meerow, D.W. Stevenson, and J. Francisco-Ortega. 2014. Plant conservation challenges in the Bahama Archipelago. Botanical Review 80:265–282.

Carter, R.L., and W.K. Hayes. 1996. *Cyclura rileyi*. The IUCN red list of threatened species 1996: e.T6033A12351578. Downloaded on 29 January 2018.

Casale, P., and A.D. Tucker. 2017. *Caretta caretta* (amended version of 2015 assessment). The IUCN red list of threatened species 2017: e.T3897A119333622. Downloaded on 29 January 2018.

Cates, D.L. 1998. Mark Catesby's Bahamian plants. Bahamas Journal of Science 5: 16–21.

Chamberlin, R.V., M.A. Cazier, W.J. Gertsch, C. Vaurie, and P. Vaurie. 1952. The centipedes (Chilopoda) of South Bimini, Bahama Islands, British West Indies. American Museum Novitates, no. 1576.

Checklist of Insecta. n.d. http://insectoid.info.

Chesser, R.T., K.J. Burns, C. Cicero, J.L. Dunn, A.W. Kratter, I.J. Lovette, P.C. Rasmussen, J.V. Remsen Jr., D.F. Stotz, B.M. Winger, and K. Winker. 2018. Check-list of North American Birds (online), http://checklist.aou.org/taxa (accessed 15 May 2018). Chicago, IL: American Ornithological Society.

Choate, P., S. Burns, L. Olsen, D. Richman, O. Pérez, M. Patnaude, and R. Pluke. 2008. A dichotomous key for the identification of the cockroach fauna (Insecta: Blattaria) of Florida. Florida Entomologist 72: 612–617.

Cigliano, M.M., H. Braun, D.C. Eades, and D. Otte. n.d. Orthoptera Species File. Version 5.0/5.0. http://Orthoptera.SpeciesFile.org (accessed 1 January 2018).

Clench, H.K. 1942. The Lycænidæ of the Bahama Islands (Lepidoptera Rhopalocera). Psyche 49(3–4): 52–60.

Clench, H.K. 1964. A synopsis of the West Indian Lycaenidae, with remarks on their zoogeography. Journal of Research on the Lepidoptera, 2(4): 247–270.

Clench, H.K. 1977. A list of the butterflies of Andros, Bahamas. Pittsburgh, PA: Carnegie Museum of Natural History.

Clough, G.C. 1972. Biology of the Bahaman Hutia, *Geocapromys ingrahami*. Journal of Mammalogy, 53(4): 807–823.

The Cornell Lab of Ornithology. n.d. https://www.allaboutbirds.org (accessed 1 January 2018). Ithaca, NY: Cornell University.

Correll, D.S. 1979. The Bahama Archipelago and its plant communities. Taxon, 28: 35–40.

Correll, D.S., and H.B. Correll. 1982. Flora of The Bahama Archipelago. Hirschberg, Germany: Strauss and Cramer.

Crab Database. 2016. https://www.crabdatabase.info (accessed 1 January 2018).

Crafton, M.C. 1986. A history of The Bahamas. 3rd edition. Ontario, Canada: San Salvador Press.

Crews, S.C., and A. Yang. 2016. Notes on the spiders (Arachnida, Araneae) of the Turks and Caicos Islands, British West Indies. Caribbean Journal of Science, 49(1): 83–90.

Crother, B.I., ed. 1999. Caribbean amphibians and reptiles. Cambridge, MA: Academic Press.

Cutts, W. 2004. Trees of The Bahamas and Florida. Oxford, UK: MacMillan Caribbean.

Darsie, R.F. Jr., D.S. Taylor, Z.A. Prusak, and T.N. Verna. 2010. Checklist of the mosquitoes of The Bahamas with three additions to its fauna and keys to the adult females and fourth instars. Journal of the American Mosquito Control Association, 26(2): 127–134.

Davis, D.R. 1975. A review of the West Indian moths of the family Psychidae with descriptions of new taxa and immature stages. Smithsonian Contributions to Zoology, No. 188.

Davis, L.R. Jr., R.K. Van der Meer, and S.D. Porter. 2001. Red imported fire ants expand their range across the West Indies. Florida Entomologist, 84(4): 735.

de Armas, L.F. 2017. Scorpions. The Antillean or West Indian fauna. http://www.scorpions etc.com/scorpions-the-antillean-or-west-indian-fauna.html (accessed 1 January 2018).

de Armas, L.F. 2017. Amblypygi. The Antillean (West Indian) fauna. http://www.scorpions etc.com/amblypygi-the-antillean-west-indian-fauna.html (accessed 1 January 2018).

Deutsch, C.J., C. Self-Sullivan, and A. Mignucci-Giannoni. 2008. *Trichechus manatus*. The IUCN red list of threatened species 2008: e.T22103A9356917. Downloaded on 29 January 2018.

Digital Commons Network. n.d. http://network.bepress.com (accessed 1 January 2018).

Discover Life. n.d. http://www.discoverlife.org (accessed 1 January 2018).

Dunkle, S.W. 1989. Dragonflies of the Florida Peninsula, Bermuda, and The Bahamas. Gainesville, FL: Scientific Publishers.

Dunkle, S.W. 1990. Damselflies of Florida, Bermuda, and the Bahamas. Scientific Publishers Nature Guide, No. 3.

Eaton, E.R., and K. Kaufman. 2007. Kaufman Field Guide to Insects of North America. New York, NY: Houghton Mifflin Harcourt.

Elliott, N.B. 1992. Wasps of The Bahamas (Hymenoptera: Scoliidae, Tiphiidae, Pompilidae, Vespidae, Sphecidae). *In* Proceedings 4th Symposium on Natural History of the Bahamas, pp. 41–48. Bahamian Field Station, San Salvador, The Bahamas.

Elliot, N.B., and H.K. Clench. 1980. Annotated list of butterflies of San Salvador, Bahamas. Journal of the Lepidopterists' Society 34(2): 120–126.

Elliott, N.B., and W.M. Elliott. 1994. Preliminary comparisons of wasp faunas of the Great and Little Bahama Banks (Hymenoptera: Tiphiidae, Scoliidae, Pompilidae, Mutillidae, Sphecidae, Vespidae). *In* Proceedings of the 5th Symposium on the Natural History of the Bahamas, pp. 29–32. Bahamian Field Station, San Salvador, Bahamas.

Elliott, N.B., and N.Y. Loudonville. 1987. Hemiptera associated with several Bahamian shrubs. *In* Proceedings of the 2nd Symposium on the Botany of the Bahamas, pp. 1–5. Bahamian Field Station, San Salvador, The Bahamas.

Elliott, N.B., and D.L. Smith. 2003. Visitors to *Pithecellobium keyense* Britt. Ex Britt. and Rose on three islands of the southern Bahamas. *In* Proceedings 9th Symposium on Natural History of the Bahamas, pp. 106–115. Gerace Research Center, San Salvador, The Bahamas.

Elliott, N., F. Kurczewski, S. Claflin, and P. Salbert. 1979. Preliminary annotated list of the wasps of San Salvador Island, The Bahamas, with a new species of *Cerceris* [*watlingensis*] (Hymenoptera: Tiphiidae, Scoliidae, Vespidae, Pompilidae, Sphecidae). Proceedings Entomological Society of Washington (USA).

Elliot, N.B., D.L. Smith, and S.G.F. Smith. 2009. Field guide to the insects of San Salvador. Gerace Research Centre, San Salvador Island, The Bahamas.

Emlen, J.T. 1977. Land bird communities of Grand Bahama Island: The structure and dynamics of an avifanua. Ornithological Monographs 24: 1–129.

Encyclopedia of Life. n.d. http://www.eol.org (accessed 1 January 2018).

Exploring Moths of the Florida Keys. n.d. https://www.keysmoths.com (accessed 1 January 2018).

Father Sanchez's Web Site of West Indian Natural History. n.d. http://www.kingsnake .com/westindian/ (accessed 10 January 2018).

Faust, L.F. 2017. Fireflies, glow-worms, and lightning bugs: Identification and natural history of the fireflies of the Eastern and Central United States and Canada. Athens, GA: University of Georgia Press.

Featured Creatures. 1996. http://entomology.ifas.ufl.edu/creatures/ (accessed January 2018). Gainesville, FL: University of Florida.

Franz, R., C.K. Dodd, and D.W. Buden. 1993. Distributional records of amphibians and reptiles from the Exuma Islands, Bahamas, including the first reports of a freshwater turtle and an introduced gecko. Caribbean Journal of Science, 29(3–4): 165–173.

Freid, E.H., J. Ortega, and B. Jestrow. 2014. Endemic seed plants in the Bahamian Archipel-ago. Botanical Review 80(3): 204–230.

Froese, R., and D. Pauly, eds. 2018. FishBase. World Wide Web electronic publication, www.fishbase.org (accessed 1 January 2018).

Fuller, J.E., K.L. Eckert, and J.I. Richardson. 1992. WIDECAST Sea Turtle Recovery Action Plan for Antigua and Barbuda (Karen L. Eckert, ed.). CEP Technical Report No. 16. UNEP Caribbean Environment Programme, Kingston, Jamaica.

Gasca-Álvarez, H.J., and B.C. Ratcliffe. 2011. An annotated checklist of the Oryctine rhi-noceros beetles (Coleoptera: Scarabaeidae: Dynastinae: Oryctini) of the Neotropical and Nearctic realms. Zootaxa 3090: 21–40.

Generic Guide to New World Scarab Beetles. n.d. http://www.museum.unl.edu/research/ entomology/Guide/Scarabs%20Gallery/index.html (accessed 1 January 2018). Univer-sity of Nebraska-Lincoln State Museum, Department of Entomology.

Gerber, G. 2004. *Cyclura carinata*. The IUCN red list of threatened species 2004: e.T6026A12317199. Downloaded on 29 January 2018.

Gillis, W.T., R. Byrne, and W. Harrison. 1975. Bibliography of the natural history of The Bahama Islands. Atoll Research Bulletin Vol. 191. Washington, DC: Smithsonian Institute.

Gittens, L.G., and M.T. Braynen. 2012. Bahamas National Report—CFMC. OSPESCA/WECAFC/CRFM working group on Queen Conch. Proceedings of the Queen Conch working group meeting, Panama City, Panama: 1–5. http://www.strombusgigas.com/ Meeting Panama/Queen Conch Meeting (23 October)/Bahamas/Bahamas national report.pdf.

Godley, B.J., A.C. Broderick, L.M. Campbell, S. Ranger, and P.B. Richardson. 2004. An assessment of the status and exploitation of marine turtles in the Turks and Caicos Islands. In An assessment of the status and exploitation of Marine Turtles in the UK Overseas Territories in the Wider Caribbean, Final Project Report for the Department of Environment, Food and Rural Affairs and the Foreign and Commonwealth Office, pp. 180–222.

Graham Reynolds Lab. The Reynolds Lab. n.d. http://www.rgrahamreynolds.info/ (accessed 1 January 2018). Asheville, NC: University of North Carolina.

Gutiérrez, E. 1995. Annotated checklist of Cuban cockroaches. Transactions of the American Entomological Society, 121: 65–85.

Hailey, A., B. Wilson, and J. Horrocks, eds. 2011. Conservation of Caribbean island herpetofaunas, Volume 1: Conservation biology and the wider Caribbean. Leiden, The Netherlands: Brill Academic Publishers.

Hailey, A., B. Wilson, and J. Horrocks, eds. 2011. Conservation of Caribbean island herpetofaunas, Volume 2: Regional accounts of the West Indies. Leiden, The Netherlands: Brill Academic Publishers.

Hallett, B. 2006. Birds of The Bahamas and the Turks and Caicos Islands. Oxford, UK: Macmillan Publishers.

Hammerton, J.L. 2001. Casuarinas in The Bahamas: A clear and present danger. Bahamas Journal of Science, 9(1): 2–14.

Harasewych, MG. 2009. *Cerion*: A web-based resource for Cerion research and identification, Version 1.2. http://invertebrates.si.edu/Cerion/ (accessed 1 January 2018).

Harvey, M.S. 2013. Pseudoscorpions of the world, version 3.0. Western Australian Museum, Perth. http://www.museum.wa.gov.au/catalogues/pseudoscorpions (accessed 15 May 2018).

Hearty, P.J. 1998. The geology of Eleuthera Island, Bahamas: A Rosetta Stone of Quaternary stratigraphy and sea-level history. Quaternary Science Reviews, 17(4–5): 333–355.

Hedges, S.B. 2002. Morphological variation and the definition of species in the snake genus *Tropidophis* (Serpentes, Tropidophiidae). Bulletin of the Natural History Museum: Zoology Series 68(2): 83–90.

Hedges, S.B. 2018. Caribherp: West Indian amphibians and reptiles (www.caribherp.org). Philadelphia, PA: Temple University.

Hedges, S.B., and C.E. Conn. 2012. A new skink fauna from Caribbean islands (Squamata, Mabuyidae, Mabuyinae). Zootaxa, 3288: 1–244.

Helgen, K.M., J.E. Maldonado, D.E. Wilson, and S.D. Buckner. 2008. Molecular confirmation of the origin and invasive status of West Indian raccoons. Journal of Mammalogy, 89(2): 282–291.

Henderson, R.W., and R. Powell. 2009. Natural history of West Indian reptiles and amphibians. Gainesville, FL: University Press of Florida.

Heselhaus, R., and M. Schmidt. 1995. Caribbean anoles. Neptune, NJ: TFH Publications.

Humann, P. 1992. Reef creatures identification, Florida, Caribbean, Bahamas. Jacksonville, FL: New World Publications.

Humann, P. 1993. Reef coral identification, Florida, Caribbean, Bahamas. Jacksonville, FL: New World Publications.

Humann, P. 1994. Reef fish identification, Florida, Caribbean, Bahamas. Jacksonville, FL: New World Publications.

Humann, P., and N. Deloach. 1994. Reef fish identification: Florida, Caribbean, Bahamas. Jacksonville, FL: New World Publications.

Hutchinson, J.M. 1999. Factors influencing the surface fauna of inland blue holes on South Andros, Bahamas. Bahamas Journal of Science, 6: 29–43.

The IUCN Red List of Threatened Species. n.d. http://www.iucnredlist.org (accessed 1 January 2018). Cambridge, UK: International Union for Conservation of Nature and Natural Resources.

Kaplan, E. 1982. A field guide to coral reefs: Caribbean and Florida. Boston, MA: Houghton Mifflin.

Kass, L.B. 2009. An illustrated guide to common plants of San Salvador Island, Bahamas, 3rd edition. Gerace Research Centre, San Salvador, The Bahamas.

Kindler, P., J.E. Mylroie, H.A. Curran, J.L. Carew, D.W. Gamble, T.A. Rothfus, M. Savarese, and S.E. Sealey. 2010. Geology of Central Eleuthera, Bahamas: A field trip guide. Gerace Research Centre, San Salvador, The Bahamas.

Kjar, D.S., and G. Sackett. 2009. Insect species reported as collected in the Bahamas. Online Database. http://bio2.elmira.edu/bahamas/. Elmira College.

Klingel, G.C. 1940. Inagua. New York, NY: Dodd, Mead.

Knapp, C.R., J.B. Iverson, and S. Buckner. 2004. *Cyclura cychlura*. The IUCN red list of threatened species 2004: e.T6035A12356382. Downloaded on 29 January 2018.

Knapp, C.R., J.B. Iverson, S.D. Buckner, and S.V. Cant. 2011. Conservation of amphibians and reptiles in The Bahamas. *In* Conservation of Caribbean Island Herpetofaunas, Volume 2: Regional accounts of the West Indies, eds. A. Hailey, B. Wilson, and J. Horrocks, pp. 53–88. Leiden, The Netherlands: Brill Academic Publishers.

Kornicker, L.S. 1963. The Bahama banks: A living fossil-environment. Journal of Geological Education, 11(1): 17–25.

Landry, C.L., B.J. Rathcke, L.B. Kass, N.B. Elliott, and R. Boothe. 2005. Flower visitors to white mangrove: A comparison between three Bahamian islands and Florida. *In* Proceedings of the 10th symposium on the Natural History of The Bahamas, pp. 84–94. Gerace Research Center, San Salvador, The Bahamas.

Langerhans Lab. n.d. http://gambusia.zo.ncsu.edu/ (accessed 1 January 2018). Evolutionary Ecology Lab, North Carolina State University.

The Levy Native Plant Preserve. n.d. http://www.levypreserve.org (accessed 1 January 2018). Eleuthera, The Bahamas.

Lewis, J.G.E. 2006. The biology of Centipedes. Cambridge, UK: Cambridge University Press.

Linley, J.R., and J.B. Davies. 1971. Sandflies and tourism in Florida and The Bahamas and Caribbean area. Journal of Economic Entomology, 64(1): 264–278.

Losos, J.B., and R.S. Thorpe. 2004. Evolutionary diversification of Caribbean Anolis lizards. *In* Adaptive speciation, eds. U. Dieckmann, M. Doebeli, J.A.J. Metz, and D. Tautz, pp. 322–324. Cambridge, UK: Cambridge University Press.

Lotts, K., and T. Naberhaus, coordinators. 2017. Butterflies and moths of North America. http://www.butterfliesandmoths.org (accessed 1 January 2018).

Maclean, W.P., R. Kellner, and H. Dennis. 1977. Island lists of West Indian Amphibians. Division of Reptiles and Amphibians, National Museum of Natural History. Washington, DC: Smithsonian Institution.

Malumphy, C., M.A. Hamilton, B.N. Manco, P.W.S. Green, M.D. Sanchez, M. Corcoran, and E. Salamanca. 2012. *Toumeyella parvicornis* (Hemiptera: Coccidae), causing severe decline of *Pinus caribaea* var. *Bahamensis* in the Turks and Caicos Islands. Florida Entomologist, 95(1): 113–119.

Manco, BN. 2008. The creepy-crawly life. http://www.timespub.tc/2008/09/the-creepy-crawly-life/ (accessed 1 January 2018). Times of the Islands: Sampling the Soul of the Turks and Caicos Islands.

Martin Lab. n.d. http://labs.bio.unc.edu/martin/ (accessed 1 January 2018). Chapel Hill, NC: University of North Carolina.

Martin, C.H., and P.C. Wainwright. 2013. A remarkable species flock of *Cyprinodon* pupfishes endemic to San Salvador Island, Bahamas. Bulletin of the Peabody Museum of Natural History, 54(2): 231–241.

McDonald, S. 2016. Seafood report: Queen Conch. Seafood Watch Seafood Reports. http://seafood.ocean.org/wp-content/uploads/2017/08/MBA_SeafoodWatch_QueenConchReport.pdf.

McNary Wood, N. 2003. Flowers of The Bahamas and the Turks and Caicos Islands. Oxford, UK: Macmillan Publishers.

Meylan, A.B. 1999. Status of the Hawksbill Turtle (*Eretmochelys imbricata*) in the Caribbean Region. Chelonian Conservation and Biology, 3(2): 177–184.

Miller, L.D., M.J. Simon, and D.J. Harvey. 1992. The butterflies (Insecta: Lepidoptera) of Crooked, Acklins and Mayaguana Islands, Bahamas, with a discussion of the biogeographical affinities of the Southern Bahamas and description of a new subspecies by H.K. Clench. Annals of the Carnegie Museum, 61(1): 1–31.

Miller, R.B., and L.A. Stange. 2011. Antlions of Hispaniola (Neuroptera: Myrmeleontidae). Insecta Mundi. Paper 694.

MolluscaBase. 2018. http://www.molluscabase.org (accessed 19th May 2018).

Montambault, J.R. 2007. Conservation status and potential of West Indian endemic bird species in a rapidly suburbanizing landscape, Middle Caicos, Turks and Caicos Islands [PhD dissertation]. University of Florida, Gainesville, Florida.

Morrison, L.W. 1998. A review of Bahamian ant (Hymenoptera: Formicidae) biogeography. Journal of Biogeography, 25(3): 561–571.

Morse, A.P. 1905. Some Bahama Orthoptera. Psyche, 12(1): 19–24.

Myers, P., R. Espinosa, C.S. Parr, T. Jones, G.S. Hammond, and T.A. Dewey. 2018. The Animal Diversity Web (online), https://animaldiversity.org (accessed 1 January 2018). University of Michigan, Museum of Zoology.

Mylroie, J.R., and J.R. Mylroie. 2007. Development of the carbonate island karst model. Journal of Cave and Karst Studies, 69(1): 59–75.

The Myriapoda (Millipedes, Centipedes) featuring the North American fauna. n.d. http://www.nadiplochilo.com (accessed 1 January 2018).

Nellis, D.W. 1997. Poisonous plants and animals of Florida and the Caribbean. Sarasota, FL: Pineapple Press.

Nickrent, D.L., W.H. Eshbaugh, and T.K. Wilson. 1988. The vascular flora of Andros Island, Bahamas. Dubuque, IA: Kendall/Hunt.

Noble, G.K., G.C. Klingel, and W.W. Coleman. 1932. The reptiles of Great Inagua Island, British West Indies. American Museum Novitates, no. 549.

North American Moth photographers group. n.d. http://mothphotographersgroup.msstate .edu (accessed 1 January 2018). Mississippi Entomological Museum. Starkville, MS: Mississippi State University.

Oliver, J.A. 1948. The anoline lizards of Bimini, Bahamas. American Museum Novitates, no. 1383.

Olson, S.L., ed. 1982. Fossil vertebrates from The Bahamas. Smithsonian Contributions to Paleobiology 48. Washington, DC: Smithsonian Institution Press.

Olson, S.L., G.K. Pregill, and W.B. Hilgartner. 1990. Studies on fossil and extant verte-brates from San Salvador (Watling's) Island, Bahamas. Washington, DC: Smithsonian Institution Press.

Palmer, R.S., ed. 1962. Handbook of North American birds. Volume 1. New Haven, CT: Yale University Press.

Palmer, R.S., ed. 1976. Handbook of North American birds. Volume 2. New Haven, CT: Yale University Press.

Palmer, R.S., ed. 1976. Handbook of North American birds. Volume 3. New Haven, CT: Yale University Press.

Patterson, J. 2002. Native trees of The Bahamas. Nassau, Bahamas: Media Enterprise.

Paulson, D. 2009. Dragonflies and Damselflies of the west. Princeton, NJ: Princeton University Press.

Paulson, D. 2012. Dragonflies and Damselflies of the east. Princeton, NJ: Princeton University Press.

Peck, S.B., and C. Beninger. 1989. A survey of insects of the Florida Keys: Cockroaches (Blattodea), mantids (Mantodea), and walkingsticks (Phasmatodea). Florida Entomol-ogist, 72(4): 612–617.

Peck, S.B., and D.E. Perez-Gelabert. 2012. A summary of the endemic beetle genera of the West Indies (Insecta: Coleoptera); bioindicators of the evolutionary richness of this Neotropical archipelago. Insecta Mundi 718. http://digitalcommons.unl.edu/ insectamundi/718.

Perez, K.E., J.R. Cordeiro, and J. Gerber. 2008. A guide for terrestrial gastropod identifica-tion. Carbondale, IL: American Malacological Society. 72 pp.

Perez-Gelabert, D.E. 2008. Arthropods of Hispaniola (Dominican Republic and Haiti): A checklist and bibliography. Magnolia Press.

Piniak, W.E.D., and K.L. Eckert. 2011. Sea turtle nesting habitat in the wider Caribbean region. Endangered Species Research, 15(2): 129–141.

Quintero, D. 1981. The amblypygid genus Phrynus in the Americas (Amblypygi, Phryni-dae). Journal of Arachnology, 9: 117–166.

Raffaele, H.A., and J. Wiley. 2014. Wildlife of the Caribbean. Princeton, NJ: Princeton University Press.

Raffaele, H.A., J. Wiley, O.H. Garrido, A. Keith, and J.I. Raffaele. 1998. A guide to the birds of the West Indies. Princeton, NJ: Princeton University Press.

Randall, J.E. 1983. Caribbean reef fishes, 2nd edition. Neptune City, NJ: TFH Publications.

Ratcliffe, B.C., and R.D. Cave. 2008. The Dynastinae (Coleoptera: Scarabaeidae) of The Bahamas with a description of a new species of *Cyclocephala* from Great Inagua Island. Papers in Entomology. Paper 105.

Ratcliffe, B.C., and R.D. Cave. 2015. The Dynastine Scarab Beetles of the West Indies: (Coleoptera: Scarabaeidae: Dynastinae). Lincoln, NE: University of Nebraska State Museum.

Rehn, J.A. 1906. The Orthoptera of The Bahamas. Bulletin of the American Museum of Natural History 22: 107.

Rein, J.O. 2017. The Scorpion files. http://www.ntnu.no/ub/scorpion-files (accessed 1 January 2018). Trondheim: Norwegian University of Science and Technology.

Rendell, N. 2011. UK Overseas Territories and Crown Dependencies: 2011 Biodiversity snapshot. Department of Environment and Coastal Resources, Turks and Caicos Islands Government.

Reynolds, R.G., and G.P. Gerber. 2012. Ecology and conservation of the Turks Island Boa (*Epicrates chrysogaster chrysogaster*: Squamata: Boidae) on Big Ambergris Cay. Journal of Herpetology, 46(4): 578–586.

Reynolds, R.G., and M.L. Niemiller. 2010. Island invaders: Introduced reptiles and amphibians of the Turks and Caicos Islands. IRCF, Reptile and Amphibians: Conservation and Natural History, 17: 117–121.

Reynolds, R.G., M.L. Niemiller, S.B. Hedges, A. Dornburg, A.R. Puente-Rolón, and L.J. Revell. 2013. Molecular phylogeny and historical biogeography of West Indian boid snakes (Chilabothrus). Molecular Phylogenetics and Evolution, 68(3): 461–470.

Reynolds, R.G., A.R. Puente-Rolón, A.J. Geneva, K.J. Aviles-Rodriguez, and N.C. Herrmann. 2016. Discovery of a remarkable new boa from the Conception Island Bank, Bahamas. Breviora, 549(1): 1–19.

Richardson, P.B., M.W. Bruford, M.C. Calosso, L.M. Campbell, W. Clerveaux, A. Formia, and K. Parsons. 2009. Marine turtles in the Turks and Caicos Islands: Remnant rookeries, regionally significant foraging stocks, and a major turtle fishery. Chelonian Conservation and Biology, 8(2): 192–207.

Richmond, N.D., and M.G. Netting. 1955. The blindsnakes (Typhlops) of Bimini, Bahama Islands, British West Indies, with description of a new species. American Museum Novitates, no. 1734.

Riley, N.D. 1975. A field guide to the butterflies of the West Indies. London, UK: Collins.

Rindge, F.H., M.A. Cazier, W.J. Gertsch, C. Vaurie, and P. Vaurie. 1952. The butterflies of the Bahama Islands, British West Indies (Lepidoptera). American Museum Novitates, no. 1563.

Robertson, D.R., and J. Van Tassell. 2015. Shorefishes of the Greater Caribbean: Online information system. Version 1.0. http://biogeodb.stri.si.edu/caribbean/en/pages (accessed 1 January 2018). Smithsonian Tropical Research Institute.

Robertson, R. 2011. Catesby's gallery. Natural History, 119: 32–37.

Robinson, W.H. 2005. Urban insects and Arachnids: A handbook of urban entomology. Cambridge, UK: Cambridge University Press.

Rodríguez-Durán, A., and T.H. Kunz. 2001. Biogeography of West Indian bats: An ecological perspective. *In* Biogeography of the West Indies: Patterns and perspectives, eds. C.A. Woods and F.E. Sergile, pp. 355–368. Boca Raton, FL: CRC Press.

Rodríguez-Robles, J.A., and R. Thomas. 1992. Venom function in the Puerto Rican racer, *Alsophis portoricensis* (Serpentes: Colubridae). Copeia, 1: 62–68.

Rolling Harbour. n.d. Birds, Wildlife & More. https://rollingharbour.com (accessed 1 January 2018). Abaco, Bahamas.

Rood, J.E. 2014. Succulent and spiny: The Bahamas' Quest for a sustainable lobster fishery [PhD dissertation]. Cambridge, MA: Massachusetts Institute of Technology.

Roody, W.C. 2003. Mushrooms of West Virginia and the Central Appalachians. Lexington, KY: University Press of Kentucky.

Rosenberg, M.S. 2014. Contextual cross-referencing of species names for fiddler crabs (Genus *Uca*): An experiment in cyber-taxonomy. PLoS ONE 9(7): e101704.

Ruckes, H. 1952. Some scutelleroid Hemiptera of the Bahama Islands, British West Indies. American Museum Novitates, no. 1591.

Salvesen, K. 2014. The Delphi Club to the birds of Abaco. Kent, UK: Delphi Club Publications.

Sanborn, A.F. 2001. Distribution of the cicadas (Homoptera: Cicadidae) of the Bahamas. Florida Entomologist, 84(4): 733.

Scheffrahn, R.H., J. Křeček, J.A. Chase, B. Maharajh, and J.R. Mangold. 2006. Taxonomy, biogeography, and notes on termites (Isoptera: Kalotermitidae, Rhinotermitidae, Termitidae) of the Bahamas and Turks and Caicos Islands. Annals of the Entomological Society of America, 99(3): 463–486.

Schwartz, A. 1965. Geographic variation in *Sphaerodactylus notatus* Baird. Revista de Biología Tropical, 13(2): 161–185.

Schwartz, A., and R.W. Henderson. 1991. Amphibians and reptiles of the West Indies: Descriptions, distributions, and natural history. Gainesville, FL: University Press of Florida.

Schwartz, A., F.I. Gonzalez, M. Rose, and R.M. Henderson. 1987. New records of butterflies from the West Indies. Journal of the Lepidopterists' Society 41(3): 145–150.

Scurlock, J.P. 1987. Native trees and shrubs of the Florida Keys. Lower Sugarloaf Key, FL: Laurel and Herbert.

Sealey, N.E. n.d. Media Enterprises. http://bahamasmedia.academia.edu/NeilSealey. Nassau, Bahamas.

Sealey, N.E. 2005. The Bahamas today, 2nd edition. Oxford, UK: MacMillan Caribbean.

Sealey, N.E. 2006. Bahamian landscape: Introduction to the geology and physical geography of The Bahamas, 3rd edition, Oxford, UK: Macmillan Publishers.

Sealey, N.E. 2011. Casuarina-induced beach erosion revisited. *In* 13th Symposium on the Natural History of The Bahamas, pp. 52–58. Gerace Research Centre, San Salvador, The Bahamas.

Selander, R.B., and J.K. Bouseman. 1960. Meloid beetles (Coleoptera) of the West Indies. *In* Proceedings of the United States National Museum, 111(3428): 197–226.

Seminoff, J.A. 2004. *Chelonia mydas*. The IUCN red list of threatened species 2004: e.T4615A11037468. Downloaded on 29 January 2018. Southwest Fisheries Science Center, U.S.

Sewlal, J.A.N., and C.K. Starr. 2011. Preliminary survey of the spider fauna of Great Inagua, Bahamas, WI. International Journal of Bahamian Studies, 17(2): 3–8.

Shattuck, G.B., ed. 1905. The Bahama Islands. Baltimore, MD: Johns Hopkins Press.

Shedd Aquarium. n.d. https://www.sheddaquarium.org/Conservation--Research/Field-Research/ (accessed 1 January 2018). Chicago, Illinois.

Shelley, R.M., and W.D. Sissom. 1995. Distributions of the scorpions *Centruroides vittatus* (Say) and *Centruroides hentzi* (Banks) in the United States and Mexico (Scorpiones, Buthidae). Journal of Arachnology, 23(2): 100–110.

Shelley, R.M., G.B. Edwards, and A. Chagas. 2005. Introduction of the centipede *Scolopendra morsitans* L., 1758, into northeastern Florida, the first authentic North American record, and a review of its global occurrences (Scolopendromorpha: Sscolopendridae: Scolopendrinae). Entomological News, 116(1): 39–58.

Sibley, D.A. 2014. The Sibley guide to birds, 2nd edition. New York, NY: Alfred Knopf.

Siebert, W.H. 1913. The Legacy of the American Revolution to the British West Indies and Bahamas: A chapter out of the history of the American loyalists (No. 1). Ohio State University.

Singing Insects of North America. n.d. http://entnemdept.ufl.edu/walker/buzz/index.htm (accessed 1 January 2018).

Slater, J.A., and R.M. Baranowski. 1990. Lygaeidae of Florida (Hemiptera: Heteroptera). Arthropods of Florida and neighboring land areas. Vol. 14. Gainesville, FL: Florida Department of Agriculture and Consumer Services, Division of Plant Industry.

Slowik, J., and D.S. Sikes. 2011. Spiders (Arachnida: Araneae) of Saba Island, Lesser Antilles: Unusually high species richness indicates the Caribbean biodiversity hotspot is woefully undersampled. Insecta Mundi, 177: 1–9.

Smith, C.A., and K.L. Krysko. 2007. Distributional comments on the Teiid lizards (Squamata: Teiidae) of Florida with a key to species. Caribbean Journal of Science, 43(2): 260–265.

Smith, D.S., L.D. Miller, J.Y. Miller, and R. Lewington. 1994. Butterflies of the West Indies and South Florida. Oxford, UK: Oxford University Press.

Smith, S.G.F., D.L. Smith, and N.B. Elliott. 2003. A first look at Acridid grasshoppers (Orthoptera) of The Bahamas. *In* Proceedings of the 9th Symposium on the Natural History of the Bahamas, pp. 121–125. Gerace Research Centre, San Salvador, The Bahamas.

Smithsonian Natural Museum of Natural History. n.d. https://naturalhistory.si.edu/ (accessed 1 January 2018). Washington, DC.

Speer, K.A., J.A. Soto-Centeno, N.A. Albury, Z. Quicksall, M.G. Marte, and D.L. Reed. 2015. Bats of The Bahamas: Natural history and conservation. Florida Museum of Natural History Bulletin, 53(3): 45–95.

St. Leger, R.G.T. 1991. The butterflies and hawk moths of the Turks and Caicos Islands. Bulletin of Amateur Entomological Society (June): 114–120.

Steadman, D.W., N.A. Albury, P. Maillis, J.I. Mead, J. Slapcinsky, K.L. Krysko, H.M. Singleton, and J. Franklin. 2014. Late-Holocene faunal and landscape change in the Bahamas. The Holocene, 24(2): 220–230.

Steadman, D.W., R. Franz, G.S. Morgan, N.A. Albury, B. Kakuk, K. Broad, S.E. Franz, K. Tinker, M.P. Pateman, T.A. Lott, and D.M. Jarzen. 2007. Exceptionally well preserved late Quaternary plant and vertebrate fossils from a blue hole on Abaco, The Bahamas. Proceedings of the National Academy of Sciences, 104(50): 19897–19902.

Steensma, J.T., N. Morken, L. Wiedman, and L. Colebrooke. 2016. A guide to birds of North Andros Island. St. Louis, MO: SHEBA Media.

Stevenson, D.W. 2010. *Zamia integrifolia*. The IUCN red list of threatened species 2010: e.T42164A10670852. Downloaded on 29 January 2018.

Stiling, P.D. 1986. Butterflies and other insects of the eastern Caribbean. London, UK: Macmillan Publishers.

Stiling, P.D. 1999. Butterflies of the Caribbean and Florida. London, UK: MacMillan Education.

Strohecker, H.F. 1953. The Gryllacrididae and Gryllidae of The Bahama Islands, British West Indies (Orthoptera). American Museum Novitates, no. 1618.

Sullivan Sealey, K., V.N. McDonough, and K.S. Lunz. 2014. Coastal impact ranking of small islands for conservation, restoration, and tourism development: A case study of The Bahamas. Ocean and Coastal Management 91: 88–101.

Terrestrial Mollusc Tool. n.d. http://idtools.org/id/mollusc/index.php (accessed 1 January 2018). Gainesville, FL: University of Florida.

Theile, S. 2001. Queen Conch fisheries and their management in the Caribbean. Brussels: TRAFFIC Europe.

Toft, C.A., and T.W. Schoener. 1983. Abundance and diversity of orb spiders on 106 Bahamian Islands: Biogeography at an intermediate trophic level. Oikos, 41(3): 411–426.

Tortoise and Freshwater Turtle Specialist Group. 1996. *Trachemys stejnegeri* (errata version published in 2016). The IUCN red list of threatened species 1996: e.T22026A97299425. Downloaded on 29 January 2018.

Tortoise and Freshwater Turtle Specialist Group. 1996. *Trachemys terrapen* (errata version published in 2016). The IUCN red list of threatened species 1996: e.T22027A97299558. Downloaded on 29 January 2018.

Turks and Caicos Islands Department of Environment and Maritime Affairs. n.d. https://www.gov.tc/dema/reptiles-and-amphibians (accessed 1 January 2018). Grand Turk, Turks and Caicos Islands.

Turnbow, R.H. Jr., and M.C. Thomas. 2008. An annotated checklist of the Coleoptera (Insecta) of The Bahamas. Insecta Mundi. Paper 347.

Turvey, S., and L. Dávalos. 2008. *Geocapromys ingrahami*. The IUCN red list of threatened species 2008: e.T9002A12949103. Downloaded on 29 January 2018.

Uetz, P., and Jiří Hošek, eds. 1996. The Reptile Database. http://www.reptile-database.org (accessed 15 May 2018).

University of The Bahamas, Gerace Research Centre, San Salvador, The Bahamas. n.d. http://www.geraceresearchcentre.com (accessed 1 January 2018).

Vaurie, P., M.A. Cazier, W.J. Gertsch, and C. Vaurie. 1952. Insect collecting in the Bimini Island group, Bahama Islands. American Museum Novitates, no. 1565.

Velazco, P., and S. Turvey. 2008. *Chilonatalus tumidifrons*. The IUCN red list of threatened species 2008: e.T14361A4435590. Downloaded on 29 January 2018.

Walker, L.N., J.E. Mylroie, A.D. Walker, and J.R. Mylroie. 2008. The caves of Abaco Island, Bahamas: Keys to geologic timelines. Journal of Cave and Karst Studies, 70(2): 108–119.

Wardle, C., L. Gape, and P. Moore (editors). 2014. Beautiful Bahama birds. Common birds of the Bahama Islands. Bahamas National Trust and BirdsCaribbean.

Warren, A.D., K.J. Davis, E.M. Stangeland, J.P. Pelham, and N.V. Grishin. 2013. Illustrated lists of American Butterflies. http://www.butterfliesofamerica.com (accessed 18 May 2018).

Westfall, M.J. 1960. The Odonata of The Bahama Islands, the West Indies. American Museum Novitates, no. 2020.

White, A.W. 1998. A birder's guide to The Bahama Islands. Colorado Springs, CO: American Birding Association.

Wieg, C. 2009. Geographic variation in the Bahamian Brown Racer *Alsophis vudii* [PhD dissertation]. Athens, OH: Ohio University.

Wild Bahamas. n.d. https://www.facebook.com/BahamianWildlife/ (accessed January 2018).

Wildscreen Arkive. n.d. https://www.arkive.org (accessed 1 January 2018).

Wood, K.M. 2003. Flowers of The Bahamas and the Turks and Caicos Islands. Oxford, UK: Macmillan Publishers.

Woodruff, D.S. 1978. A remarkably diverse group of West Indian land snails. Malacologia, 17(2): 223–239.

Woodruff, D.S., and S.J. Gould. 1980. Geographic differentiation and speciation in *Cerion*—a preliminary discussion of patterns and processes. Biological Journal of the Linnaean Society, 14(3–4): 389–416.

Woods, C.A., and F.E. Sergile, eds. 2001. Biogeography of the West Indies: Patterns and perspectives. Boca Raton, FL: CRC Press.

The World of Stick and Leaf Insects. n.d. http://www.phasmatodea.com (accessed 1 January 2018).

World Spider Catalog. 2018. World Spider Catalog. Natural History Museum Bern. http://wsc.nmbe.ch, version 19.0 (accessed 12 May 2018).

Yokoyama, M. 2013. The incomplete guide to the wildlife of Saint Martin. http://www.sxmwildlife.com/publications/wildlife-of-st-martin-2nd-edition/ (accessed 1 January 2018).

Photo Credits

Introduction

All photos by J. M. Wunderle, with the exception of the following:

Bahama Cave Fish (*Lucifuga spelaeotes*). Keith Pamper.

Cats pose a threat to native species. Dave Currie.

Fungi

All photos by D. Jean Lodge, with the exception of the following:

Clathrus crispus. Scott Johnson.

Phellinus gilvus. Nick Legon.

Hexagonia hydnoides. Camilla Adair.

Lentinus sp. Kurt Miller.

Gymnopus neotropicus. Tim Baroni.

Cantharellus coccolobae. Father Alejandro Sanchez.

Plants

All photos by Ethan Freid, with the exception of the following:

Amyris elemifera. Ron Lance.

Invertebrates

All photos by Dave Currie, with the exception of the following:

Hemitrochus sp. J. M. Wunderle.

Orthalicus undatus. J. M. Wunderle.

Bulimulus sp. Matthew L. Niemiller, Ph.D.

Drymaeus multilineatus. Mark Yokoyama.

Sarasinula plebeia. Scott Johnson.

Gecarcinus lateralis. Robert Perger, Colección Boliviana de Fauna. La Paz, Bolivia.

Uca leptodactyla. Wilfredo R. Rodriguez H.

Aratus pisonii. Alan Cressler.

Heteropoda venatoria. Mark Yokoyama.

Antillochernes sp. Mark Yokoyama.

Sphendononema guildingii. Scott Johnson.

Libellula needhami. Ancilleno O. Davis.

Erythrodiplax umbrata, female. Mark Yokoyama.

Erythemis simplicicollis, male. D. Gordon E. Robertson.

Celithemis eponina, female or juvenile male. Jeff Pippen.

Lestes spumarius, female. Lic. Alfredo D. Colón Archilla (www.alfredocolo .zenfolio.com).

Ischnura hastata, male. Jeff Pippen.

Ischnura ramburii, male. Jeff Pippen.

Nehalennia minuta, male. Denis Gaschignard.

Stagmomantis carolina, female eating male. Thomas M. Mitchell.

Periplaneta americana. Mark Yokoyama.

Reticulitermes flavipes, workers. Steve Nanz.

Apis mellifera. Mark Yokoyama.

Polistes bahamensis. Janson Scott Jones.

Pepsis sp., female attacking Bahamian Tarantula (*Cyrtopholis bonhotei*). Scott Johnson.

Danaus plexippus plexippus. Steve Nanz.

Heliconius charithonia. Mark Yokoyama.

Eunica tatila. Father Alejandro Sanchez.

Papilio polyxenes, male. Donald W. Hall, University of Florida.

Papilio aristodemus ponceanu, male. Jaret C. Daniels, Ph.D., Florida Museum of Natural History.

Ascalapha odorata, female. J. M. Wunderle.

Automeris io lilith, female. Janson Scott Jones.

Cautethia grotei grotei. Ancilleno O. Davis.

Hemaris thysbe. Jeff Pippen.

Perigonia lusca. David Fine (www.keysmoths .com).

Pseudosphinx tetrio. David Fine (www .keysmoths.com).

Pseudosphinx tetrio, caterpillar. David Fine (www.keysmoths.com).

Xylophanes tersa. Valerie Bugh.

Efferia sp. Lic. Alfredo D. Colón Archilla (www.alfredocolo.zenfolio.com).

Culicoides furens. Roxanne Connelly, UF/IFAS/FMEL.

Aedes aegypti. James Gathany, PHIL.

Lucilia sp. Steve Nanz.

Tabanus sp. Steve Nanz.

Strategus tapa, male. Lic. Alfredo D. Colón Archilla (www.alfredocolo.zenfolio.com)

Blapstinus sp. Valerie Bugh.

Melanophila notata. Tony DeSantis.

Stenodontes chevrolati, female. Scott Johnson.

Stenodontes chevrolati, male. Graham Savage.

Placosternus difficilis. J. M. Wunderle.

Lagocheirus araneiformis. Mark Yokoyama.

Harpalus pensylvanicus. Steve Nanz.

Cicindela (Ellipsoptera) marginata. Jeff Pippen.

Nemognatha punctulata. Steve Nanz.

Chalcolepidius silbermanni. J. M. Wunderle.

Photinus sp. Steve Nanz.

Chondrocera laticornis. Valerie Bugh.

Jadera haematoloma. Valerie Bugh.

Loxa viridis. Valerie Bugh.

Euthyrhynchus floridanus. Valerie Bugh.

Sphyrocoris obliquus. Valerie Bugh.

Zelus longipes. Valerie Bugh.

Diceroprocta bonhotei, nymph. J. M. Wunderle.

Neonemobius cubensis. Lyle Buss, Entomology and Nematology Department, University of Florida.

Vella fallax, adult. Margy Green (www.margygreen.com).

Family Myrmeleontidae, larva. Margy Green (www.margygreen.com).

Fish

Cyprinodon brontotheroides, male. Christopher Martin.

Cyprinodon brontotheroides, female. Christopher Martin.

Cyprinodon desquamator, male. Christopher Martin.

Cyprinodon desquamator, female. Christopher Martin.

Gambusia hubbsi, two males. Brian Langerhans, Ph.D.

Gambusia sp., female. Brian Langerhans, Ph.D.

Megalops atlanticus. Paul Davill Photography.

Albula vulpes. Brian Gratwicke.

Makaira nigricans. Pat Ford.

Epinephelus striatus. Melinda S. Riger.

Amphibians

Eleutherodactylus planirostris. Matthew L. Niemiller, Ph.D.

Eleutherodactylus rogersi. Scott Johnson.

Osteopilus septentrionalis, at night. J. M. Wunderle.

Osteopilus septentrionalis, during the day. Dave Currie.

Hyla squirella. Janson Scott Jones.

Gastrophryne carolinensis. Janson Scott Jones.

Rana grylio. Janson Scott Jones.

Rana clamitans. Steve Nanz.

Lithobates sphenocephalus. Janson Scott Jones.

Rhinella marina. J. M. Wunderle.

Reptiles

Trachemys terrapen. Dave Currie.

Trachemys stejnegeri malonei. Joe Burgess.

Trachemys scripta. Dave Currie.

Chelonia mydas mydas. Jason R. Folt.

Eretmochelys imbricata imbricata. Jason R. Folt.

Caretta caretta caretta. Adam Rees.

Dermochelys coriacea. Joe Burgess.

Pholidoscelis auberi thoracica, male. Dave Currie.

Pholidoscelis maynardii maynardii. Joe Burgess.

Pholidoscelis maynardii uniformis. Joe Burgess.

Spondylurus caicosae. Joe Burgess.

Spondylurus turksae. R. Graham Reynolds, University of North Carolina Asheville.

Leiocephalus carinatus virescens, male. Dave Currie.

Leiocephalus greenwayi, male. Dave Currie.

Leiocephalus inaguae, male. Joe Burgess.

Leiocephalus inaguae, female. Scott Johnson.

Leiocephalus loxogrammus parnelli, male. Dave Currie.

Leiocephalus loxogrammus parnelli, female. Dave Currie.

Leiocephalus punctatus. Dave Currie.

Leiocephalus psammodromus, male. Joe Burgess.

Leiocephalus psammodromus, female. Joe Burgess.

Anolis angusticeps, dark coloration. Dave Currie.

Anolis angusticeps, light coloration. Jason R. Folt.

Anolis brunneus. Dave Currie.

Anolis brunneus. Dave Currie.

Anolis brunneus. Dave Currie.

Anolis distichus. Dave Currie.

Anolis distichus. Dave Currie.

Anolis fairchildi. R. Graham Reynolds, University of North Carolina Asheville.

Anolis equestris. Janson Scott Jones.

Anolis sagrei, male. Dave Currie.

Anolis sagrei, male, dewlap extended. Dave Currie.

Anolis sagrei, female. Dave Currie.

Anolis scriptus. Matthew L. Niemiller, Ph.D.

Anolis smaragdinus. Jason R. Folt.

Cyclura cychlura cychlura, male. Joe Burgess.

Cyclura cychlura cychlura, female. Joe Burgess.

Cyclura carinata carinata, male. Joe Burgess.

Cyclura carinata carinata, female. Joe Burgess.

Cyclura rileyi rileyi, male. Dave Currie.

Cyclura rileyi rileyi, female. Dave Currie.

Iguana iguana. Mark Yokoyama.

Hemidactylus garnotii. J. D. Willson.

Hemidactylus mabouia, at night. Dave Currie.

Hemidactylus mabouia, during day. Dave Currie.

Tarentola americana warreni. Joe Burgess.

Aristelliger barbouri. Aaron H. Griffing, Marquette University, Milwaukee, Wisconsin USA.

Aristelliger hechti. Joe Burgess.

Sphaerodactylus argus. Joe Burgess.

Sphaerodactylus caicosensis, male. Matthew L. Niemiller, Ph.D.

Sphaerodactylus caicosensis, female. Matthew L. Niemiller, Ph.D.

Sphaerodactylus copei, male. Scott Johnson.

Sphaerodactylus copei, female. Scott Johnson.

Sphaerodactylus corticola. Joe Burgess.

Sphaerodactylus inaguae, male. Joe Burgess.

Sphaerodactylus mariguanae. Joe Burgess.

Sphaerodactylus nigropunctatus, male. R. Graham Reynolds, University of North Carolina Asheville.

Sphaerodactylus nigropunctatus, female. Joe Burgess.

Sphaerodactylus notatus. Joe Burgess.

Sphaerodactylus underwoodi, female. Joe Burgess.

Sphaerodactylus underwoodi, male. Joe Burgess.

Cubophis vudii. Dave Currie.

Cubophis vudii. Dave Currie.

Cubophis vudii, hood showing. Scott Johnson.

Pantherophis guttatus. R. Graham Reynolds, University of North Carolina Asheville.

Chilabothrus exsul. Scott Johnson.

Chilabothrus chrysogaster chrysogaster, spotted morph. Joe Burgess.

Chilabothrus chrysogaster chrysogaster, striped morph. Joe Burgess.

Chilabothrus strigilatus strigilatus. Dave Currie.

Chilabothrus strigilatus strigilatus with White-crowned Pigeon (*Patagioenas leucocephala*). Dave Currie.

Chilabothrus argentum. R. Graham Reynolds, University of North Carolina Asheville.

Chilabothrus schwartzi, juvenile. Joe Burgess.

Tropidophis curtus. R. Graham Reynolds, University of North Carolina Asheville.

Tropidophis canus. Joe Burgess.

Tropidophis greenwayi, dark color. Matthew L. Niemiller, Ph.D.

Tropidophis greenwayi, light color. Matthew L. Niemiller, Ph.D.

Tropidophis greenwayi eating Silver Key Anole (*Anolis scriptus*). Matthew L. Niemiller, Ph.D.

Epictia columbi. Dave Currie.

Typhlops biminiensis. Jason R. Folt.

Typhlops lumbricalis. Jason R. Folt.

Typhlops platycephalus. Joe Burgess.

Ramphotyphlops braminus. Dave Currie.

Birds

Dendrocygna arborea. Anthony Levesque.

Spatula discors, female and male. Bruce Hallett.

Spatula clypeata, female and male. Bruce Hallett.

Anas platyrhynchos, male. Dave Currie.

Anas platyrhynchos, female. Dave Currie.

Anas bahamensis. Dave Currie.

Anas acuta, male. Bruce Hallett.

Anas acuta, female. Bruce Hallett.

Aythya collaris, male. Bruce Hallett.

Aythya collaris, female. Bruce Hallett.

Aythya affinis, male. Bruce Hallett.

Aythya affinis, female. Bruce Hallett.

Lophodytes cucullatus, male and female. Bruce Hallett.

Mergus serrator, female and male. Steve Nanz.

Oxyura jamaicensis jamaicensis, male. Bruce Hallett.

Oxyura jamaicensis jamaicensis, female. Bruce Hallett.

Colinus virginianus, female and male. Tom Sheley.

Phasianus colchicus, male. Dave Currie.

Phoenicopterus ruber. Bruce Hallett.

Phoenicopterus ruber, flock. Bruce Hallett.

Tachybaptus dominicus dominicus. Gerlinde Taurer.

Podilymbus podiceps. Bruce Hallett.

Columba livia. Bruce Hallett.

Patagioenas leucocephala. Dave Currie.

Streptopelia decaocto. Bruce Hallett.

Columbina passerina. Dave Currie.

Geotrygon chrysia. Andrew Dobson.

Leptotila jamaicensis. Bruce Hallett.

Zenaida asiatica. Bruce Hallett.

Zenaida aurita. Craig Nash.

Zenaida macroura. Dave Currie.

Coccyzus minor. Dave Currie.

Coccyzus merlini bahamensis. Dave Currie.

Crotophaga ani. Dave Currie.

Chordeiles gundlachii. Dave Currie.

Calliphlox evelynae, male. Dave Currie.

Calliphlox evelynae, male, in flight. Dave Currie.

Calliphlox evelynae, female. Dave Currie.

Calliphlox evelynae, female, in flight. Dave Currie.

Calliphlox lyrura, male. D. Neil McKinney.

Chlorostilbon ricordii, male. Bruce Hallett.

Chlorostilbon ricordii, female. Keith Salvesen.

Rallus longirostris coryi. Dave Currie.

Porzana carolina. Bruce Hallett.

Porphyrio martinicus. Bruce Hallett.

Gallinula galeata. Bruce Hallett.

Fulica americana. Bruce Hallett.

Aramus guarauna. Bruce Hallett.

Himantopus mexicanus. Dave Currie.

Haematopus palliatus. Dave Currie.

Pluvialis squatarola, winter plumage. Bruce Hallett.

Pluvialis squatarola, partial breeding plumage. Dave Currie.

Charadrius nivosus. Bruce Hallett.

Charadrius wilsonia. Dave Currie.

Charadrius semipalmatus. Dave Currie.

Charadrius melodus. Bruce Hallett.

Charadrius vociferus. Bruce Hallett.

Numenius phaeopus. Bruce Hallett.

Arenaria interpres. Bruce Hallett.

Arenaria interpres, partial breeding plumage. Dave Currie.

Calidris alba. Dave Currie.

Calidris minutilla. Bruce Hallett.

Calidris pusilla. Bruce Hallett.

Limnodromus griseus. Bruce Hallett.

Actitis macularius, winter plumage. Bruce Hallett.

Actitis macularius, spring plumage. Bruce Hallett.

Tringa flavipes. Bruce Hallett.

Tringa semipalmatus. Bruce Hallett.

Tringa semipalmatus. Bruce Hallett.

Tringa melanoleuca. Bruce Hallett.

Leucophaeus atricilla, breeding plumage. J. M. Wunderle.

Leucophaeus atricilla, juvenile. Bruce Hallett.

Larus delawarensis. Dave Currie.

Larus argentatus. Bruce Hallett.

Larus fuscus, adult. Bruce Hallett.

Larus marinus, adult. Bruce Hallett.

Larus marinus, juvenile. Dave Currie.

Anous stolidus. Dave Currie.

Onychoprion fuscatus. Dave Currie.

Onychoprion anaethetus. Dave Currie.
Sternula antillarum. Dave Currie.
Gelochelidon nilotica. Bruce Hallett.
Gelochelidon nilotica, in flight. Alex Hughes.
Sterna dougallii. Benjamin Hack.
Thalasseus maxima, winter plumage. Dave
    Currie.
Thalasseus maxima, summer plumage, in
    flight. Dave Currie.
Thalasseus sandvicensis, winter plumage.
    Dave Currie.
Phaethon lepturus. Dave Currie.
Phaethon lepturus, in flight. Dave Currie.
Puffinus lherminieri. Bruce Hallett.
Puffinus lherminieri, in flight. Bruce Hallett.
Fregata magnificens, male, in flight. Dave
    Currie.
Fregata magnificens, female, in flight. Bruce
    Hallett.
Fregata magnificens, displaying males. Mark
    Yokoyama.
Sula leucogaster leucogaster, in flight. Bruce
    Hallett.
Sula leucogaster leucogaster. Bruce Hallett.
Sula sula sula, white morph. Bruce Hallett.
Sula sula sula, dark morph. Bruce Hallett.
Phalacrocorax brasilianus. Tom Sheley.
Phalacrocorax auritus auritus. Bruce Hallett.
Phalacrocorax auritus auritus, juvenile. Bruce
    Hallett.
Pelecanus occidentalis, adult. Bruce Hallett.
Pelecanus occidentalis, juvenile. Bruce Hallett.
Pelecanus occidentalis, in flight. Bruce Hallett.
Ixobrychus exilis. Bruce Hallett.
Ardea herodias. Dave Currie.
Ardea alba. Mark Yokoyama.
Egretta thula. Mark Yokoyama.
Egretta caerulea, adult. Dave Currie.
Egretta caerulea, juvenile. Bruce Hallett.
Egretta tricolor. J. M. Wunderle.
Egretta rufescens, dark morph. Dave Currie.
Egretta rufescens, white morph. Tom Sheley.
Bubulcus ibis. Dave Currie.
Butorides virescens bahamensis, adult. Dave
    Currie.
Butorides virescens bahamensis, juvenile.
    Dave Currie.
Nycticorax nycticorax, adult. Bruce Hallett.

Nycticorax nycticorax, juvenile. Bruce
    Hallett.
Nyctanassa violacea. Dave Currie.
Eudocimus albus, adult. Bruce Hallett.
Platalea ajaja, adult. Elwood D. Bracey.
Cathartes aura. Dave Currie.
Pandion haliaetus ridgwayi, resident
    subspecies. Dave Currie.
Pandion haliaetus carolinensis, North
    American subspecies, in flight. Dave
    Currie.
Buteo jamaicensis jamaicensis. Bruce Hallett.
Buteo jamaicensis jamaicensis, in flight. Bruce
    Hallett.
Tyto alba. Bruce Hallett.
Athene cunicularia floridana. Bruce Hallett.
Megaceryle alcyon. Bruce Hallett.
Melanerpes superciliaris nyeanus, female.
    Dave Currie.
Melanerpes superciliaris nyeanus, male. Dave
    Currie.
Sphyrapicus varius, adult. Bruce Hallett.
Picoides villosus maynardi, male. Bruce
    Hallett.
Falco sparverius sparveroides, light morph.
    Dave Currie.
Falco sparverius sparveroides, dark morph.
    Dave Currie.
Falco columbarius. Becky Marvil.
Falco peregrinus. Bruce Hallett.
Amazona leucocephala bahamensis. Keith
    Salvesen.
Contopus caribaeus bahamensis. Dave Currie.
Myiarchus sagrae lucaysiensis. Dave Currie.
Tyrannus dominicensis. Bruce Hallett.
Tyrannus caudifasciatus bahamensis. Dave
    Currie.
Vireo griseus. Bruce Hallett.
Vireo crassirostris crassirostris. Bruce Hallett.
Vireo flavifrons. Bruce Hallett.
Vireo altiloquus. Bruce Hallett.
Corvus nasicus. Nikolaj Mølgaard Thomsen.
Tachycineta cyaneoviridis. Bruce Hallett.
Tachycineta cyaneoviridis, in flight. Bruce
    Hallett.
Sitta pusilla insularis. Bruce Hallett.
Polioptila caerulea. Dave Currie.
Turdus plumbeus plumbeus. Dave Currie.

Dumetella carolinensis. Bruce Hallett.
Margarops fuscatus. Ancilleno O. Davis.
Mimus gundlachii. Dave Currie.
Mimus polyglottos polyglottos. Bruce Hallett.
Sturnus vulgaris, breeding plumage. Dave Currie.
Anthus rubescens. Bruce Hallett.
Bombycilla cedrorum. Bruce Hallett.
Passer domesticus, male. Dave Currie.
Passerculus sandwichensis. Bruce Hallett.
Spindalis zena, male. Dave Currie.
Icterus northropi. Daniel C. Stonko.
Agelaius phoeniceus, male. Keith Salvesen.
Agelaius phoeniceus, juvenile male. Bruce Hallett.
Molothrus bonariensis, female. Bruce Hallett.
Molothrus bonariensis, male. Bruce Hallett.
Seiurus aurocapilla. Bruce Hallett.
Helmitheros vermivorum. Bruce Hallett.
Parkesia noveboracensis. Bruce Hallett.
Vermivora cyanoptera, male. Bruce Hallett.
Mniotilta varia, female. Bruce Hallett.
Protonotaria citrea, male. Bruce Hallett.
Oreothlypis peregrina, male. Bruce Hallett.
Geothlypis formosa, male. Bruce Hallett.
Geothlypis rostrata coryi, male. Dave Currie.
Geothlypis rostrata, female. Bruce Hallett.
Geothlypis trichas, male. Bruce Hallett.
Geothlypis trichas, female. Bruce Hallett.
Setophaga citrina, male. Bruce Hallett.
Setophaga ruticilla, male. Bruce Hallett.
Setophaga ruticilla, female. Bruce Hallett.
Setophaga kirtlandii, spring male. Dave Currie.
Setophaga tigrina, spring male. Bruce Hallett.
Setophaga tigrina, first fall/winter plumage. Bruce Hallett.
Setophaga americana, fall male. Bruce Hallett.
Setophaga americana, female. Bruce Hallett.
Setophaga magnolia, fall male. Bruce Hallett.
Setophaga petechia flaviceps, male. Dave Currie.
Setophaga striata, spring male. Bruce Hallett.
Setophaga caerulescens, male. Bruce Hallett.
Setophaga caerulescens, female. Bruce Hallett.
Setophaga palmarum palmarum, winter plumage. Bruce Hallett.

Setophaga pityophila, male. Bruce Hallett.
Setophaga pinus achrustera, male. Tom Reed.
Setophaga coronata, winter plumage. Bruce Hallett.
Setophaga dominica, male. Bruce Hallett.
Setophaga flavescens. Bruce Hallett.
Setophaga discolor, male. Bruce Hallett.
Setophaga virens, male. Bruce Hallett.
Pheucticus ludovicianus, male. Bruce Hallett.
Pheucticus ludovicianus, female. Bruce Hallett.
Passerina caerulea, male. Bruce Hallett.
Passerina caerulea, juvenile. Bruce Hallett.
Passerina cyanea, male. Bruce Hallett.
Passerina ciris, male. Bruce Hallett.
Passerina ciris, female. Bruce Hallett.
Coereba flaveola bahamensis. Bruce Hallett.
Tiaris canorus, male. Bruce Hallett.
Tiaris canorus, female. Bruce Hallett.
Tiaris bicolor, male. Bruce Hallett.
Melopyrrha violacea violacea. Dave Currie.

Mammals

Macrotus waterhousii (top right). MerlinTuttle.org.
Macrotus waterhousii (left). Nancy Albury.
Monophyllus redmani. MerlinTuttle.org.
Artibeus jamaicensis (top right). J. Angel Soto-Centeno, Ph.D., Rutgers University, Newark.
Artibeus jamaicensis (top left). Nancy Albury.
Brachyphylla nana. MerlinTuttle.org.
Erophylla sezekorni. MerlinTuttle.org.
Chilonatalus tumidifrons (top right). J. Angel Soto-Centeno, Ph.D., Rutgers University, Newark.
Chilonatalus tumidifrons (top left). Nancy Albury.
Nyctiellus lepidus. J. Angel Soto-Centeno, Ph.D., Rutgers University, Newark.
Eptesicus fuscus (middle left). MerlinTuttle.org.
Eptesicus fuscus (middle right). Nancy Albury.
Lasiurus borealis. MerlinTuttle.org.
Tadarida brasiliensis. MerlinTuttle.org.
Geocapromys ingrahami (middle left). Mark C. Erdos.
Geocapromys ingrahami (middle right). Ethan Freid.

Rattus norvegicus. Glenn Vermeersch.
Rattus rattus. Dave Currie.
Mus musculus domesticus. Dave Currie.
Procyon lotor. Ancilleno O. Davis.
Sus scrofa (bottom right). Dave Currie.
Sus scrofa (top left). Dana Lowe.
Trichechus manatus. Bahamas Marine
    Mammal Research Organisation.
Trichechus manatus. Bahamas Marine
    Mammal Research Organisation.
Tursiops truncatus. Steve Nanz.

Index